Lecture Notes of the Institute for Computer Sciences, Social Informatics and Telecommunications Engineering 547

T0220826

The LNICST series publishes ICST's conferences, symposia and workshops.
LNICST reports state-of-the-art results in areas related to the scope of the Institute.
The type of material published includes

- Proceedings (published in time for the respective event)
- Other edited monographs (such as project reports or invited volumes)

LNICST topics span the following areas:

- General Computer Science
- E-Economy
- E-Medicine
- Knowledge Management
- Multimedia
- Operations, Management and Policy
- Social Informatics
- Systems

Lin Yun · Jiang Han · Yu Han
Editors

Advanced Hybrid Information Processing

7th EAI International Conference, ADHIP 2023
Harbin, China, September 22–24, 2023
Proceedings, Part I

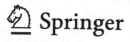 Springer

Editors
Lin Yun
Harbin Engineering University
Harbin, China

Jiang Han
Harbin Engineering University
Harbin, China

Yu Han
Harbin Engineering University
Harbin, China

ISSN 1867-8211 ISSN 1867-822X (electronic)
Lecture Notes of the Institute for Computer Sciences, Social Informatics
and Telecommunications Engineering
ISBN 978-3-031-50542-3 ISBN 978-3-031-50543-0 (eBook)
https://doi.org/10.1007/978-3-031-50543-0

This Springer imprint is published by the registered company Springer Nature Switzerland AG
The registered company address is: Gewerbestrasse 11, 6330 Cham, Switzerland

Paper in this product is recyclable.

Preface

We are delighted to introduce the proceedings of the 7th edition of the European Alliance for Innovation (EAI) International Conference on Advanced Hybrid Information Processing (ADHIP 2023). This conference brought together researchers, developers and practitioners around the world who are leveraging and developing advanced information processing technology. This conference aimed to provide an opportunity for researchers to publish their important theoretical and technological studies of advanced methods in social hybrid data processing, and their novel applications within this domain.

The technical program of ADHIP 2023 consisted of 108 full papers. The topics of the conference were novel technology for social information processing and real applications to social data. Aside from the high-quality technical paper presentations, the technical program also featured three keynote speeches. The three keynote speakers were Cesar Briso from Technical University of Madrid, Spain, Yong Wang from Harbin Institute of Technology, China, and Yun Lin from Harbin Engineering University, China.

Coordination with the steering chairs, Imrich Chlamtac, Shuai Liu and Yun Lin was essential for the success of the conference. We sincerely appreciate their constant support and guidance. It was also a great pleasure to work with such an excellent organizing committee team for their hard work in organizing and supporting the conference. In particular, the Technical Program Committee, led by our TPC Co-Chairs, Yun Lin, Ruizhi Liu and Shan Gao completed the peer-review process of technical papers and made a high-quality technical program. We are also grateful to the Conference Manager, Ivana Bujdakova, for her support and to all the authors who submitted their papers to the ADHIP 2023 conference.

We strongly believe that the ADHIP conference provides a good forum for all researchers, developers and practitioners to discuss all technology and application aspects that are relevant to information processing technology. We also expect that the future ADHIP conferences will be as successful and stimulating as indicated by the contributions presented in this volume.

Yun Lin

Organization

Organizing Committee

General Chair

Yun Lin   Harbin Engineering University, China

General Co-chairs

Zheng Dou   Harbin Engineering University, China
Yan Zhang   University of Oslo, Norway
Shui Yu   University of Technology Sydney, Australia
Joey Tianyi Zhou   Institute of High-Performance Computing, A*STAR, Singapore
Hikmet Sari   Nanjing University of Posts and Telecommunications, China
Bin Lin   Dalian Maritime University, China

TPC Chairs and Co-chairs

Yun Lin   Harbin Engineering University, China
Guangjie Han   Hohai University, China
Ruolin Zhou   University of Massachusetts Dartmouth, USA
Chao Li   RIKEN-AIP, Japan
Guan Gui   Nanjing University of Posts and Telecommunications, China
Ruizhi Liu   Harbin Engineering University, China

Sponsorship and Exhibit Chairs

Yiming Yan   Harbin Engineering University, China
Ali Kashif   Manchester Metropolitan University, UK
Liang Zhao   Shenyang Aerospace University, China

Local Chairs

Jiang Hang	Harbin Engineering University, China
Yu Han	Harbin Engineering University, China
Haoran Zha	Harbin Engineering University, China

Workshops Chairs

Nan Su	Harbin Engineering University, China
Peihan Qi	Xidian University, China
Jianhua Tang	Nanyang Technological University, Singapore
Congan Xu	Naval Aviation University, China
Shan Gao	Harbin Engineering University, China

Publicity and Social Media Chairs

Jiangzhi Fu	Harbin Engineering University, China
Lei Chen	Georgia Southern University, USA
Zhenyu Na	Dalian Maritime University, China

Publications Chairs

Weina Fu	Hunan Normal University, China
Sicheng Zhang	Harbin Engineering University, China
Wenjia Li	New York Institute of Technology, USA

Web Chairs

Yiming Yan	Harbin Engineering University, China
Zheng Ma	University of Southern Denmark, Denmark
Jian Wang	Fudan University, China

Posters and PhD Track Chairs

Lingchao Li	Shanghai Dianji University, China
Jibo Shi	Harbin Engineering University, China
Yulong Ying	Shanghai University of Electric Power, China

Panels Chairs

Danda Rawat	Howard University, USA
Yuan Liu	Tongji University, China
Yan Sun	Harbin Engineering University, China

Demos Chairs

Ao Li	Harbin University of Science and Technology, China
Guyue Li	Southeast University, China
Changbo Hou	Harbin Engineering University, China

Tutorials Chairs

Yu Wang	Nanjing University of Posts and Telecommunications, China
Yi Zhao	Tsinghua University, China
Qi Lin	Harbin Engineering University, China

Technical Program Committee

Zheng Dou	Harbin Engineering University, China
Yan Zhang	University of Oslo, Norway
Shui Yu	University of Technology Sydney, Australia
Joey Tianyi Zhou	A*STAR, Singapore
Hikmet Sari	Nanjing University of Posts and Telecommunications, China
Bin Lin	Dalian Maritime University, China
Yun Lin	Harbin Engineering University, China
Guangjie Han	Hohai University, China
Ruolin Zhou	University of Massachusetts Dartmouth, USA
Chao Li	RIKEN-AIP, Japan
Guan Gui	Nanjing University of Posts and Telecommunications, China
Zheng Ma	University of Southern Denmark, Denmark
Jian Wang	Fudan University, China
Lei Chen	Georgia Southern University, USA
Zhenyu Na	Dalian Maritime University, China
Peihan Qi	Xidian University, China
Jianhua Tang	Nanyang Technological University, Singapore

Contents – Part I

Mobile Education, Mobile Monitoring, Behavior Understanding and Object Tracking

Wireless Communication for Social Information Processing, Artificial Intelligence Technology

Personalized Recommendation Method of Online Career Guidance Curriculum Resources Based on Collaborative Filtering

Juanjuan Zou[✉]

Chongqing Vocational Institute of Engineering, Jiangjin 402260, China
17726637816@163.com

Abstract. Aiming at the problems of low recommendation efficiency and inaccurate recommendation results of existing resource recommendation algorithms, this paper proposes a personalized recommendation method design of online career guidance curriculum resources based on collaborative filtering. Firstly, analyze the principle of personalized recommendation of course resources, then establish a user social network model, and calculate the similarity based on user interest preferences and course resource ratings. Finally, based on this, complete the design of personalized recommendation methods for online employment guidance course resources. The feasibility of the proposed method was demonstrated through comparative experiments. The test results showed that the MAE value of the proposed method was between 0.7 and 0.78, the average recommendation time was less than 13.3 ms, and the F-value was higher than 0.95, which is superior to the comparative method. The recommendation efficiency is higher and the recommendation results are more accurate, indicating good application value.

Keywords: Collaborative Filtering · Employment Guidance Courses · Curriculum Resources · Personalized Recommendations

1 Introduction

Nowadays, the job market is becoming increasingly competitive, and many graduates or job seekers feel confused when facing the employment problem. They don't know how to choose the industry, the company and how to improve their abilities. In order to better guide job seekers, many universities and job centers provide a variety of career guidance course resources, which cover the content of resume making, career planning, interview skills, workplace psychology and other aspects, so it has good reference value for job seekers [1]. However, for different job seekers, due to the differences in learning background, personal hobbies, job hunting direction and other factors, each person needs different course recommendation, but manual screening and recommendation requires a lot of time and energy [2]. In the face of such huge course resources, how to provide the most suitable course recommendation for job seekers becomes more complicated.

L. Yun et al. (Eds.): ADHIP 2023, LNICST 547, pp. 3–15, 2024.
https://doi.org/10.1007/978-3-031-50543-0_1

As proposed in reference [3], a personalized recommendation system for course resources based on data mining is proposed. It utilizes 22 servers to form a server module, and uses an S3C3210x processor for processing to form a hardware system design. The software mainly uses the Orange toolbox to mine course resource data, and then uses hybrid recommendation technology to achieve personalized recommendation. A personalized recommendation algorithm for large-scale open online course resources is proposed in reference [4]. It first clusters online course resources, then generates candidate queues for course resource recommendation. Through column and set operations, the queue length with behavior height that overlaps with the user's own information is obtained. Finally, course resource recommendation is implemented based on the TF-IDF method. A multi task feature recommendation algorithm integrating Knowledge graph is proposed in literature [5], which is based on multi task feature learning and embeds Knowledge graph in tasks; By establishing higher-order connections between potential features and entities through cross compression units between tasks, a recommendation model is established, achieving precise recommendation of course resources based on learners' goals, interests, and knowledge levels. After verification, all three methods have problems with low recommendation efficiency and low recommendation quality. Therefore, further research should be conducted on them.

Collaborative filtering is an algorithm widely used in the field of personalized recommendation. Its principle is to find users with similar preferences and views through the analysis of historical user behavior data, and then recommend similar content to target users according to their evaluation or selection records. The online resource recommendation system based on collaborative filtering algorithm has been widely used in education, e-commerce and other fields, and has achieved good recommendation results, with high user satisfaction.

Based on this, this study will improve the collaborative filtering algorithm, establish the social network model, and realize the calculation of user similarity from two aspects, so as to enhance the individuation and accuracy of the recommendation algorithm, better grasp the user needs, and improve the efficiency of resource utilization. At the same time, this study will also explore data processing and model optimization problems, improve the running speed and stability of the algorithm, and further promote the utilization rate and quality of online career guidance course resources. In conclusion, this study aims to provide personalized and intelligent employment guidance services for job seekers and make positive contributions to talent training and career development.

2   Principles of Personalized Recommendation of Course Resources

Collaborative filtering is based on a collective wisdom. Users with similar interests may have preferences for similar projects. The collaborative filtering algorithm first needs to find a set of neighbor users with similar interests to the target object in the system, and then predict and judge whether to recommend according to the preference of the neighbor users in the set for the predicted items. The most important steps are: collecting user preferences, finding neighboring users or similar items, and calculating recommendations. The user preference information in the process of collecting user preferences comes from user data, mainly including basic user information, user learning data, user behavior data, et al.The earliest collection of user preference information is mainly based on user display data, such as rating, voting, forwarding, favorites, comments, etc., which is called display rating. Later, with the development of technology and the need for recommendation accuracy, recommendation based on data mining technology is widely studied and used. Users' implicit data are obtained through click rate, purchase behavior, page stay time, etc., and weighted into users' score on resources, which is called implicit score. The former is more intuitive and can accurately reflect users' preferences for resources, while the latter indirectly obtains resource scores through relevant data to make up for the sparse data caused by the lack of real scores. However, its disadvantage is that these data cannot reflect users' real preferences in some cases, resulting in low accuracy of scores. Finding similar neighbors and items is the core step of collaborative filtering recommendation. Common similarity algorithms include cosine similarity, Euclidean distance and Pearson correlation coefficient.

The collaborative filtering recommendation algorithm does not consider the content information of items in the recommendation process, so it overcomes the problem of extracting the content features of video, audio and other types of items in the content-based recommendation algorithm, so that the system needs less information when making recommendations and is easier to obtain. At the same time, the recommendation system is not limited to text items, so it is more universal.

However, collaborative filtering algorithm also has some shortcomings, that is, in the actual recommendation system, the number of items purchased or evaluated by users is often very small in a massive number of projects. In this case, the quality of the set of near neighbor users sought for target users based on similarity will be greatly reduced. And this inaccurate set of neighbor users will affect the accuracy of the results produced when the system makes predictions for the target users later. When a user has just registered as a new user, there is no history of the new user in the system, so it is not possible to find the user's nearest neighbor set based on its evaluation of the project through similarity calculation.

On this basis, the algorithm is improved and the personalized recommendation method of online career guidance course resources is researched.

3 Establish a User Social Network Model

Traditional collaborative filtering algorithms only rely on users' historical scores for recommendation. Firstly, the global calculation of similarity among users leads to a large number of redundant calculations, which increases the computing burden and response time of the system. Secondly, when there are illegal users in the system deliberately imitate the browsing records of some real users of the system, resulting in the illusion that illegal users and real users are very similar, and then deliberately score high on certain resources to make certain resources appear in the recommendation results of real users of the system, this kind of support attack will reduce the robustness of the system [6]. To solve the above problems, this paper proposes to establish a user social network model to screen out the target user trust user set.By utilizing users' social information to establish a social network model, a trust set is established based on the model. Transforming the previous global similarity calculation into local similarity calculation in the trust set can effectively reduce system complexity and improve real-time performance. At the same time, the trust set can filter out unrelated malicious users, ensuring that user recommendations are not maliciously misled and increasing the robustness of the system. Social network models can effectively increase the accuracy of the nearest neighbor set, thereby alleviating the impact of data sparsity.

There are two main variables in the user social network model. One is trust capacity, which represents the degree of trust that the target object can propagate to its friends and friends of friends; The second is trust traffic, where the target object allocates the level of trust based on the strength of their social relationships with other users [7].

The establishment of user social network model is mainly divided into three steps:

Step 1: First, set the target user as the start node. Assume that the target user fully trusts the initial trust capacity set by himself is 1.

Step 2: A person will ask the people around him one by one when consulting his friends. Based on this idea, breadth-first search method is adopted to allocate the trust capacity for the social circle of the target user (starting node). The formula for calculating the trust capacity of subsequent nodes is as follows:

$$Cap_{suc} = Cap_{pre} \times w_{ij} \tag{1}$$

Among them, Cap_{suc} represents the trust capacity of the current node, Cap_{pre} represents the trust capacity of the previous node, and w_{ij} represents the social intensity coefficients of users i and j [8]. The calculation formula is as follows:

$$w_{ij} = \begin{cases} SoSim_{ij}, \, inter_{ij} = 0 \\ \dfrac{2 \times inter_{ij} + SoSim_{ij}}{inter_{max}^i + SoSim_{ij}}, \, inter_{ij} \neq 0 \end{cases} \tag{2}$$

where, $inter_{ij}$ is the number of interactions between user i and j, and $inter_{max}^i$ is the maximum number of interactions between user i and other users. $SoSim_{ij}$ represents the social circle similarity of users i and j. When the interaction times of user i and j are 0, $SoSim_{ij}$ is used to represent w_{ij}, and the calculation formula is as follows:

$$SoSim_{ij} = \begin{cases} 1, \, \dfrac{JaSim_{ij}}{2 \times avg(JaSim)} \geq 1 \\ \dfrac{JaSim_{ij}}{2 \times avg(JaSim)} \end{cases} \tag{3}$$

Among them, $JaSim_{ij}$ is the degree of overlap of friend circles, $avg(JaSim)$ is the average degree of overlap of friend circles for all users in the system [9]. $SoSim_{ij}$ represents the degree of overlap between users i and j's circle of friends. If the subsequent nodes are not the direct circle of friends of the target object, but rather the non direct relationship of friends, then according to the idea of the Ford Fulkerson algorithm, a maximum path can be found to transfer the trust relationship and obtain the trust capacity of the subsequent nodes.

Step 3: Collect the trust capacity of all users in the network model and establish the trust user set of target users.

Let CD store the trust capacity propagated by all nodes in the social network model of user u through the initial node, where the node whose value is greater than threshold

T can enter the candidate set, and CDF is the candidate set of users. TL is the set of trusted users of user u. The calculation process is as follows:

(1) Set node u as the starting node;
(2) From u's local social network adjacency matrix D, find out the user who can join the candidate user set. The user's trust capacity needs to be greater than the threshold T set by the system, and has not been selected into the candidate user set CDF and the trusted user set TL [10].
(3) Find the largest user v in CDF by comparing the trust capacity, and then place it in TL;
(4) Subtract 0.1 from user v's value in CD, and then set $[u, v]$ to 1 in D;
(5) Take user v as the starting node and repeat step (2), and then delete user v from CDF [11];
(6) If CDF is empty, the trusted user set of user u is output. Otherwise, repeat step (3) to obtain the trust set of target user u after the algorithm ends. In the following text, use an improved similarity calculation method to calculate the similarity between users and u in the trust set, and then obtain the nearest neighbor set [12].

3. User similarity calculation.

This article adopts a combination of interest preference model and resource scoring method for neighbor calculation. Firstly, it collects neighbor users based on the user interest preference model, and secondly, it collects neighbor users based on user learning records:

1) Similarity calculation based on user interest preferences
   a. User knowledge level similarity
      The similarity of knowledge level of user u and user v is defined as $S(u, v)$, and $S(u, v)$ is 1 when the range of knowledge level of the two users is consistent. $S(u, v)$ is 0 when two users' knowledge ranges are inconsistent [13]. Where, S_u represents the knowledge ability level of user u, and S_v represents the ability level of user v, then the knowledge level similarity $S(u, v)$ of user u and user v can be calculated as follows:

$$S(u, v) = \begin{cases} 1, S_u = S_v \\ 0, S_u \neq S_v \end{cases} \tag{4}$$

   b. Interest tag similarity
      The similarity of interest labels between user u and user v is defined as $I(u, v)$, with user u's interest label set $I_u = \{I_{u1}, I_{u1}, \cdots, I_{um}\}$ and user v's interest label set $I_v = \{I_{v1}, I_{v1}, \cdots, I_{vn}\}$, where the same number of labels has a value of k, then the similarity of interest labels $I(u, v)$ between user u and user v is calculated using the following formula:

$$I(u, v) = \frac{k}{m + n + k} \tag{5}$$

   c. Curriculum preference similarity
      The similarity of career guidance course preferences of user u and user v is defined as $J(u, v)$, user u prefers course set $J_u = \{I_{u1}, I_{u1}, \cdots, I_{um}\}$, user v prefers

course set $J_v = \{I_{v1}, I_{v1}, \cdots, I_{vn}\}$, where the number of the same course is h, then the calculation formula of course preference similarity $I(u, v)$ of user u and user v is as follows:

$$I(u, v) = \frac{h}{m + n + h} \tag{6}$$

d. Employment Guidance Learning Scope Preference Similarity

The similarity of the employment guidance learning range between user u and user v is represented by $L(u, v)$, where user u has a learning range set of $L_u = \{L_{u1}, L_{u1}, \cdots, L_{um}\}$ and user v has a learning range preference set of $L_v = \{L_{v1}, L_{v1}, \cdots, L_{vn}\}$, where the same number of range values are r[14, 15]. The formula for calculating the course preference similarity $L(u, v)$ between User a and User b is as follows:

$$L(u, v) = \frac{r}{m + n + r} \tag{7}$$

Similarly, the similarity of user preference for career guidance teachers is calculated by the same principle. The similarity value of user teacher preference is defined as $D(u, v)$, then the user similarity $SoSim_{ij}$ based on the user interest preference model is shown as follows:

$$SoSim_{ij\text{model}} = \alpha S(i, j) + \beta I(i, j) + \chi J(i, j) + \delta L(i, j) + \varepsilon D(i, j) \tag{8}$$

Among them, $\alpha, \beta, \chi, \delta, \varepsilon$ is the weight factor, and its sum is 1.

2) similarity calculation based on the user resources score:

Step 1: Assume that the user rating data contains m user set and n career guidance course resources, and generate a "user-1 rating" matrix according to the user's learning record. $V(m, n)$ represents the rating of resources as shown in Table 1.

Table 1. User-course resource scoring matrix

	Career guidance Course Resources 1	Career guidance Course Resources 2	Career guidance Course Resources 3	...	Career guidance Course Resources n
User 1	$V_{1,1}$	$V_{2,1}$	$V_{3,1}$	...	$V_{m,1}$
User 2	$V_{2,1}$	$V_{2,2}$	$V_{2,3}$	...	$V_{2,n}$
...	...	...	...	...	...

Step 2: Construct a data structure of "Employment Guidance Course Resource One User" and obtain a set of user neighbors who have studied the same course, as shown in Table 2.

Table 2. Employment guidance curriculum resources a user matrix

	User 1	User 2	User 3		User n
Course Resources 1	$U_{1,1}$	$U_{1,2}$	$U_{1,3}$	...	$U_{1,n}$
Course Resources 2	$U_{2,1}$	$U_{2,2}$	$U_{2,3}$	...	$U_{2,n}$
...	...	...	...	...	...
Course Resources m	$U_{m,1}$	$U_{m,2}$	$U_{m,3}$	...	$U_{m,n}$

Step 3: Based on the co rated course, a modified cosine similarity algorithm is used to calculate the similarity values between users. The average scores of users i and j in the co rated items are R_i, R_j, respectively. The calculation formula is as follows:

$$SoSim_{ij\text{score}} = \frac{\sum_{k \in I_{ij}} \left(R_{i,k} - \overline{R}_i\right)\left(R_{j,k} - \overline{R}_j\right)}{\sqrt{\sum_{k \in I_i} \left(R_{i,k} - \overline{R}_i\right)^2}\sqrt{\sum_{k \in I_j} \left(R_{j,k} - \overline{R}_j\right)^2}} \tag{9}$$

Because the cosine similarity algorithm does not consider the problem of user rating differences, but only distinguishes in the direction, it is possible to calculate a high similarity even if there are obvious differences between two resources. The modified cosine algorithm compensates for this shortcoming by subtracting the average user rating of the resource. Finally, the result values of the two calculation methods are summed. The rule of calculation is to superimpose the two similarity values of the same user. If the user has no result value in one of the calculation methods, the similarity value of the user in this dimension is 0 by default. The final similarity of the user is defined as $SoSim_{ij\text{final}}$, λ is the weight factor, then the calculation formula of the result is as follows:

$$SoSim_{ij\text{final}} = \lambda SoSim_{ij\text{model}} + (1 - \lambda)SoSim_{ij\text{score}} \tag{10}$$

After calculating the results, $SoSim_{ij\text{final}}$ is sorted by size, and the top N items are selected as sub items of the nearest neighbor user set in descending order. Finally, recommendations are made based on the resource prediction score of the nearest neighbor. The search and calculation of recent neighbors is the focus and key of this article's research. The selection of neighbor users has a significant impact on the final recommendation project, as these resources are all sourced from their neighbors. Accurate neighbor relationships determine the accuracy of recommendation results.Based on the calculation of similar neighbors based on user interest preference model, this paper summarizes students' interest preference from attributes, behaviors, students' learning ability, project attributes and other aspects, and captures students' interest preference from multiple perspectives. For new users, no scoring resource can be used for recommendation calculation through interest preference model to solve the cold start problem. For the problem of data sparsity, the calculation based on the user's interest preference model is a supplement to the calculation of resource score. The combined method of the two will improve the quality and reliability of the user's neighbors and provide accurate data resources for the final recommendation.

4   Personalized Recommendation of Online Employment Guidance Course Resources

Through the above extended neighbors based on users' interests and preferences and direct neighbors based on course resource scoring, predict the unused and unrated items of the target users in the final neighbor user resource project, and select the first item N with the highest predicted score as the recommended item to the users. The calculation method is as follows:

$$P_{u,j} = \overline{R}_u + \frac{\sum_{v \in s(u)} \left(R_{v,j} - \overline{R}_v\right) \cdot SoSim(u, v)}{\sum_{v \in s(u)} |SoSim(u, v)|} \tag{11}$$

During the period, $P_{u,j}$ represents the predicted score of user u on unrated resources, $s(u)$ represents a set of similar neighboring users for user u, $\overline{R}_u$ represents the average score of user u's rated resources, and $\overline{R}_v$ represents the average score of neighboring user v's rated resources.

To sum up, complete the design of personalized recommendation method for online career guidance curriculum resources based on collaborative filtering.

5   Experimental Analysis

5.1   Experimental Environment Settings

This research experiment is carried out based on two laptops, and the experimental environment is set up as shown in Table 3.

Table 3. Experimental Environment Settings

Index	Parameter
Computer	Two sets
Memory	4 GB
Processor	Inter(R) CORE (TM)-i5-7200U@2.60 GHz
Operating system	Windows 10
The server	Apache/2.4.26(Win32)
Database	MySQL

In order to test the performance of Collaborative filtering recommendation based on user interest preferences proposed in this study, 40 testers were selected to test the system. These 40 students are students from educational technology majors, and the test process is as follows (Fig. 1)

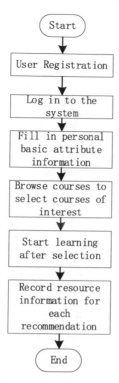

Fig. 1. Test Flow Chart

In this test, a total of 40 students studied and rated 200 career guidance course resources, with a total of 250 scoring records. The main purpose of this test is to test the recommendation effectiveness of the system's recommendation methods. Due to limitations in the experimental environment and conditions, the quality of the testers did not achieve the expected results, and the number of testers was relatively limited. However, the overall recommendation effect was relatively obvious, achieving the expected value.

5.2 Selection of Experimental Indicators

In this test, mean absolute error (MAE) was selected as one of the evaluation indexes to evaluate the performance of the method. MAE determines whether the recommendation is accurate by comparing the degree of deviation. Deviation degree refers to the error between the predicted score of the recommendation algorithm and the actual score of users. The lower MAE value is, the smaller the error between the predicted score and the actual score is, and the more accurate the recommendation is. Conversely, the higher the MAE value, the greater the error and the worse the recommendation effect. When the number of resources scored by the user is defined as n, the system predicts that the user's score on resource i is X_i, and the actual score on resource i is y_i, then the calculation formula of MAE index is shown as follows:

$$MAE_i = \frac{\sum_{j=1}^{n} |x_i - y_i|}{n} \qquad (12)$$

Introduce the comprehensive evaluation index F-measure to evaluate the accuracy and recall of the testing system. The range of F-measure is between 0 and 1, and the larger its value, the better the performance of the classifier and the stronger the predictive ability of the model. When F-measure is equal to 1, it indicates that the model's prediction is very accurate and can perfectly match the required results; When F-measure is equal to 0, it indicates that the predicted value of the model is completely useless and cannot be evaluated for its effectiveness. The specific calculation formula is as follows:

$$F = \frac{2PR}{P + R} \qquad (13)$$

In the formula, F represents the F-measure value, P represents the accuracy rate of recommendation results of employment guidance course resources, and R represents the recall rate of recommendation results of employment guidance course resources.

5.3   Analysis of Experimental Results

Based on the above, literature [2] system and literature [3] system were selected as comparative methods to conduct personalized recommendation of online employment guidance course resources together with the proposed method. Multiple tests were conducted, and the changes in MAE values of the three methods as the algorithm iterated are compared as shown in Fig. 2.

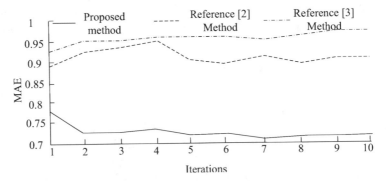

Fig. 2. Changes in MAE values

As shown in Fig. 2, the MAE values of the three methods are also constantly changing with the change of the number of iterations, but the trend is not large. The MAE values obtained by the proposed method are relatively stable, but decrease with the increase of the number of iterations, and their values are stable between 0.7 and 0.78, while the other two methods are stable, both higher than 0.89. Moreover, its value varies greatly and does not decrease as the number of iterations increases, so the proposed method has higher recommendation accuracy. Next, the recommendation efficiency of the three methods is compared, and the results are shown in Table 4.

Table 4. Average Recommended Time

Iterations	Average recommended time of the proposed method/ms	Reference [2] Method Average Recommended Time/ms	Reference [3] Method Average Recommended Time/ms
1	13.2	44.2	46.2
2	11.1	45.5	43.3
3	12.4	44.8	44.2
4	11.2	44.9	42.5
5	11.9	42.3	42.8
6	12.3	45.6	45.2
7	12.6	43.3	43.2
8	11.4	44.2	44.6
9	13.3	44.6	43.9
10	12.6	44.6	45.6

According to Table 4 above, it can be seen that the recommendation time of the proposed method is the shortest, which is all less than 13.3 ms, while the recommendation time of the method in literature [2] is at least 42.3 ms, and the recommendation time of the method in literature [2] is at least 42.5 ms. Therefore, the recommendation efficiency of the proposed method is higher and it has greater application value.

Next, the F value of the three methods changes with the iteration of the algorithm, as shown in Fig. 3.

As shown in Fig. 3, the F value of the system established by the proposed method is all higher than 0.95, while the method in literature [2] and the method in literature [3] are both lower than 0.85. Therefore, the recommended performance of the proposed method resources is better and it has greater application value.

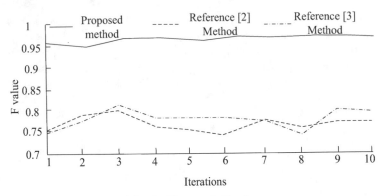

Fig. 3. Changes in F-value

6 Conclusion

The development and popularization of internet technology have greatly enriched the types and channels for people to obtain information, and also brought great convenience to people's work, study, and life. However, these vast amounts of data have already exceeded the limits that individuals can accept and have caused the famous "information overload" problem, which seriously affects the efficiency of users to obtain effective information. As a result, personalized recommendation systems have emerged. This research proposes a personalized recommendation method of online career guidance curriculum resources based on collaborative filtering to solve the problem of poor recommendation effect of contemporary graduates' career guidance curriculum resources. In order to realize the personalized recommendation of career guidance course resources, the author first establishes the user social network model and then calculates the user similarity from two aspects. The results show that the MAE values of the proposed methods are stable, ranging from 0.7 to 0.78, the average recommendation time is less than 13.3ms, and the F values are higher than 0.95, which is better than the comparison methods and has greater application value. Although this study has already achieved certain research results, there are still some shortcomings in the exploration of curriculum resources for employment indicators, which is also the next research direction.

References

1. Chen, Y., Dai, Y., Han, X., et al.: Dig users' intentions via attention flow network for personalized recommendation. Inf. Sci. **547**, 1122–1135 (2021)
2. Liu, J., Yang, Z., Li, T., et al.: SPR: similarity pairwise ranking for personalized recommendation. Knowl.-Based Syst. **239**, 107828 (2022)
3. Yang, Z., Liu, X.: Design of personalized recommendation system for ideological and political theory resources based on data mining. Tech. Autom. Appl. **42**(01), 93–96 (2023)
4. Jia, Z., Bi, H., Bai, T.: Design of personalized recommendation algorithm for large open online course resources. Tech. Autom. Appl. **41**(08), 146–149 (2022)
5. Wu, H., Xu, X., Meng, F.: Knowledge graph-assisted multi-task feature-based course recommendation algorithm. Comput. Eng. Appl. **57**(21), 132–139 (2021)

6. Bao, T., Xu, L., Zhu, L., et al.: Successive point-of-interest recommendation with personalized local differential privacy. IEEE Trans. Veh. Technol. **9**, 66371–66386 (2021)
7. Yang, H., Gao, H.: Personalized content recommendation in online health communities. Ind. Manag. Data Syst. **122**(2), 345–364 (2022)
8. Liu, Z., Wang, L., Li, X., et al.: A multi-attribute personalized recommendation method for manufacturing service composition with combining collaborative filtering and genetic algorithm. J. Manuf. Syst. **58**, 348–364 (2021)
9. Zhou, L., Wang, C.: Research on Recommendation of Personalized exercises in English learning based on data mining. Sci. Program. **14**, 1–9 (2021)
10. Fang, C., Lu, Q.: Personalized recommendation model of high-quality education resources for college students based on data mining. Complexity, 1–5 (2021)
11. Liu, F., Tian, F., Li, X., Lin, L.: A collaborative filtering recommendation method for online learning resources incorporating the learner model. CAAI Trans. Intell. Syst. **16**(06), 1117–1125 (2021)
12. Yang, X.: Personalized recommendation system of library bibliography based on collaborative filtering. Microcomput. Appl. **37**(09), 169–171+175 (2021)
13. Gao, F., Chen, D., Yan, T.: Based on content recommendation and collaborative filtering algorithm achieving personalized evaluation. J. Anhui Univ. (Nat. Sci.) **46**(02), 22–29 (2022)
14. Wang, S., Sun, X., Gao, Y., Sun, B.: Personalized product recommendation method based on neural collaborative filtering. Inf. Technol. **355**(06), 143–147 (2021)
15. Hao, Z., Liao, X., Wen, W., Cai, R.: Collaborative filtering recommendation algorithm based on multi-context information. Comput. Sci. **48**(03), 168–173 (2021)

Tool Condition Monitoring and Maintenance Based on Deep Reinforcement Learning

Yong Ge[(✉)], Guangyi Zhao, and Zhihong Wang

School of Electrical Engineering, Anhui Technical College of Mechanical and Electrical Engineering, Wuhu 241002, China
geyong7894@163.com

Abstract. Tool status monitoring requires collecting a large amount of data to complete analysis, and different types of tools may exhibit different wear and failure modes during processing, making tool status monitoring more difficult. Therefore, a tool condition monitoring method based on deep reinforcement learning is proposed. The feature of tool wear is extracted by wavelet packet analysis. Introduce regression network into deep reinforcement learning network, and complete tool condition monitoring by combining regression algorithm with deep reinforcement learning network algorithm. Finally, specific suggestions for tool status maintenance are provided. To verify the effectiveness of the proposed method, comparative experiments were designed. The results show that the accuracy of tool condition monitoring is high, the monitoring decision coefficient can be maintained above 0.95, and the mean absolute percentage error is smaller.

Keywords: Deep Reinforcement Learning · Tool Status · Monitoring Methods · Regression Algorithm

1 Introduction

With the development of modern manufacturing, the requirements for product accuracy and quality are becoming increasingly high. As a key component in machining, the state of cutting tools has a significant impact on the entire machining process. Therefore, studying tool condition monitoring and maintenance can improve processing efficiency and product quality, reduce processing costs and mechanical failures, and have important industrial application value. At present, the technology of tool condition monitoring and maintenance has been widely applied. For example, using sensors, digital signal processing, artificial intelligence and other technologies, real-time monitoring and diagnosis of tool wear, fracture, temperature, vibration and other conditions can be realized, and timely warning and maintenance can be carried out to improve the utilization and life of machine tools and tools, shorten downtime and production cycle, and reduce maintenance costs and losses. However, existing tool condition monitoring and maintenance technologies still face some challenges and shortcomings, such as low sensor accuracy and reliability, immature data processing methods, and insufficient standardization and intelligence. Further research and improvement are needed to meet the needs of different processing fields.

© ICST Institute for Computer Sciences, Social Informatics and Telecommunications Engineering 2024
Published by Springer Nature Switzerland AG 2024. All Rights Reserved
L. Yun et al. (Eds.): ADHIP 2023, LNICST 547, pp. 16–28, 2024.
https://doi.org/10.1007/978-3-031-50543-0_2

Yang et al. [1] proposed a recognition method of tool wear state based on one-dimensional depth convolution automatic encoder, selecting the effective value of three-phase current of motor under different working conditions and doing normalization. Utilize a one-dimensional deep convolutional autoencoder to perform unsupervised training on the processed values and extract feature information. Using sample labels for secondary supervised recognition of different wear states of cutting tools. Dong et al. [2] proposed a woodworking tool wear condition monitoring method based on discrete wavelet transform and genetic BP neural network. The main shaft power signals under different spindle speeds, milling depths and tool wear conditions are collected, their approximate coefficients are extracted by discrete wavelet transform, sample data sets are established, and they are input into BP neural network model training, and optimized by genetic algorithm to achieve accurate monitoring of woodworking tool wear conditions. Wu et al. [3] proposed an indirect monitoring method for tool wear based on the spindle current signal and particle swarm optimization support vector machine model. The main shaft signal characteristic parameters related to tool wear are extracted as the input feature vector, and the parameters are optimized through the particle swarm optimization algorithm, and the model is established to complete the tool state detection.

Deep learning is a machine learning method that simulates the information processing mechanism of human brain neural networks. It is built on the basis of artificial neural networks and gradually constructs high-level abstractions from local features to global semantics through multi-level nonlinear transformations. However, there are certain shortcomings in current deep learning, such as the need for a large amount of data for training, so data acquisition and organization are very difficult tasks. Due to the use of a series of neurons in deep learning, and the complex interactions between each neuron, the interpretability of deep learning models is poor, making it difficult to distinguish which features ultimately determine the output of the model [4]. Therefore, this article proposes a deep reinforcement learning method, which has the ability to handle continuous actions and state spaces, and performs better than traditional reinforcement learning methods.

2   Feature Extraction of Tool Wear

The number of sensor output signals collected on site by machine tools is not only large, but most of them do not have availability. If the collected signal data is directly used as a prediction basis, it will not only prolong the training time but also increase hardware costs. Therefore, it is necessary to extract the signal characteristics of tool wear from the initial signal data. Tool wear signal includes high and low frequency bands. The frequency resolution of high frequency band information is small, which usually causes information loss. In order to make the high-frequency part of the signal more refined and the high-frequency information more complete, the wavelet packet analysis method [5, 6] is used to divide the frequency band into multiple levels, and the signal characteristics are used as the reference basis for selecting the frequency spectrum and frequency band to accurately extract the tool wear characteristics.

Given the high frequency band D and low frequency band A of tool wear signal, wavelet packet S divides the frequency band into layer n, so the decomposition expression of wavelet packet is as follows:

$$S = A^n n + DA^n n + ADAn + D^{n-1} An$$
$$+ A^{n-1} Dn + DADn + AD^{n-1} n + D^n n \tag{1}$$

In the process of decomposing high-frequency information, if the following equation relationship exists between Hilbert space $L^2(R)$ and wavelet function closure W_j, it indicates that the main basis of wavelet packet decomposition is scale factor j:

$$L^2(R) = \sum_j^n W_j \tag{2}$$

Let the decomposition level be 0 to simplify the calculation complexity of wavelet function[7, 8], then the following simplified formula is obtained:

$$\begin{cases} u_1(x) = \sum_j h_j u_0\left(\sqrt{2}x\right), \{h_j\} \in L^2(R) \\ u_0(x) = \sum_j g_j u_0\left(\sqrt{2}x\right), \{g_j\} \in L^2(R) \end{cases} \tag{3}$$

If the solved wavelet functions $u_1(x)$ and $u_0(x)$ degenerate to scale function and wavelet basis function respectively, then there is a double scale equation set as shown in the following formula:

$$\begin{cases} \phi(x) = \sum_j h_j \phi\left(\sqrt{2}x\right) \\ \varphi(x) = \sum_j g_j \varphi\left(\sqrt{2}x\right) \end{cases} \tag{4}$$

In conclusion, the characteristics of tool wear are constructed by using the two scale equations:

$$\begin{cases} u_{L^2}^{j,2n}(x) = \sum_j h_{j-2L^2} d_j^{2j+1,n^2} \\ d_{L^2}^{j,2n+1}(x) = \sum_j g_{j-2L^2} d_j^{2j+1,n^2} \end{cases} \tag{5}$$

$$d_{L^2}^{j+1,n}(x) = \sum_j \left(h_{L^2-2j} d_j^{j,2n^2} + g_{L^2-2j} d_j^{j,2n^2+2} \right) \tag{6}$$

3   Tool Condition Monitoring Based on Deep Reinforcement Learning

In the past, tool status detection was mostly achieved through deep learning methods. Although tool status monitoring can also be achieved, a large number of candidate areas are needed for status monitoring operations, which undoubtedly reduces the efficiency of tool status monitoring. Compared with deep learning methods, deep reinforcement learning has significant efficiency advantages in tool condition monitoring, and has been widely used in tool condition monitoring for a period of time [9, 10]. However, deep reinforcement learning algorithms only perform region search operations based on the specifications of the current candidate region, which reduces the accuracy of state monitoring to a certain extent. Introducing regression networks into deep reinforcement learning networks can significantly improve the accuracy of tool state monitoring. Based on this, this article introduces regression networks into deep reinforcement learning networks and uses motion state monitoring methods based on regression and deep reinforcement learning networks to complete tool state monitoring.

A regression network has been added to the existing network of deep reinforcement learning. The newly added reinforcement learning network consists of a VGG network for feature extraction, a DQN network responsible for performing path search, and a regression network capable of performing regression operations on candidate regions. After introducing the regression network, joint optimization is carried out on the regression network and DQN network in the following two ways to output better tool condition monitoring results.

(1) Loss function. Generally, root mean square error and smoothL1 are regarded as the loss function of DQN and regression network respectively. Since the network based on regression and deep reinforcement learning only adds regression network on the basis of the original deep reinforcement learning network, correspondingly, as long as the DQN and the loss function of regression network are combined together, the loss function of the learning network in this paper can be obtained by weighting. The solution process can be described by the formula:

$$A(s, a, t) = \frac{1}{N_{dpn}} \sum_j (y_i - Q(s_j, a))^2 + \lambda \frac{1}{N_{reg}} \sum_j R\left(t_j - t_j^*\right) \tag{7}$$

In the above equation, the sample index value is represented by j; Output from DQN network to y_j; The weighted parameters and the size of regression losses are described by λ and $R\left(t_j - t_j^*\right)$; The number of input samples and expected output values of the DQN network are described by N_{dpn} and $Q(s_j, a)$ respectively; SmoothL1 loss function is represented by R; The candidate area coordinates and actual area coordinates t and t_j^* after parameterization meet the following equation:

$$\begin{cases} t = (t_d, t_y, t_w, t_h) \\ t_j^* = \left(t_d^*, t_y^*, t_w^*, t_h^*\right) \end{cases} \tag{8}$$

Use b to represent the coordinates of the candidate region, satisfy $b = (d, y, w, h)$, and perform parameterization operations on it. The parameterization process can be expressed as:

$$\begin{cases} t_x = \frac{d - d_a}{w_a}, t_y = \frac{y - y_a}{h_a} \\ t_w = \log\left(\frac{w}{w_a}\right), t_h = \lg\left(\frac{h}{h_a}\right) \end{cases} \tag{9}$$

$$\begin{cases} t_x^* = \frac{d^*-d_a}{w_a}, t_y^* = \frac{y^*-y_a}{h_a} \\ t_w^* = \log\left(\frac{w^*}{w_a}\right), t_h^* = \lg\left(\frac{h^*}{h_a}\right) \end{cases} \tag{10}$$

In the above equation, the center points and width/height of the candidate regions output by the regression network and obtained by the DQN network are described by d, y, w, h and d_a, y_a, w_a, h_a respectively; The center point and width height of the actual target area are described by d^*, y^*, w^*, h^* and respectively.

(2) Model training. In order to enable the DQN network to better complete path search work and perform action search and decision operations using probabilities ε and $1 - \varepsilon$, the value of ε is set to 1 in this process. Use *epoch* to represent the training cycle. As *epoch* increases, ε will continue to decrease until its value decreases to 0.1.

In order to ensure the accuracy of regression network training, I_OU with a target area and a real area higher than a certain threshold is sent to the regression network for network training operations.

For a tool image, define the entire tool image as the initial candidate area, perform size normalization on it, normalize it into a tool image with size, and then send it to VGG for image feature extraction. Then, action search and decision-making are carried out with probabilities ε and $1 - \varepsilon$. After the action is completed, new image candidate regions and rewards will be obtained accordingly. Then, normalization operations will continue on the newly obtained image candidate regions [11], which will be sent to VGG to obtain the new state. Repeat the above process continuously. When the search step reaches the upper limit or the action is completed, use a regression network to adjust the image candidate regions reasonably to obtain the final tool status monitoring results.

4   Tool Wear Status Maintenance Suggestions

Tool wear refers to the gradual loss of sharpness and functionality of a tool due to factors such as friction, abrasion, and chemical reactions during its usage. Tool wear can lead to various hazards, and here are some common ones:

① Decreased machining quality: Tool wear can result in increased surface roughness, larger dimensional deviations, and irregular shapes. When a tool loses its original sharpness and geometry, it cannot effectively cut the material, leading to burrs, cracks, or other defects. This directly affects the appearance and functionality of the products, reducing machining quality.

② Reduced production efficiency: Worn-out tools require more time and energy to complete the same machining tasks. The increased friction generated during cutting with worn-out tools slows down the machining speed, thus reducing production efficiency. Additionally, worn-out tools need to be replaced or repaired more frequently, leading to increased downtime and maintenance costs.

③ Shortened tool life: Tool wear significantly shortens the lifespan of a tool. When a tool becomes too worn to perform effective cutting, it needs to be replaced. Frequent tool changes not only increase costs but also disrupt production schedules and workflow. Moreover, frequent tool changes pose challenges for tool inventory management.

④ Increased safety hazards: Worn-out tools can generate high temperatures, splattering metal chips, or other hazardous substances during cutting. These substances can cause harm to operators, such as burns, cuts, or eye injuries. Additionally, worn-out tools are prone to breakage or detachment, which can create safety hazards in the workplace.

⑤ Increased costs: Tool wear not only increases the costs of tool replacement and repair but also raises production costs. Ineffective cutting by worn-out tools can result in material waste and extended machining time. Furthermore, worn-out tools require more frequent maintenance and adjustments, adding to operational costs.

In summary, tool wear poses significant hazards to machining quality, production efficiency, tool lifespan, work safety, and costs. Therefore, mitigating the hazards associated with tool wear becomes crucial.

When a tool is used for a period of time, wear is inevitable. How to maintain the good condition of the tool to ensure its good cutting performance and lifespan is a key issue. The following are suggestions for maintaining tool wear status.

① Timely tool replacement: For tools under high pressure, high speed and other processing conditions, timely replacement is very important. Once the tool experiences wear and cracks, it must be replaced in a timely manner. Timely tool replacement can ensure the cutting performance and lifespan of the tool [12, 13].

② Clean the tool: The tool is prone to contamination such as oil and dust during processing, which can lead to increased wear of the tool. So it is necessary to regularly clean the cutting tools, keep them dry and tidy, and use appropriate lubricants to reduce friction and wear.

③ Tool maintenance [14]: Pay attention to the tool maintenance during use, regularly apply anti rust oil and cutting fluid, and do other necessary maintenance work to ensure the normal operation of the tool.

④ Choose the correct cutting conditions: While selecting the machining tool, it is also necessary to choose the correct cutting conditions based on the specific processing conditions, such as cutting speed, feed speed, etc. The correct cutting conditions can not only improve cutting performance, but also extend the service life of the tool [15].

⑤ Shockproof measures: when using high-speed rotating cutting tools, vibration and noise will often occur due to their high-speed operation, which will lead to increased tool wear and reduced processing quality. Effective shockproof measures should be taken to reduce tool wear and ensure machining accuracy.

In short, for enterprises, maintaining the condition of cutting tools is very important. It can significantly improve the efficiency of machine use, reduce production downtime, reduce production costs, and improve the overall production efficiency and competitiveness of the enterprise.

5   Experimental Design and Result Analysis

Set the parameters as shown in Table 1 to construct a deep reinforcement learning model. On the CK6180 large CNC lathe provided by a certain CNC machine tool factory, a simulation experimental environment was established using three way dynamic piezoelectric force measuring equipment, kistler high-precision charge amplifier, acquisition card, and wear measurement microscope to simulate the status monitoring results of 20 1mm titanium alloy cutting tools (Fig. 1 and Table 2).

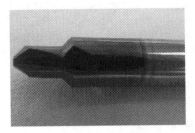

Fig. 1. Schematic diagram of hard alloy cutting tools

Table 1. Experimental setup details

Set Content	Details
Cutting	The cutting speed is 16 m per minute and the cutting depth is 11 μ m
Laser	The diameter of the light spot and the distance between the light spot and the tool tip are set to 0.9 mm and 5 mm respectively. Set the incident angle to 61° and the laser power to 351 W
Tool structure	Set the tip arc to 0.9 mm
Measuring instrument	DVM5000 HD Ultra Depth of Field Microscope
Ultrasonic	The frequency range of ultrasonic vibration is 36 kHz, with an amplitude of 5 μ m

Table 2. Setting of parameters related to deep reinforcement learning

Indicator Name	Specific parameters
Number of network structure layers	4
Number of input layer nodes	16
Number of hidden layer nodes	16
Number of output layer nodes	10
Training frequency	800
Training error	0.0001
Learning rate	0.1
Momentum factor	0.2

5.1 Monitoring Accuracy

According to the preset experimental plan, the overall monitoring accuracy of different methods was compared using a holistic analysis method during the analysis process of the experimental indicators. The specific experimental results are shown in Fig. 2.

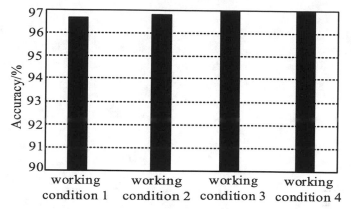

Fig. 2. Experimental Results of Tool Status Monitoring Accuracy

From Fig. 2, it can be determined that from an overall perspective, the wear monitoring accuracy of the methods in the following four working conditions is between 96% and 97%, and the monitoring ability and sensitivity of the methods in the article to tool condition are relatively ideal.

5.2 Test Using the Mean of Determination Coefficient and Absolute Percentage Error

To better demonstrate the application advantages of the method, the determination coefficient R^2 and the mean absolute percentage error $MAPE$ are used as evaluation indicators to demonstrate the universality of the method from different levels. The determination coefficient R^2 represents the fitting level between the output wear value and the actual wear value. The closer the R^2 value approaches 1, the better the fitting level between the two, and the more accurate the monitoring results. The calculation process is as follows:

$$R^2 = 1 - \frac{\sum (\psi_i - \hat{\psi}_i)^2}{\sum (\psi_i - \overline{\psi})^2} \tag{11}$$

Among them, $\overline{\psi}$ represents the mean of all true values, ψ_i is the true value calculated for the i-th time, and $\hat{\psi}_i$ is the monitoring value calculated for the i-th time.

MAPE can effectively highlight the true situation between the wear monitoring value and the actual wear value, and use it as a cost function to statistically analyze the operational performance of the three methods, recorded as:

$$MAPE = \sum_{i=1}^{n_w} \left| \frac{\psi_i - \overline{\psi}}{\psi_i} \right| \times \frac{100}{n_w} \qquad (12)$$

Among them, n_w is the number of calculated iterations.

The number of experiments is set to 700, and the mean coefficient of determination for every 100 experiments is used as the analysis object. The comparison methods are the tool state monitoring method based on deep convolutional automatic encoder proposed in reference [1] (referred to as deep convolutional automatic encoding method) and the tool state monitoring method based on discrete wavelet transform and genetic BP neural network proposed in reference [2] (referred to as discrete wavelet transform and genetic BP neural network method). The comparison results of tool wear output value determination coefficients of the three methods are shown in Fig. 3.

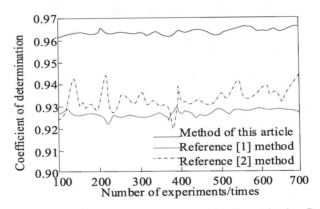

Fig. 3. Comparison Results of tool wear Output Value Determination Coefficient

Observing Fig. 3, it can be seen that under the same experimental operation, the determination coefficient value of the method in this paper is always higher than that of the deep convolutional automatic coding method and the discrete wavelet transform and genetic BP neural network method. The value of the research method is always stable above 0.95, while the curve fluctuation amplitude of the two literature methods is relatively large, indicating that their computational stability is not strong and they are prone to deviation from the true value.

The trend of cost function changes for the three methods is shown in Fig. 4.

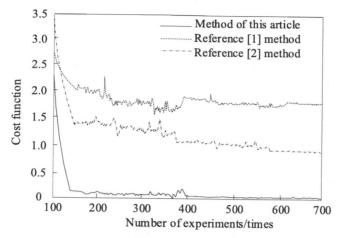

Fig. 4. Comparison of the Trend of Cost Function Changes

According to Fig. 4, as the number of iterations increases, the cost function value gradually decreases, and the mean absolute percentage error between the wear detection value and the actual value decreases, ultimately tending to converge smoothly. The two reference methods may generate gradient dispersion effects during calculation, resulting in high cost function values and slow optimization rates; Compared with this method, this method has the fastest convergence rate, can complete high-quality tool wear monitoring tasks in the shortest time, has strong generalization ability, and has more significant application advantages.

5.3 Application Accuracy Testing of the Method in This Paper Under Different Turning Environments

(1) Conventional turning environment

In the turning environment, the tool wear is measured by the method in this paper when the tool passes through different cutting paths. The results are shown in Table 3.

From the analysis of Table 3, it can be seen that in the conventional turning environment, the measurement results of the tool wear of the method in this paper are highly consistent with the actual value when the tool passes through different cutting paths. The measurement error of the method in this paper is 0.01 mm when the cutting path is 301 m, but the measurement accuracy still meets the requirements of tool wear measurement.

(2) Laser ultrasonic composite ultra precision turning environment

Divide the details of tool wear into four parts: the width, area, perimeter, and depth of the wear range, and then randomly extract the cutting paths in sequence of 151 m, 451 m, 573 m, 301 m, 73 m, and 523 m. The measurement results of this method are shown in Table 4.

Table 3. Measurement results of this method

Cut path/m	Actual value/mm	The measurement value of the method in this article/mm	Error value/mm
73	0.03	0.03	0
151	0.11	0.011	0
223	0.12	0.12	0
501	0.15	0.15	0
573	0.14	0.14	0
451	0.17	0.17	0
523	0.19	0.19	0
301	0.21	0.22	0.01

Table 4. Measurement results of this method in laser ultrasonic composite ultra precision turning environment

Cut path/m	Maximum width of the wear range/mm	Actual value/mm	Maximum area of wear range/	Actual value/mm	Perimeter of wear range/mm	Actual value/mm	Maximum depth in the z-direction of the wear range/mm	Actual value/mm
73 m	0.43	0.43	0.22	0.22	1.95	1.95	0.53	0.53
151 m	0.43	0.43	0.21	0.22	1.94	1.95	0.41	0.421
573 m	0.43	0.43	0.22	0.22	1.96	1.95	0.55	0.53
451 m	0.43	0.43	0.23	0.22	1.95	1.95	0.44	0.42
301 m	0.43	0.43	0.22	0.22	1.92	1.95	0.41	0.41

In the laser ultrasonic composite ultra precision turning environment, the measurement accuracy of the method in this paper for tool wear is 0.99. Therefore, the method in this paper is applicable to the measurement task of laser ultrasonic composite ultra precision turning tool wear.

6 Conclusion

The increasingly popular CNC machine tools have greatly improved processing efficiency and quality, and vigorously promoted the automation process in the industrial production field. However, during mechanical processing, due to factors such as continuous processing at a single station, lagging or overshoot of tool holders at multiple stations, and varying positions of tool tips during tool clamping, the tools used are bound to experience varying degrees of wear.

Once the tool wear reaches a certain degree, it will lead to machine tool failure and even workpiece scrap, and the machining will be interrupted for a long time, which will increase the time cost and production cost. Therefore, in order to ensure the continuous operation of machine tool processing, improve production efficiency and machine tool utilization, and create economic benefits, a tool condition monitoring and maintenance method based on deep reinforcement learning is designed. By using wavelet packet analysis method to process tool signals and extract tool wear features. The regression network is introduced into the Deep reinforcement learning network to learn the mapping relationship between tool wear characteristics and tool status, so as to complete tool status monitoring.

The experimental results of this study indicate that the research method has a high level of accuracy in monitoring tool status, and the monitoring determination coefficient can be maintained above 0.95. The mean absolute percentage error is close to 0, indicating better application performance. According to the experimental results, the proposed method can achieve accurate monitoring of tool status. Future research will further enhance the performance and application scope of tool condition monitoring and maintenance methods, explore their applicability and effectiveness in different fields, and contribute to intelligent manufacturing and industrial automation.

References

1. Yang, G.W., Li, H.K., Zhang, M.L., et al.: Tool condition monitoring method based on ODCAE. J. Vibr. Shock **40**(21), 223–233+274 (2021)
2. Dong, W.H., Hu, Y., Tian, G.J., et al.: Woodworking tool wear condition monitoring based on discrete wavelet transformation and genetic algorithm - BP neural network. J. Central South Univ. Forestry Technol. **41**(06), 157–166 (2021)
3. Wu, Y.: Monitoring cutting tool wear based on spindle current signal multi-feature fusion. Manuf. Technol. Mach. Tool **717**(03), 44–48 (2022)
4. Awaisi, K.S., Abbas, A., Khattak, H.A., et al.: Deep reinforcement learning approach towards a smart parking architecture. Clust. Comput. **26**(1), 255–266 (2023)
5. Zhang, Y., Xie, X., Li, H., et al.: An unsupervised tunnel damage identification method based on convolutional variational auto-encoder and wavelet packet analysis. Sensors **22**(6), 2412 (2022)
6. Yang, J., Zhou, C.: A fault feature extraction method based on LMD and wavelet packet denoising. Coatings **12**(2), 156 (2022)
7. D'Amato, V., D'Ecclesia, R., Levantesi, S.: ESG score prediction through random forest algorithm. CMS **19**(2), 347–373 (2022)
8. Yu, Z., Wang, Z., Jiang, Q., et al.: Analysis of factors of productivity of tight conglomerate reservoirs based on random forest algorithm. ACS Omega **7**(23), 20390–20404 (2022)
9. Mohamed, A., Hassan, M., M'Saoubi, R., et al.: Tool condition monitoring for high-performance machining systems—a review. Sensors **22**(6), 2206 (2022)
10. Patange, A.D., Jegadeeshwaran, R., Bajaj, N.S., et al.: Application of machine learning for tool condition monitoring in turning. Sound Vibr. **56**, 127–145 (2022)
11. Liu, Y., Guo, L., Gao, H., et al.: Machine vision based condition monitoring and fault diagnosis of machine tools using information from machined surface texture: a review. Mech. Syst. Sig. Process. **164**, 108068 (2022)
12. Gao, Z., Hu, Q., Xu, X.: Condition monitoring and life prediction of the turning tool based on extreme learning machine and transfer learning. Neural Comput. Appl. **34**(5), 3399–3410 (2022)

13. Wu, Y., Ma, X.: A hybrid LSTM-KLD approach to condition monitoring of operational wind turbines. Renew. Energy **181**, 554–566 (2022)
14. Wang, C.A., Wang, H.X., Huang, Z.X.: Simulation analysis on influencing factors of tool break in copper micro-milling. Comput. Simul. **39**(04), 201–204+217 (2022)
15. Cheng, M., Jiao, L., Yan, P., et al.: Intelligent tool wear monitoring and multi-step prediction based on deep learning model. J. Manuf. Syst. **62**, 286–300 (2022)

Research on Control of Virtual and Real Drive System of Intelligent Factory Robot Based on Digital Twin

Yong Ge$^{(\boxtimes)}$, Yechao Shen, and Zhihong Wang

School of Electrical Engineering, Anhui Technical College of Mechanical and Electrical Engineering, Wuhu 241002, China
geyong7894@163.com

Abstract. Traditional virtual and real robot drive systems use management methods, but due to the impact of the production environment, the system's command response speed is slow, which affects the control effect. To address this issue, this article proposes a control method for the virtual and actual drive system of intelligent factory robots based on digital twins. This method achieves high-speed interaction of the main control data of the driving system by setting up a data interaction scenario between the virtual and actual driving systems, and controlling the connections between the logical units and various modules. At the same time, by constructing control models of virtual and actual driving systems for robots, and combining virtual and actual driving interaction scenarios, the position and attitude conversion data of intelligent engineering robots are obtained. When sending instructions, smooth switching mode is used to control the virtual and actual driving systems of intelligent factory robots to achieve fast and dynamic response, and combined with smooth switching function to ensure the stable performance of the overall driving system. Through comparative experiments, the superiority of the driving system control method has been proven, which can be applied in practical life.

Keywords: Digital twins · Smart factory · Robot · Virtual reality driven · Control methods

1 Introduction

In the virtual and real driving process of the robot, because the robot and the driver are in different physical spaces, and the working environment of the robot is a dynamic and unstructured environment, there are certain safety hazards and operational risks [1] in the virtual and real driving process of the robot. In addition, most of the existing virtual and real driving technologies of robots adopt the master-slave control mode. The operation process is cumbersome and requires the driver to have some professional knowledge, so the human-machine interaction is poor. How to improve the safety of robot virtual and real driving and how to improve the human-machine interaction of

L. Yun et al. (Eds.): ADHIP 2023, LNICST 547, pp. 29–45, 2024.
https://doi.org/10.1007/978-3-031-50543-0_3

robot virtual and real driving are the technical problems that need to be solved urgently. With the development of workshop digitalization, the demand for interconnection and intercommunication of various equipment in the manufacturing workshop continues to increase, so industrial systems, Internet of Things data analysis, data interaction and other technologies are gradually applied in modern manufacturing workshops [2]. At present, SCADAS is widely used in industrial enterprises. Through SCADAS, the operation status of equipment in the workshop can be monitored and the workshop can be feedback controlled interactively with MES. Now, due to the difference between the workshop equipment suppliers and the automation level, the communication protocols between these equipment are usually unable to communicate, which leads to the difficulty of workshop data interaction. For the data interaction of workshop equipment, in addition to analyzing and accessing the communication protocol used by the respective equipment, the monitoring and control of workshop equipment can also be realized by the way of embedded equipment collecting control signals.

Digital twin technology provides a new idea for virtual and real robot driving. The digital twin technology is used to establish the digital twin model of the robot and its working scene at the remote end. The communication link is used to connect the digital twin model at the remote end with the robot at the local end. The driver plans the movement path and posture of the robot at the local end through human-computer interaction equipment at the remote end, the local sensor feeds back the real-time status of the robot and its working scene to the driver, realizing remote control of the robot and real-time monitoring of the changes in the working scene [3]. The application of digital twin technology in virtual and real robot drive can overcome the restrictions of environment and space on robot application. Robots can be used in harsh environments such as high temperature, high radiation or inaccessible environments, instead of human beings to perform dangerous and complex driving tasks, improve driving safety and greatly expand the driving space [4].

Therefore, this paper uses digital twin technology to design the control method of virtual and real drive system of intelligent factory robot, and uses virtual and real drive interaction technology and target detection technology to plan the path and posture of robot, effectively improving the interactivity and safety of robot drive. This paper investigates the contradictory relationship between the dynamic response and steady-state performance of the controller, and proposes a new controller design method to solve this problem through smooth switching function. This method can use a signal controller to achieve fast tracking when the error is large, and an energy controller to ensure steady-state performance when the error is small. At the same time, the paper also introduces the advantages of sliding mode Variable structure control algorithm, and solves the steady-state chattering problem of sliding mode control by introducing smooth switching control. This study is innovative in improving the rapid response ability and steady-state performance of the system.

2 Design of Control Method for Virtual and Real Drive System of Intelligent Factory Robot Based on Digital Twin

2.1 Build Data Interaction Scenarios of Virtual and Real Driving Systems

This paper controls the connection between the logic unit and each module of the virtual and real drive system to ensure the high-speed interaction of the main control data of the drive system. At present, the main control virtual and real drive modes of common virtual and real drive system implementation schemes mainly include the following three types: FPGA virtual and real drive mode, which is a connection mode that can be used to configure the logic units and modules inside the virtual and real drive mode through Verilog HDL hardware description language to achieve custom configuration of circuit logic [5]. DSP virtual and real driving mode is a virtual and real driving mode specially designed for digital signal processing, which is mainly used to analyze, transform, filter, detect, modulate and demodulate the collected data. ARM virtual real drive mode is a microprocessor based on RSIC (Reduced Instruction Set), which reduces the difficulty of hardware design by using RSIC. In addition, the Cortex architecture series of ARM processors have rich interfaces suitable for general industrial control scenarios. Among them, FPGA is the most closely connected with hardware, and the virtual and real drive system using FPGA as the main control can achieve ultra-high speed data interaction. However, because of this feature, FPGA has no good versatility, and the flexibility of program design is poor. As DSP is a virtual and real driving mode designed for signal processing, it has strong data processing capability but needs highly optimized code, so it can only be developed using the development tools provided by the manufacturer [6]. In order to improve the universality of the system and reduce the difficulty of hardware design, this paper selects the ARM processor of Cartex architecture series as the main control virtual and real driving mode of the virtual and real driving system. In view of the difficulty of expansion of the existing virtual reality drive system, this paper has built a virtual reality drive interaction scenario, as shown in Fig. 1 below.

As shown in Fig. 1, the virtual reality driven interaction scenario built in this paper is divided into three layers, namely the perception layer, the network layer, and the application layer. The perception layer is responsible for collecting the signals of intelligent factory equipment, converting the physical signals into digital signals, transmitting them to the data processing gateway, and receiving data from the data interaction gateway to drive the equipment at the corresponding point [7]. The network layer is mainly responsible for the transmission and interaction of data between the physical entity world data of the data interaction system and the information simulation world. The application layer mainly analyzes the collected data, generates and issues decisions. The whole drive process takes ARM drive as the core of data processing and transmission, and uses a variety of point expansion modules to enhance the point where the system can collect data. In order to meet the needs of distributed drive management and control, WIFI chips and remote modules are added to realize a wide range of intelligent factory data interaction through wireless network communication. The system communicates with the upper computer through Ethernet to transmit data, upload the collected data and receive the control instructions of the upper computer. In the drive system, the network layer and the application layer communicate through Ethernet, and the server and client only

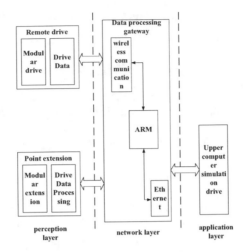

Fig. 1. Schematic diagram of virtual reality driven interaction scene

communicate with each other. You can make full use of the hardware performance at both ends of the architecture by reasonably allocating data processing tasks to the server and client. The front end is only responsible for displaying and transmitting request information through the browser, and the main system logic is from the server side. This architecture greatly increases the compatibility of the system, and the upper computer only needs to install a browser to realize the client. However, this architecture has higher requirements for server performance [8] because it places business processing logic in the server. Because in the virtual and real drive system designed in this paper, the data processing gateway is used as the server, and the ARM driver of Cortex architecture series is used as the master drive mode for the purpose of reducing costs and facilitating use, the performance is limited. Moreover, the purpose of this virtual and real driving system design is to collect and drive physical equipment, and the decision-making level is not the content of this system design. In this paper, the input circuit and output circuit are designed at one point, and then external switches or data latches are used to ensure that only one circuit works normally at the same time. By flexibly adjusting the functions of the virtual and real drive system, the data interaction of the production smart factory is realized.

2.2 Building Control Model of Virtual and Real Driving System of Robot Based on Digital Twins

In this paper, the virtual reality driven interactive scene is used to obtain the position and pose conversion data of intelligent engineering robots, which makes the whole control process more reasonable. In order to improve the human-machine interaction and safety of virtual and real driving robot, this paper designs a robot remote teaching system based on digital twin technology. First, the virtual robot pose is planned using the virtual and real driving method of robot augmented reality based on object detection. In the virtual and real driving digital twin model of robot, whether the virtual robot collides and interferes with the working scene or personnel is detected. If there is no collision and interference, the robot control command is generated and sent to the physical robot to control the movement of the physical robot [9]. The digital twin model is shown in Fig. 2 below.

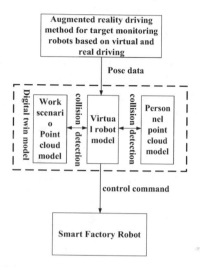

Fig. 2. Schematic diagram of digital twin control model

As shown in Fig. 2, the framework is mainly composed of physical unit and digital twin unit. The physical unit is composed of physical robot and working scene, robot control cabinet and RGB-D sensor, etc. The digital twin unit includes robot digital model unit and robot virtual and real drive unit. The physical unit uses the RGB-D sensor to obtain the geometric parameters and RGB-D images of the robot and its working scene, and directly reads the joint data of the robot in the current position and posture through ROS, and sends the acquired real-time data to the digital model unit; The digital model unit analyzes and processes the obtained geometric data, joint data, RGB-D image data, position and posture data, and establishes the digital twin model of the robot and its working scene. The virtual and real driving unit of the robot presents the pose data, twin model and collision detection results in real time in the computer, providing visual basis for the teaching staff to plan the pose of the robot [10]. The teaching staff observes in real time through the teaching platform, and uses the virtual reality drive of the object detection robot proposed in this paper to plan the virtual robot pose, sends the planned pose data to the collision detection model of the digital model unit, detects the collision interference between the robot and the object or person in the working scene during the teaching process, and generates control commands when there is no collision interference, control physical robot movement through ROS. According to the motion of the robot, the axial acceleration, angular velocity and other parameters of the robot are analyzed, and the formula is as follows:

$$\alpha(t) = \alpha \tag{1}$$

$$\varphi(t) = \frac{1}{2}\alpha t^2 \tag{2}$$

$$\omega(t) = -\alpha t + \alpha t^2 \tag{3}$$

In Eqs. (1–3), $\alpha(t)$ Is the angular acceleration; α Is a constant value; $\varphi(t)$ Is the angle; t Is the time; $\omega(t)$ Is the angular velocity. Segmental angular acceleration $\alpha(t)$ Is a constant value. Angular velocity $\omega(t)$ In the first section, the neutral line increases to the maximum value, and then in the second section, it decreases linearly to the static state. Rotation angle passed in section I and II $\varphi(t)$ Increase by parabola function. This type of running contour can achieve the shortest positioning time. By the preset ending angle φ_{Max} And the corresponding time point t can calculate the required angular acceleration constant or angular deceleration constant. For the sake of simplicity, the temporary transition phase that accelerates the formation and elimination of angular impact will not be considered [11]. The analysis shows that the ending angle is known φ_{Max} And the corresponding time t, the required angular acceleration or angular deceleration can be calculated. The maximum action range and maximum speed of the robot are shown in Table 1 below.

Table 1. Maximum Action Range and Maximum Speed of Robot

project	Maximum action range	Maximum speed (rad/s)
J1	$\pm 180°$	3.20
J2	$-60°{-}120°$	3.08
J3	$\pm 180°$	5.23
J4	$-60°{-}80°$	4.15
J5	$\pm 180°$	6.54
J6	$\pm 135°$	6.54
J7	$\pm 360°$	10.47

As shown in Table 1, $\varphi_{Max} = 180° = \pi$, $\omega_{Max} = 3.20$ rad/s, it is known that the angular velocity curve has the same area in two stages, then:

$$\varphi_{Max} = 2\left[\frac{\alpha}{2}\left(\frac{t}{2}\right)^2\right] \tag{4}$$

In Eq. (4), φ_{Max} Is the maximum angle range of the end angle. The drive model of intelligent factory robot is simplified. The length of the connecting rod axes J1, J2, J3, J4, J5, J6, J7 is ln, the radius is rn, and the mass is mn. There is a servo motor and driver at each node. The rotation direction of each shaft is as follows: J1, J7 and J4 rotate around the shaft itself, J2, J3, J5 and J6 rotate around the mechanical arm J2, J3, J5 and J6 respectively, and the angle of Ji is θi, the counterclockwise rotation direction is the positive direction, and the horizontal included angle of the robot drive is:

$$\theta_3' = \frac{\pi}{2} - \theta_2 - \theta_3 \cos\theta_7 \tag{5}$$

$$\theta_5' = \theta_3' - \theta_5 \cos\theta_4 \tag{6}$$

In formulas (5, 6), θ_3' Is the angle between J3 and the horizontal line; θ_2 Is the corner of J2; θ_3 Is the corner of J3; θ_7 Is the corner of J7; θ_5' Is the angle between J5 and the horizontal line; θ_5 Is the corner of J5; θ_4 It is the corner of J4. In this paper, the robot drive is divided into different modules, and different robots are controlled by different drive systems. Digital twin modeling is carried out in the No. 2 drive system. The RGB D image of the working scene is obtained by using the RGB D sensor for 3D reconstruction. The conversion matrix between the RGB-D sensor coordinate system and the robot base mark is obtained by hand eye calibration [12]. The robot virtual model is fused with the point cloud model of the working scene to complete the digital twin modeling. Build an augmented reality drive platform in No. 1 drive system. The RGB sensor acquires the RGB image of the working scene, completes the augmented reality registration according to the conversion matrix between the RGB sensor coordinate system and the robot base, and superimposes the virtual robot onto the actual working scene. In the No. 1 driving system, the robot augmented reality driving method of object detection proposed in this paper is used to plan the position and pose of the virtual robot. The planned pose data is sent to the No. 2 drive system for collision detection, including collision detection between the robot and the work scene based on the octree model, and human-machine minimum distance detection based on human skeleton node recognition. The pose data without collision interference will be generated into robot operation instructions to control the physical robot to execute driving tasks.

2.3 Control the Smooth Switching Mode of Virtual and Real Driving of Intelligent Factory Robots

With the help of digital twin technology, this paper controls the robot to stably complete the task of the drive system, and controls the smooth switching mode, so that the robot can quickly respond dynamically while the virtual and real drive systems issue commands, and combines the smooth switching function to ensure the overall stable performance of the robot drive. The dynamic response and steady state performance of a single controller are often contradictory, and the controller with fast dynamic response is often accompanied by poor steady state performance [13], and vice versa. In this paper, the controller is designed from the perspectives of signal and energy, and the smooth switching function is designed, so that the robot control system can use the signal controller to track quickly when the error is large, and the energy controller to ensure the steady state performance when the error is small. The smooth switching function plays a role in smooth transition. At the same time, it has the rapidity of signal control and the stability of energy control. Sliding mode variable structure control is a typical signal control with fast response because of its simple algorithm and fast response speed. A new sliding mode control is proposed in this paper, which further improves the fast response of the system [14]. The chattering problem of sliding mode control in steady state is always difficult to solve, and smooth switching control also provides another solution to eliminate chattering. Hamiltonian control studies the variation of system energy. After the system reaches steady state, the energy dissipation is given priority, which can make the system have good steady state performance. The smooth switching control scheme is shown in Fig. 3 below.

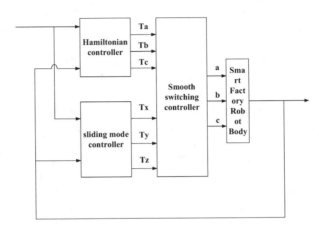

Fig. 3. Schematic diagram of smooth switching control scheme

As shown in Fig. 3, Ta, Tb and Tc are the output commands obtained by the Hamilton controller; Tx, Ty and Tz are the output commands obtained by designing the sliding mode controller from the perspective of signal; a, b and c are the input torque of the robot. Because of its simple algorithm structure, fast dynamic response, strong robustness to unknown disturbances and parameter deviations, sliding mode control improves the control effect of the system through the approach law. Exponential approach law, power approach law and double power approach law are important control indicators. The double power approach law is described as:

$$S = -k_1|s|^\alpha \text{sgn}s - k_2|s|^\beta \text{sgn}s \tag{7}$$

In Eq. (7), S Is the law of double power approach; $k_1 \sim k_2$ Is the distance between the sliding surface and the driving system; sgn$s$ Is the symbolic function of the sliding surface; β Is a constant value. In this paper, the scale function is introduced σ, when σ when the value is different [15], the control effect of the drive system is different. The smooth switching function is expressed as:

$$Q(s) = \begin{cases} \exp(-\left(\frac{|S|-1}{\sigma}\right)^2), & |S| > 1 \\ 1, & |S| \leq 1 \end{cases} \tag{8}$$

In Eq. (8), $Q(s)$ It is a smooth switching function. The design of virtual and real drive system of intelligent factory robot is similar to that of permanent magnet synchronous motor, and the equivalent magnetic circuit method is generally used to analyze and design the motor in engineering. The equivalent magnetic circuit method is simple and can meet the design requirements after several iterations. However, in the analysis and processing, the permanent magnet should be considered as a magnetic source to participate in the calculation of the equivalent magnetic circuit. At the same time, due to the diversity of the rotor magnetic circuit structure and the complexity of the magnetic field distribution, it is difficult to describe the real magnetic field situation only by relying on a few concentrated parameters. Therefore, the magnetic circuit calculation can only be used to estimate the initial scheme and compare similar structures, and cannot obtain some key coefficients such as magnetic leakage coefficient, pole arc coefficient, etc. In order to improve the accuracy of the design, this paper uses the field circuit combination method to design the virtual and real drive system of intelligent factory robots. The field circuit combination method is to use the electromagnetic field numerical calculation (finite element method) to obtain the magnetic leakage coefficient, calculate the pole arc coefficient and other parameters that cannot be obtained in the calculation of the magnetic circuit method, and then combine the magnetic leakage coefficient, calculate the pole arc coefficient and other parameters into the equivalent magnetic circuit to modify, improve the accuracy of the calculation results and reduce the dependence on empirical data, therefore, the field circuit combination method has the advantages of the equivalent magnetic circuit method and the finite element method, and can effectively combine the magnetic field and magnetic circuit. During the operation of the robot drive system, there are not only the input and output of signals, but also the increase and decrease of mechanical energy. Therefore, PMSM and robot can be regarded as the devices for signal transformation and energy transformation. Based on the point of view of signal transformation, the controlled object is regarded as a signal transformation device that converts the input signal into the output signal. The control principle is to quickly converge the error signal to 0. The controller based on this control concept is called the signal controller. The PCH control method studies the relationship between the energy transformation of the system and the change of mechanical energy from the perspective of the energy transformation of the system. Its control principle is that the total energy of the digital twin technology makes the system meet the requirements of control objectives through damping injection and other rules. The controller based on this control concept is called energy controller. After being controlled, the virtual and real drive system can detect signals in time, quickly adjust the position and speed of the robot, and has good dynamic performance. The drive system management and control method based on digital twin technology can minimize the loss of the system, realize energy optimization control when the system is in steady state, and have good steady state indicators. In this paper, feedback linearization control is used to control the virtual and real drive systems. The key point is to accurately convert the nonlinear system into a linear system using the existing methods, so that the linear system method can be used to solve the control problems faced by the nonlinear system.

3 Experiment

In order to verify whether the control method of the drive system designed in this paper has practical value, this paper conducts an experimental analysis of the above methods. The conventional control method of virtual and real drive system of intelligent factory robots based on point cloud model, the conventional control method of virtual and real drive system of intelligent factory robots based on convolutional neural network, and the control method of virtual and real drive system of intelligent factory robots based on digital twins designed in this paper are used to control robots and select the best control scheme. The specific experimental preparation process and the final experimental results are shown below.

3.1 Experiment Preparation

In the virtual reality drive system of the robot, the data processing gateway is mainly used as the server to receive and send data, and can access additional point signals through wired and wireless ways. The data processing module is based on the independent control chip to expand the interface of the gateway to achieve the purpose of accessing data to the virtual and real driving system of the robot. The main function of the driver module is to improve the driving ability of the signal expansion board to drive high-power equipment, and play an isolation role to prevent high-power equipment from burning the data processing module. The remote terminal is a client that transmits and receives data with the gateway through wireless communication technology, and also adds a wired communication module, which can further increase the number of access points by connecting the expansion board card. In industrial equipment, the most important thing to access control is PLC, whose power supply is 24 V. In order to facilitate the installation of the gateway, a voltage conversion module is designed.24 V direct power supply is used to realize voltage conversion through 24 V–5 V step-down circuit. This paper selects XL2596S-ADJE1 as the voltage conversion chip, which can set the voltage output through the voltage stabilizing feedback circuit to improve the stable 5V3A output, which is the most recommended power supply mode for the data processing gateway. The voltage conversion chip has a control pin (ON/OFF) through which the output of the chip can be controlled. Therefore, the design adds a normally closed switch on this pin to enable it to provide a switching function in addition to the 24 V–5 V function, which facilitates the quick restart of the data processing gateway when the gateway fails to work normally due to errors. Under the condition of normal operation of the virtual and real drive system, this paper configures the parameters of the experimental environment, as shown in Table 2 below.

Table 2. Parameters of Experimental Environment

name	Model/version
Hardware	
operating platform	Ubuntu 18.04
CPU	Intel(R)xeon(R) CPU E5-2650 V4@2.20 GHz
GPU	TITAN xp 1288Mi B
Graphics card	NVIDIA-SMI432.00
Software	
CUDA	10.2
CUDNN	7.6.5
Pytorch	1.8.0
Pycharm	2018.3.2
Python	3.7

As shown in Table 2, this experiment uses the server training and experimental yolov5 model, and uses the Pytorch deep learning framework to train and experiment the yolov5 model. Convert the weight file obtained from the deep learning framework training, and then call it through the DNN module of Opencv to ensure that the control scene of the entire virtual and real robot drive system is more urgent to meet the actual industrial needs. The virtual reality driven data interaction system aims to comprehensively collect the data of the equipment in the production workshop and realize the digital abstraction process of the workshop. The control signal transmission between workshop equipment is electrical signal, and the current is mostly when the equipment transmits analog signal. Due to the complex working environment of the workshop, there are a large number of high-voltage lines on the site, and the voltage of various electromagnetic sensors is vulnerable to interference, which leads to the distortion of the collected data and affects the operation of the equipment. Using current signal as analog control signal can better ensure the accuracy of control. The range of equipment signals is variable. The range of electrical signals output by different equipment produced by different manufacturers is different, and their corresponding physical meanings are also different. Common ranges are 0–5 V, 0–10 V, 0–20 mA, etc. If you want to collect data from devices with different ranges, you must also use those that support various ranges. The number of equipment is large and scattered. With the increase of digitalization of manufacturing workshops, the number of sensors is increasing. It is necessary to collect physical quantities of many points at the same time. And because the processing equipment such as machine tools, robots, etc. are dangerous, the equipment must be installed with a suitable safety distance, which leads to the distribution of the data to be collected in the whole workshop is far away. This paper takes the intelligent factory inspection robot as an example to analyze its structure, as shown in Fig. 4 below.

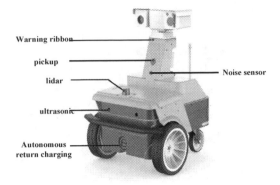

Warning ribbon

pickup

lidar

Noise sensor

ultrasonic

Autonomous
return charging

Fig. 4. Structure diagram of intelligent factory patrol robot

As shown in Fig. 4, the structure of the patrol robot includes not only basic components such as warning ribbons, pickups, laser radars, ultrasound, autonomous return charging, noise sensors, but also components such as pan tilt, gas detectors, antennas, light strips, and interface panels. In the process of robot driving, there are relatively many virtual and real signals, and there should be enough digital and analog acquisition channels, at least 64 digital channels and 8 analog channels, to ensure the data interaction effect of the driving system. The digital channel of the data interaction system needs to support two modes of high-level trigger and low-level trigger. The analog quantity channel of data interaction system supports 0–5 V and 0–20 mA analog quantity. In the virtual reality drive system of the robot, the data processing gateway is used as the server, and the ARM processor of Cortex architecture series is used as the main control chip for the purpose of reducing cost and facilitating use, so the performance is limited. Moreover, the virtual and real driving system of the robot is designed to collect physical equipment and drive physical equipment, and the decision-making level is not the design content of the system. In this way, the C/S architecture puts the business processing logic on the client side, which more meets the design goal of the system. Because the C/S architecture only needs to transmit interactive data, when the amount of data is large, the communication pressure of using the C/S architecture is less, which can ensure the real-time performance of the system data. Therefore, this paper chooses to use C/S structure as the data interaction structure between system layers, and uses B/S structure to assist in monitoring the working state of wireless terminals. In this paper, during the driving process of the intelligent factory inspection robot, the camera imaging is used to transform with the coordinate system. The conversion relationship is shown in Fig. 5 below.

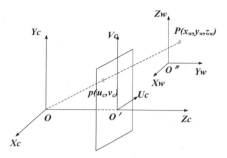

Fig. 5. Schematic diagram of the relationship between imaging and coordinate system conversion

As shown in Fig. 5, O is the camera concern for robot imaging; O′ is the center of the image plane; O″ is the center of robot base coordinate system; P (uc, vc) is the actual position coordinate of the robot drive; P (xw, yw, zw) is the three-dimensional coordinate driven by the robot after the conversion of the base coordinate system. The virtual and real driving technology of industrial robots is different from the traditional servo system. It does not need to use mechanical transmission components such as belts (pulleys), gears or cams. External loads can be directly connected with torque direct drive motors. The use of virtual and real driving technology has many advantages, which can improve the accuracy and repeatability. Compared with a "high-precision" planetary gear, the repetition error of a standard virtual reality driven rotating motor is less than 1 arc second. The backlash of the planetary gear can reach 1 min of arc, which may cause the load to move 1 min of arc for the absolutely stable drive motor. Therefore, the precision of direct drive motor structure is 60 times higher than that of traditional motor structure. After the virtual and real driving technology improves the precision, the machine can produce higher quality products. While the end actuator of the virtual robot model follows the attitude teaching device, the attitude teaching device acquires the attitude of the attitude teaching device in the northeast sky coordinate system through the inertial sensor module, transmits the attitude data generated by the attitude teaching device to the computer through the Bluetooth module, and controls the attitude of the robot end actuator, the teaching staff adjusts the posture of the virtual robot terminal by adjusting the posture of the posture teaching device, and observes and adjusts in real time through the AR visual interface to achieve the expected posture.

3.2 Experimental Results

Under the above experimental conditions, this paper makes the intelligent factory robot move to any position in the factory, and records the required and actual coordinates of the robot. Compare the coordinates of the conventional control method of virtual and real drive system of intelligent factory robots based on point cloud model, the coordinates of the conventional control method of virtual and real drive system of intelligent factory robots based on convolutional neural network, and the coordinates of the control method of virtual and real drive system of intelligent factory robots based on digital twins designed in this paper. The closer to the reference coordinates, the better the control effect of robot virtual and real drive system. The experimental results are shown in Table 3 below.

Table 3. Experimental Results

Drive times	Coordinate/mm	Coordinates/mm of conventional control method for virtual and real drive system of intelligent factory robot based on point cloud model	Coordinates of conventional control method for virtual and real drive system of intelligent factory robot based on convolutional neural network/mm	Coordinates/mm of control method for virtual and real drive system of intelligent factory robot based on digital twin designed in this paper
1	reference resources	(262.0, 313.6, 687.0)		
	measure	(262.5, 313.8, 685.0)	(261.0, 312.6, 687.8)	(262.0, 313.6, 687.0)
2	reference resources	(383.4, −260.6, 323.1)		
	measure	(381.4, −265.6, 326.1)	(382.4, −260.5, 321.1)	(383.4, −260.6, 323.1)
3	reference resources	(314.4, −386.8, 641.1)		
	measure	(318.4, −387.8, 646.1)	(312.4, −384.8, 641.6)	(314.4, −386.8, 641.1)
4	reference resources	(379.5, −268.3, 254.8)		
	measure	(379.1, −268.0, 249.8)	(376.5, −268.8, 252.8)	(379.5,-268.3,254.8)
5	reference resources	(469.0, −23.4, 441.4)		
	measure	(469.9, −28.4, 440.4)	(470.8, −25.4, 441.4)	(469.0,-23.4,441.4)

As shown in Table 3, the smaller the difference between the robot driving position coordinates and the actual position coordinates, the better the control effect will be in the process of virtual and real driving control of the robot. If other conditions are consistent, after using the conventional control method of virtual and real drive system of intelligent factory robot based on point cloud model, the reference coordinate of the robot is (262.0313.6687.0), and the actual position coordinate of the robot is (262.5313.8685.0). In this three-dimensional coordinate, there is a difference of 2.0 mm from the reference coordinate. The error between the actual position coordinates and the reference position coordinates of the other four times is about ±5.0 mm, which shows that the control effect of the robot is poor and cannot meet the control requirements of the virtual and real driving system of the robot. After using the conventional control method of virtual and real drive system of intelligent factory robot based on convolutional neural network, the actual position coordinate of the first drive was (261.0312.6687.8), which had errors (1.5, 1.2, - 2.8) with the reference position coordinate. The other four drive coordinates also had errors of varying degrees with the reference position coordinate, which affected the robot control effect and needed further improvement. However, after using the virtual and real drive system control method of intelligent factory robot based on digital twins designed in this paper, the actual position coordinate of the robot driven for the first time is (262.0313.6687.0), which is consistent with the reference position coordinate in height. The other four drives are the same as the reference position coordinate, which can ensure the control effect of the robot, it meets the control requirements of the virtual and real drive system of intelligent factory robots.

Further analyzing the performance of the method proposed in this article, the two datasets were randomly divided into three groups, with the driving control speed as the comparison indicator. The test results of the three methods are compared as shown in Table 4

Table 4. Drive Control Speed/ms

group	Control Method for Virtual Reality Drive System of Intelligent Factory Robot Based on Point Cloud Model	Control Method of Virtual and Real Drive System for Intelligent Factory Robot based on Convolutional Neural Network	The virtual and real drive system of the intelligent factory robot designed based on the digital twin brothers
1	37	57	12
2	39	59	13
3	45	62	11
4	55	66	14
5	56	68	12

According to the analysis of Table 4, the space segmentation speed of the method in this paper is the fastest, with an average of 12.83 ms, while the driving control speed of the control method of the virtual reality drive system of intelligent factory robots based on the point cloud model and the control method of the virtual and real drive systems of intelligent factory robots based on Convolutional neural network are slower, with an average of 48.33 ms and 62.83 ms, respectively. This shows that the method in this paper is applied to drive control, It can achieve efficient management and control of industrial robots, which is conducive to rapid production.

4 Conclusion

Since the birth of the first generation of robots, robots have been widely used in machinery manufacturing, automobile and ship manufacturing, electronic and electrical processing industry, rubber and other raw material processing industry, medicine and food processing industry and other fields. At present, robots have been widely used in welding and grinding, machining and assembly of mechanical parts and other fields. Most of the traditional robot driving programming methods use the drive box to import the robot running program or manually control the robot to perform the driving task. Although this robot driving method can make the driver visually observe and adjust the robot driving posture, the driving efficiency is low. In the actual production environment, because the driver and the robot are not necessarily in the same physical space, and are restricted by the uncertainty and complexity of the working environment, the traditional robot cannot complete the driving task independently; In addition, due to the poor environmental awareness of existing intelligent robots, there are potential safety hazards in the process of autonomous task execution, which makes it difficult to achieve safe and reliable autonomous programming of robots. Therefore, it is necessary for drivers to monitor the status of robots and changes in their working scenes in real time through devices such as vision sensors, and remotely control the robots to execute driving tasks. Therefore, this paper uses digital twin technology to design the control method of virtual and real drive system of intelligent factory robots. From the driving scene, control model, control switching mode and other aspects, we can better drive the robot and truly improve the control effect of the robot.

References

1. Kl, A., Pr, B., Csk, C., et al.: Real time exploited BLDC motor drive BLDC motor drive in lab view virtual instrumentation environment. Mater. Today Proc. (3) (2021)
2. Long, D., Xu, R., Liu, J., et al.: Enterprise service remote assistance guidance system based on digital twin drive. Mob. Inf. Syst. **2021**(8), 1–9 (2021)
3. Visnic, B.: Real-time processors help drive the zonal E/E REVOLUTION. Automot. Eng. **6**, 9 (2022)
4. Wang, P., Luo, M.: A digital twin-based big data virtual and real fusion learning reference framework supported by industrial internet towards smart manufacturing. J. Manuf. Syst. **58**(1), 16–32 (2021)
5. Saito, K., Akagi, H.: A real-time real-power emulator for a medium-voltage high-speed electrical drive: discussion on mechanical vibrations. IEEE Trans. Ind. Appl. 1 (2021)
6. Jiang, Y., Li, M., Guo, D., et al.: Digital twin-enabled smart modular integrated construction system for on-site assembly. Comput. Ind. **136**(6), 103594 (2022)
7. Fan, L., Li, H.J.: Research on optimization of frequency conversion harmonic based on three-phase power supply. Comput. Simul. **40**(1), 6 (2023)
8. Qing, Z., Liguo, Z., Yulin, D., et al.: From real 3D modeling to digital twin modeling. Acta Geodaetica et Cartographica Sinica **51**(6), 1040–1049 (2022)
9. Yang, F., Feng, T., Xu, F., et al.: Collaborative clustering parallel reinforcement learning for edge-cloud digital twins manufacturing system. China Commun. **19**(08), 138–148 (2022)
10. Semeraro, C., Lezoche, M., Panetto, H., et al.: Digital twin paradigm: a systematic literature review. Comput. Ind. **130**, 103469 (2021)
11. Aheleroff, S., Xu, X., Zhong, R.Y., et al.: Digital twin as a service (DTaaS) in industry 4.0: an architecture reference model. Adv. Eng. Inform. **47**, 101225 (2021)
12. Bhatti, G., Mohan, H., Singh, R.R.: Towards the future of smart electric vehicles: digital twin technology. Renew. Sustain. Energy Rev. **141**, 110801 (2021)
13. Jafari, M., Kavousi-Fard, A., Chen, T., et al.: A review on digital twin technology in smart grid, transportation system and smart city: challenges and future. IEEE Access (2023)

14. Sasikumar, A., Vairavasundaram, S., Kotecha, K., et al.: Blockchain-based trust mechanism for digital twin empowered industrial internet of things. Futur. Gener. Comput. Syst. **141**, 16–27 (2023)
15. Uhl, A., Schmidt, A., Hlawitschka, M.W., et al.: Autonomous liquid-liquid extraction operation in biologics manufacturing with aid of a digital twin including process analytical technology. Processes **11**(2), 553 (2023)

Information Check and Control System of Substation Telemotor Based on Computer Vision

Wenwei Li[✉]

Qinzhou Power Supply Bureau of Guangxi Power Grid Co., Ltd., Qinzhou 535019, China
zhangyongwei942@163.com

Abstract. Traditional verification methods require a large number of telecontrol signals as the basis, and it is difficult to detect the addresses of all regulatory information objects, resulting in inaccurate verification control results. For this purpose, this study designed a computer vision based substation remote motor information verification control system. In the hardware part of the system, a verification view module is constructed to present the data defined in the data model, a verification signature module is constructed to achieve information transfer between the verification side and the scheduling management side, and a verification network scheduling module is constructed to correct the monitoring point information on the work side. In the system software section, computer vision technology is used to identify the remote motor information of the substation. At the same time, encryption technology is introduced to prevent unauthorized users from stealing important information tables. Finally, design a scheduling master station simulation system to implement the motion machine information verification process and generate verification reports. According to the test results, it can be seen that the information configuration of the system is reasonable and the obtained information is valid, indicating that the effectiveness of the verification control is good.

Keywords: Computer vision · Substation remote motor · Information verification · Control system

1 Introduction

China's power grid companies have gradually started to implement the pilot construction of intelligent substations in 2009. The modeling and management of intelligent substation information model, model simulation, source end maintenance and other technologies are constantly developing, and intelligent substations are gradually entering the stage of comprehensive development. Intelligent substation automation system is widely used to achieve unified modeling and data exchange based on ICE standard. However, in the process of implementing intelligent substations, non-standard information models or misconfiguration will lead to communication errors in substations or problems in interactive operations. In the process of checking the monitoring information of intelligent substation, the shutdown of the data communication network, that is,

© ICST Institute for Computer Sciences, Social Informatics and Telecommunications Engineering 2024
Published by Springer Nature Switzerland AG 2024. All Rights Reserved
L. Yun et al. (Eds.): ADHIP 2023, LNICST 547, pp. 46–63, 2024.
https://doi.org/10.1007/978-3-031-50543-0_4

the telecontrol device, as the station control layer equipment, plays a key role [1]. The upper part can transmit the substation monitoring information in real time through the dispatching data network to ensure that the dispatching centers at all levels can monitor, control and adjust the substation, and the lower part can ensure that the equipment information can be obtained normally through the station control layer network, which is the key guarantee to realize the integrated operation of substation regulation. Therefore, how to improve the verification efficiency of substation telecontrol information between the main station end and the plant station end has become one of the difficulties in the engineering commissioning of intelligent substation monitoring system [2].

At present, the efficiency of checking the substation telecontrol information between the main station end and the plant station end is relatively low, and the method of manually operating point by point according to the monitoring information table is still used. The plant station end contacts the main station end by telephone to verify whether the monitoring information is correctly delivered. Checking the substation telecontrol information in this way causes a lot of waste of resources. At the same time, the verification of telecontrol information is often in the late stage during the commissioning stage of the entire intelligent substation project, which will inevitably increase the pressure on the substation commissioning and on-site commissioning personnel. Therefore, it is necessary to design an efficient and convenient telecontrol information quick verification system to improve the efficiency of monitoring information verification.

The traditional model verification only pays attention to the consistency test of the communication model, and ignores the data problems in the process of engineering use. In order to promote the standardization of engineering applications, the State Grid Corporation of China has proposed a secondary equipment model to meet the monitoring requirements of engineering application standards. Based on this, the rapid verification of the operation information of intelligent substation is realized, so as to effectively solve the problems in the verification process, and effectively improve the standardization level of the use of intelligent substation model engineering and the efficiency of model automation verification.

Based on the above analysis, this paper designs the information check and control system of substation remote motor based on computer vision.

2   Overall Structural Design of the System

The simulation master station in this system uses a secure and stable Linux operating system, obtains the IED model through the SCD file, and imports the telecontrol information point table into the simulation master station through Excel format. The analog master station communicates with the monitoring background and the telecontrol device through the MMS communication protocol, and realizes the automatic inspection of telecontrol information [3] by receiving the IEC 104 signal transmitted by the telecontrol device. The simulation master station can automatically record the action time, reset time and manual acceptance confirmation time (acceptance time) of the remote signaling point, and can monitor 104 "four remote" information messages in real time, and automatically generate an acceptance report after the test is completed.

In view of the problems existing in the operation of the monitoring background of the substation remote motor information check and control system, an independent network protection monitoring unit is set in the background to realize the effective monitoring of the background data protection and remote control acquisition equipment, and improve the reliability and security of equipment operation [4, 5].

The network topology in the intelligent substation telecontrol information quick check station is shown in Fig. 1.

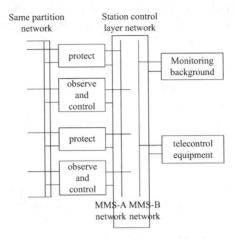

Fig. 1. Overall Structure Diagram of the System

The network protection monitoring unit consists of network protection monitoring software and network protection monitoring unit. Its function is to monitor the status, alarm, data analysis and recording of network layer equipment, effectively monitor the operation status of background data protection and remote control acquisition equipment, find problems in time, and provide accurate information for power grid faults through analysis and processing, prevent failure [6]. The designed telecontrol information quick check system is an open loop system, so the substation monitoring information can flow between the protection and monitoring device, the central switch at the station control layer, the telecontrol device, the dispatching data network, and the simulation master station.

The workflow of the telecontrol information quick check system is as follows: the "four remote" information of the relay protection device is submitted at the station terminal according to the point number of the ammeter; The data communication network shuts down the actual "four remotes" information of relay protection devices in the receiving station, and forwards it to the analog master station in the form of IEC104 message; After receiving the data forwarded by the communication network in the substation after shutdown, the analog master station automatically forms the master station record data table and checks with the monitoring background record [7]. This structure can effectively monitor the background data protection and remote control acquisition equipment, and has the characteristics of simple operation and strong practicability. It can ensure the normal operation of the background data protection and remote control acquisition equipment, greatly improving the working efficiency of operators, and also improving the reliability of substation equipment operation.

3   System Hardware Structure Design

The information checking system designed in this paper mainly includes the master station simulation system, motion devices, power protocol analyzer, and the scheduling of equipment in the substation bay.

The bay level protection device of the substation collects the data in the IED equipment, and uses the IEC communication protocol to send the operation data to the moving device [8]. The simulation system of the dispatching master station creates communication with the moving device, obtains the real-time data of the moving device by using IEC protocol, simulates the dispatching instructions issued, and provides the IEC message parsing function, and realizes the forwarding information test, software and hardware performance test, and information interaction function test [9] for the moving device of the tested object. The simulation system of the dispatching master station also has the ability of IEC communication. The power protocol analyzer uses IEC communication to connect the protection device, the simulation system of the dispatching master station and the measurement and control device, analyzes and compares the messages of the two in real time, and uses the information correspondence between the operation and movement of the automatic monitoring station terminal to enable the information collation to carry out closed-loop management, ensure the reliability and accuracy of the calibration results.

3.1   Check View Module

The check view module refers to transmitting the remote machine information data of the substation to the remote machine information check converter through the remote machine information integration protection device and storing it in the memory. The check view module is an abstraction of this type of data, and applications can use the

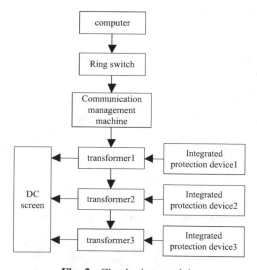

Fig. 2. Check view module

view module to process this data. For the design of remote machine information check view module, see Fig. 2.

Users can use the view to operate the application and realize interaction with the program. The view has a visual interface to display the data defined in the data model. Users can use the view to operate on the data and return the data operation results of the data model to the user. The user uses the view module to send an operation command to the communication management machine, and uses the remote machine information check converter to complete the transformation processing of the data model and definition according to the corresponding programming logic. At the same time, the transformation result is fed back to the user with the view [10].

3.2 Verification Signature Module

The verification signature module is an important component of the substation remote machine information verification control system. It is mainly used to sign and authenticate the verification results, ensuring the traceability of the verification process and the integrity of the data.

In the verification signature module, after each remote machine information verification, the system will automatically generate a verification report or record, which includes detailed verification results, timestamp, and relevant operator information. This information will be stored in the system's database for future reference and auditing.

Next, the verification signature module will generate a digital signature for the verification report or record, and can also provide signature verification functionality. By using a public key to decrypt a digital signature, the authenticity and integrity of the verification results can be verified, and any tampering or errors during the verification process can be confirmed.

Overall, the verification signature module plays an important role in the information verification control system of the substation's remote control system. It ensures the credibility of verification results and data integrity through digital signature and signature verification, while also providing traceability of the verification process. This provides important support for the safety management and audit of substations.

The design purpose of the verification signature module is to use the monitoring information table transmitted by the information transmission unit between the remote machine of the telecontrol substation and the communication management machine to exchange information between the customized sheet (that is, the monitoring information table) generated at the substation side and the dispatching management system, and obtain the verification signature information through the encryption and decryption unit. The module structure is shown in Fig. 3.

The check signature module mainly includes information exchange, information transmission, information encryption and decryption units. In this module, the transmission part mainly generates the monitoring information table at the substation side, which contains various monitoring information, equipment operation status points, equipment operation parameter points, equipment measurement parameter points, etc., it also includes equipment information (such as GIS equipment, SF6 transformer, circuit breaker hydraulic mechanism, protection device, measurement and control device, intelligent terminal, etc.), status information (such as accident information, knife switch

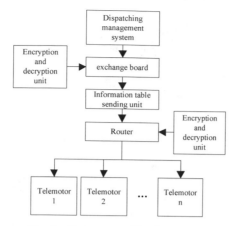

Fig. 3. Signature verification module

position information, working information), etc. The monitoring information table is an important information of the signature verification module. It is obtained through the exchange board information of the communication management machine, and then through encryption and decryption. It is generally in PDF, WORD, and EXCEL formats. The data generated from the substation end is transmitted to the remote control motor through the router, so that it has the automatic identification function and ensures the reliable transmission of information.

The information of the signature verification module is basically stored in the signature information table, and the data is sent to the signature system through the router.

3.3 Check Network Dispatching Module

The substation remote machine information check control system used is used in the power information intranet. Figure 4 shows the check network dispatching module structure.

The network automation team is configured in the master station machine room. The leaders of substation operation, monitoring acceptance, automation and dispatching use the power information network to access and use this system. The network of the supervision leadership office can realize centralized management of system monitoring point information. The system is put into use as the information of the substation monitoring point, and the automatic check control system is isolated and dispatched by using the horizontal isolation device. The network automation team completes the change confirmation of the monitoring points of the horizontal isolation network, uses an image server to complete the mapping of data in this network, and corrects abnormal data through the management of monitoring information.

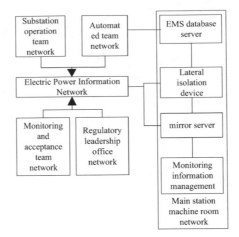

Fig. 4. Check network scheduling module

4 System Software Design

Computer vision is an interdisciplinary field, which focuses on extracting advanced information from images through computers, and seeks to automate tasks completed by human vision systems.

4.1 Information Identification Based on Computer Vision

The verification network scheduling module is an important component of the substation remote control information verification control system. It is mainly responsible for verifying and scheduling the remote motor information, ensuring the accuracy and consistency of the remote motor status, parameters, and indicator lights in the substation.

In the verification network scheduling module, it is first necessary to establish a verification rule library that contains various standards and requirements for remote machine information. These rules can be defined based on national standards, industry norms, and the requirements of the substation itself. The verification rule library can include the legal range of values for the state of the remote machine, the correctness of parameter settings, the on/off state of the indicator light, and so on.

Next, in the verification network scheduling module, computer vision technology will be used to automatically identify and analyze the remote machine information on the substation equipment. Through image processing and pattern recognition algorithms, information such as the status, parameters, and indicator lights of the remote motor can be extracted and compared with the verification rule library.

The verification network scheduling module can also achieve real-time monitoring and alarm functions for remote machine information. Once it is found that the remote machine information is inconsistent with the standards in the verification rule library or there are abnormal situations, the system will promptly issue an alarm to remind the operator to pay attention and take corresponding processing measures.

In addition, the verification network scheduling module can also record and store verification results, and generate corresponding reports and statistical information. These data can be used for analyzing and evaluating the operational status of equipment, helping to optimize the operation and maintenance management of substations, and providing support for decision-making.

In order to accurately identify the normal and abnormal information of substation remote motor, the information target of substation remote motor must be highlighted as much as possible. And computer vision technology can identify substation remote motor information with high accuracy and consistency, without being affected by human subjective factors. It can also record and analyze the identification results, generate detailed reports and statistical information. Overall, the application of computer vision technology to identify substation remote control information can improve work efficiency, reduce errors, improve system reliability, and provide more data support and decision-making basis for the operation and management of substations.

In the process of information processing, it is necessary to process the grayscale of the color image of the remote motor of the substation. The collected image is RGB color space image, and the RGB value is between 0 and 255. The so-called "grayscale" is to convert RGB images into black and white images. Since the obtained image is an 8-bit grayscale image, it is very different from a pure black and white image. Its grayscale index has 256 levels. Therefore, the grayscale value is also expressed as the brightness value: 0 is the darkest and 255 is the lightest. The pixel value in the original image is converted into the grayscale space grayscale parameter by the weighted method, and the formula is:

$$\varepsilon_i = r \times R_i + g \times G_i + b \times B_i \tag{1}$$

In formula (1), r, g, b Respectively represents the parameter values of red, green and blue colors converted into gray; R_i, G_i, B_i Respectively represent the i the coordinate pixel values in the three figures. Through this function, the image can be converted from a color image to a gray image. The transmission line is prominent after graying, and can be used as the correct input data for subsequent edge gradient fitting.

In the process of image processing, the gradient obtained is different from the edge fitting gradient. The gradient reflects the change of image pixels. Assuming that the image to be processed only contains 0 and 1 pixels, where 0 is the background and 1 is the line edge, fit the pixel with 1 pixel into a straight line, and calculate the slope after fitting, that is, the gradient. The slope after fitting is as follows:

$$\lambda = \tan^{-1} \frac{I_y}{I_x} \tag{2}$$

In formula (2), I_x, I_y Represent the horizontal and vertical coordinates of the pixel points of the fitting point respectively.

The calculation result of formula (2) is the basis for judging the normal information and abnormal information of the substation remote motor. When $\lambda > 1$ It means that the image collected by the remote motor of the substation is white pixels, that is, the information collected in this case is normal information; When $0 < \lambda \leq 1$ It means that the image collected by the remote motor of the substation is black pixel, that is, the information collected in this case is abnormal information.

4.2   Encryption and Decryption Before Verification Information Table

Because encryption technology is a relatively safe way for data transmission in the network, this kind of security defense measure that users can control is particularly popular in the power system, that is, it can provide comprehensive security protection for important information at a low cost. In the field of encryption technology, the data type that is encoded and protected after the original data (i.e. plaintext) is encrypted by the encryption device (hardware, software) and the key (public key or private key) is called ciphertext. The method of recovering the ciphertext to the original plaintext by decryption is called decryption, which is the reverse processing of encryption, as shown in Fig. 5.

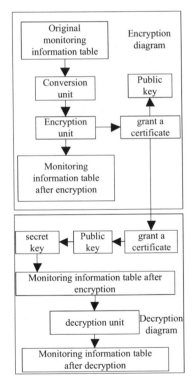

Fig. 5. Encryption and decryption in front of the check information table

Due to the importance of the monitoring information table, in this system, especially before sending the monitoring information table, encryption measures are taken for the sent monitoring information table, and then the received monitoring information table is decrypted. AES algorithm is used for encryption. In the AES algorithm, the byte replacement operation of the S-box is mainly completed by the S-box. The S-box of AES is a structure of inverse multiplication transformation, which is the only nonlinear transformation method. After the inverse multiplication transformation of the S-box, the results are as follows:

$$S[c] = h(\alpha(c)) \qquad (3)$$

In formula (3), h Indicates affine operation; α Represents the multiplication inverse operation; c Indicates clear text. The implementation of S-box is to decompose the elements of a large area into elements of a small area, and then carry out the multiplication inverse operation in the large area after a series of square, multiplication and other operations. In combination with formula (3), the inverse multiplication operation is transformed, and the transformation result is:

$$S^{-1}[c] = h^{-1}\left(\alpha^{-1}(c)\right) \qquad (4)$$

In formula (4), h^{-1} Represents an affine transformation.

Combining formula (3) with formula (4), we can get the results of inverse S-box multiplication and inverse multiplication, thus realizing S-box transformation in AES algorithm. With the support of S-box transformation structure, row displacement processing adopts a method based on multiplication, addition and matrix multiplication, and each column of data is regarded as a polynomial in a finite area. In the algorithm process, plaintext is byte based, and the corresponding components rearrange the bytes in the register to achieve line displacement under the S-box transformation structure.

For the column mixing operation, the data is multiplied by the same multiplication coefficient. The row mixing operation is completed by increasing and looping, and is saved in the register. When the calculation is completed, its output value will be transmitted to the serial and conversion device, which effectively ensures the security of data transmission and achieves the purpose of preventing selected ciphertext attacks. The above principles and methods are used for encryption and decryption. The original monitoring information table (clear text) is converted into the format required by the user through the conversion unit, and encrypted through the encryption unit. The encrypted key is obtained by the user by issuing a certificate. The encrypted ciphertext is sent and decrypted into its reverse.

4.3   Information Verification Process Design

The movement machine information verification process is realized through the simulation system of the dispatching master station, as shown below:

Step 1: Realize the configuration of the movement forwarding table and communication channel, create a verification platform according to the simulation system of the dispatching master station, and detect the channel conditions, alarm information and movement information.

Step 2: simulate the dispatching master station, receive the telemetry value of the moving device, and use it in the manual point-to-point and multi-point testing of the moving device;

Step 3: Real time monitoring, archiving, recording, analysis and comparison of communication messages to automatically detect the correspondence between the output of the station end and the response of the master station, so as to effectively achieve information closed-loop management.

Step 4: display the wrong message classification, analyze the fault tolerance of code missing, realize the diagnosis and analysis of message and data exceptions, and realize fault location and traceability. According to the abnormal information diagnosis process, the best threshold is selected to determine the best clustering result. The larger the optimal threshold, the higher the accuracy of bad data identification, the more information clusters, and the greater the possibility of miscalculation; On the contrary, the smaller the optimal threshold, the lower the accuracy of bad data identification, the fewer the number of information clusters, and the smaller the possibility of miscalculation. Under the condition of ensuring both diagnostic accuracy and stable clustering, the best threshold is fully considered to determine the change rate of clustering number:

$$P_j = \frac{n_j - n_{j+1}}{\delta_j - \delta_{j+1}} \tag{5}$$

In formula (5), j Represents the number of clusters; n_j, n_{j+1} Respectively represent the j times and times $j+1$ the number of sub cluster data; δ_j, δ_{j+1} Respectively represent the j times and times $j+1$ Secondary clustering threshold. Based on this situation, set the information diagnostic criteria as follows:

$$\begin{cases} \delta = \max(\delta_j) \\ P_j = \min(P_j) \end{cases} \tag{6}$$

According to the identification criteria j Sub cluster threshold δ_j Is the optimal threshold. Under this threshold, the information transmission process has high identification accuracy and stable clustering effect.

Step 5: In order to improve the accuracy of verification, mask technology is used. It is a very effective method to protect the encryption system from attacks. It uses randomly generated numbers (masks) to mask sensitive data. Since this operation is completed based on mask data, sensitive data will not be easily disclosed. In mask compensation, because of the direct exposure of mask parameters, it is vulnerable to energy shocks. The mask compensation operation first calculates the middle value of a mixed output mask type. In addition to the tail wheel, it can be described by the following formula:

$$\mu(Ma) = E(\phi(\vartheta(Ma))) \tag{7}$$

In formula (7), E Represents a mask group; ϕ Represents the middle value of the mask; $\vartheta(Ma)$ indicates the replacement mask value. On this basis, using the design idea of the algorithm, the mask value needed for the next round is obtained. In each mask value, the median value of the updated mask type is corrected by the middle value of the previous mask:

$$\phi' = \mu(Ma) \oplus Ma \tag{8}$$

The compensation calculation methods for masks are different for different rounds. In the first round of mask compensation algorithm, the middle value of the mask type obtained by column mixing operation will be generated on this basis. This method can effectively avoid the danger of leakage of the median value of the mixed input mask type in the first round or its Hamming distance, thus enhancing the anti attack capability of the side channel.

Step 6: Automatically record the verification process, save the modification record, and generate a verification report.

5   Performance Test and Analysis

In order to verify the feasibility of the computer vision based information check and control system for substation remote motor, the following tests are designed.

5.1   Verification Data Flow Design

The intelligent station describes the model, instance configuration, communication parameters and signal contact information (such as GOOSE connection) of all IEDs in the station through the SCD file. Each IED completes the real-time exchange of monitoring information with each station control layer equipment (including telecontrol device) by mapping to the manufacturing message specification. Dispatching automation information acquisition is the basis of power grid monitoring and operation, which mainly needs to go through the three-level equipment of information acquisition unit, telecontrol device and dispatching master station. The acquisition unit is responsible for collecting remote signaling, telemetry and other information in the station, and sending it to the background, telecontrol devices and other station control layer equipment. The telecontrol device collects the information sent by each acquisition unit and forwards it to the dispatching master station. The relationship is shown in Fig. 6.

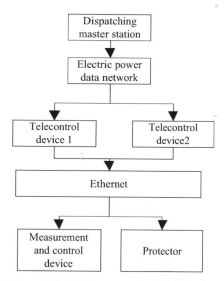

Fig. 6. Data Flow Diagram of Substation Remote Motor

Under the technical system of intelligent substation, the telecontrol device collects the information message of interval measurement and control device, protection device, etc. with MMS message, and the master station receives the data uploaded by the telecontrol device in 104 protocol message format, realizing four general functions of remote signaling, remote control, telemetry and remote regulation. Usually, there are tens of thousands of information in the intelligent station, and the main station business does not need all of them. The designer will select one or several information in the station and merge them one to one or more to one to form a regulation information point, which will be identified by the address of the unique master station information object (generally natural number). All regulation information points will be summarized to form a regulation information table of the whole station. The regulation information table specifies the information to be transmitted to the master station by the telecontrol device and its corresponding number. Table 1 shows the main contents of the regulation information table. For the convenience of subsequent discussion, the station information code is assigned to each SCD information.

Table 1. Contents of regulation information table and corresponding station information

Information address	Describe	In station index name	Information Code
35	5011 switch control circuit	CB5011 .A1m1	W
36	Phase A closing position of switch 5011 of Shuanglan Line	CB5011 .A1m2	D
37	Phase A closed position of switch 5011 of Shuanglan Line	CB5011 .A1m3	C
38	SF6 pressure is low	CB5011 .A1m4	A
39	High SF6 pressure	CB5011 .A1m5	S
40	The switch is pressed without timeout	CB5011 .A1m6	H
41	Switch pressurization timeout	CB5011 .A1m7	K

Technicians need to complete the configuration of telecontrol devices to form the corresponding relationship between the information index name in the station and the address of the information object in the master station. During operation, the telecontrol device receives and analyzes the action information in the station. Once the information corresponds to the configured index name, the action information to the dispatching master station is generated, and 104 protocol message is used to inform the master station of the action of the corresponding control information point, with SOE time attached.

5.2   Conversion Results of Telecontrol Configuration

According to the flow direction of automation information, the time stamp of action signal is formed by the process layer or bay layer device according to the acquisition time in the signal acquisition link to reflect the signal occurrence time. The time information is identified and recorded in the telecontrol device, and transmitted to the master station together with the signal point in 104 protocol (SOE mode).

On this basis, the idea of multi attribute evidence theory in fault tolerance processing is used for reference to change the singleness of judgment quantity in work, and the correlation between attributes is artificially established and used for subsequent judgment. That is, the SOE time information bit is used to synchronously transmit the address of the control information point to the master station, so as to help the master station personnel judge whether the telecontrol device has performed the correct forwarding. The specific scheme is as follows:

Use simulation equipment (such as MMS communication simulation software, etc.) to send MMS communication message of information action in a station. Part of time data in SOE time scale in the message is forced to be consistent with the information address of the corresponding control information point. Then, after the correct forwarding of the telecontrol device, the data corresponding to the second and millisecond bits of the SOE time of the information received by the dispatching master station should be consistent with the information address of the master station. In case of inconsistency, it can be judged as the remote control device conversion error, as shown in Fig. 7.

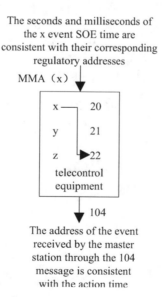

The seconds and milliseconds of
the x event SOE time are
consistent with their corresponding
regulatory addresses

The address of the event
received by the master
station through the 104
message is consistent
with the action time

Fig. 7. Results of correct conversion

As shown in Fig. 7, in the regulation information table in the left box, the x information in the station corresponds to the information address of No. 22 of the master station. Send the x event action signal through the simulation equipment. If the project configuration is correct, the telecontrol device will judge that the information corresponds to the regulation information point of the telecontrol master station with address 22 after receiving the message, and send the "22 event" information to the master station with 104 protocol message. If the project configuration is wrong, the x event is mapped to the information addresses 20 and 21, and the telecontrol device will send the message "20 and 21 event occurs".

In actual work, the simulation equipment can successively send out all the action information in the station to be checked in the order of 104 address number, and the master station side will record all the action information sent by the telecontrol device. This process only takes tens of minutes if one information point is counted every half second. Next, the master station personnel only need to observe that the SOE time data of the received action information is consistent with the address of the information object, and the number of action information and the address distribution of the information object are consistent with the expectation to judge whether the conversion of the telecontrol device is correct. If the software is used for automatic judgment, it can be completed instantaneously.

5.3 Test Results and Analysis

Use end-to-end verification methods, secondary equipment model verification methods, and computer vision based verification methods to compare and analyze the remote conversion results, and compare them with Fig. 7 to verify the accuracy of information verification. If the configuration results are reasonable, the information verification is effective, otherwise it is invalid. The comparison results of the three methods are shown in Fig. 8.

It can be seen from Fig. 8 that the configuration results of the end-to-end verification method and the secondary equipment model verification method are unreasonable, and the information obtained is invalid, indicating that the verification control effect is poor. The use of computer vision based verification methods results in reasonable configuration, correct information verification, and effective information obtained indicates that the verification control effect is good.

The seconds and milliseconds of
the x event SOE time are
consistent with their corresponding
regulatory addresses

MMA（x）

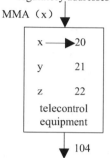

104

The address of the event
received by the master
station through the 104
message is inconsistent
with the action time

(a) End to end verification method

The seconds and milliseconds of
the x event SOE time are
consistent with their corresponding
regulatory addresses

MMA（x）

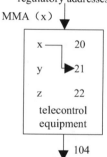

104

The address of the event
received by the master
station through the 104
message is inconsistent
with the action time

(b) Verification method of secondary equipment model

Fig. 8. Comparison and analysis of verification results of three methods

The seconds and milliseconds of
the x event SOE time are
consistent with their corresponding
regulatory addresses

MMA (x)

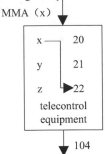

The address of the event
received by the master
station through the 104
message is consistent
with the action time

(c) Verification method based on computer vision

Fig. 8. (*continued*)

6 Conclusion

This article designs a computer vision based substation remote motor information verification and control system. The system consists of a verification view module, a verification signature module, and a verification network scheduling module, forming the hardware environment of the system. Based on this, computer vision technology is used to identify the remote motor information of the substation. After encryption and decryption processing, closed-loop management and clustering processing are used to achieve the verification process of the information.

Despite achieving certain results in this study, there are still some challenges and room for improvement. For example, the robustness of the system in complex environments still needs to be further improved.

In summary, the computer vision based substation remote machine information verification and control system provides an innovative solution for substation operation and maintenance. We believe that in future development, this system will be further improved and promoted, and will bring greater convenience and benefits to the operation and management of substations.

References

1. Nassu, B.T., Marchesi, B., Wagner, R., et al.: A computer vision system for monitoring disconnect switches in distribution substations. IEEE Trans. Power Deliv. **37**(2), 833–841 (2021)

2. Li, Y.S., Gao, B., Xu, L., et al.: An anomaly detection method for digital substation abnormal data based on fusion of difference sequence variance and CPS. Power Syst. Clean Energy **37**(02), 30–41 (2021)
3. Zhang, K.X., Tan, L., Xu, M.H., et al.: Research on intelligent substation simulation modelling technology based on data acquisition technology. IOP Conf. Ser. Earth Environ. Sci. **55**(9), 5206–5210 (2020)
4. Wang, Z.H., Tu, Y.X., Wang, H.Y., et al.: Interactive data monitoring method of substation regulation based on full communication link. J. Nanjing Univ. Sci. Technol. **46**(04), 451–459 (2022)
5. Yan, M.Q., Yang, Y.J., Zhao, F.: A data stream anomaly detection method based on an improved OCSVM smart substation. Power Syst. Protect. Control **50**(06), 100–106 (2022)
6. Jin, X.B., Zheng, W.Z., Kong, J.L., et al.: Deep-learning forecasting method for electric power load via attention-based encoder-decoder with Bayesian optimization. Energies **14**(6), 1596–1603 (2021)
7. Raveendran, S., Chandrasekhar, A.: Inspecting and classifying physical failures in MEMS substrates during fabrication using computer vision. Microelectron. Eng. **254**, 111–116 (2022)
8. Zeng, Y.X., Su, W., et al.: Power data storage and sharing method based on blockchain and data lake. Electr. Power Eng. Technol. **41**(03), 48–54 (2022)
9. Jimenez, V.A., Will, A.: A new data-driven method based on Niching genetic algorithms for phase and substation identification. Electr. Power Syst. Res. **35**(10), 199–209 (2021)

Intelligent Fusion Method for College Students' Psychological Education Score Data Based on Improved Bp Algorithm

Liang Zhang[✉] and Yu Zhao

Changchun University of Finance and Economics, Changchun 130000, China
zliang1510@163.com

Abstract. In order to solve the problem of poor fusion of college students' psychological education score data, an intelligent fusion method for college students' psychological education score data based on improved BP algorithm is proposed. Innovatively adopting genetic algorithms to optimize the weights and thresholds of BP neural networks; Standardize the hierarchical data using matrix transformation and preprocess the fused data using the Grobeis criterion; Extract multi-source heterogeneous level data features, establish an intelligent fusion model for college students' psychological education level data based on improved BP neural network, and achieve intelligent fusion of college students' psychological education level data. The experimental results show that this method has lower latency and higher packet switching rate in data fusion, proving its good fusion performance in data fusion.

Keywords: Improving BP Algorithm · Psychological Education Scores · Data Intelligent Fusion

1 Introduction

As a special group in society, college students are facing various pressures and increasing mental health problems, leading to the emergence of extreme events. Therefore, the mental health status of college students, as well as the psychological intervention models and methods, have become a research hotspot. The use of mental health education content to guide the mental health of college students and identify the existing problems is an effective method to solve mental health problems. With the continuous expansion of enrollment in universities, current university education is transitioning from "elite education" to "mass education". Due to the increasing number of college students facing psychological problems, mental health education for college students is gradually being valued by major universities. Psychological health education for college students is not limited to the study of theoretical knowledge, just like other subjects. Instead, targeted teaching of psychological health knowledge and psychological counseling work are carried out based on the psychological characteristics of this special group of college students. For example, conducting separate psychological counseling or organizing

© ICST Institute for Computer Sciences, Social Informatics and Telecommunications Engineering 2024
Published by Springer Nature Switzerland AG 2024. All Rights Reserved
L. Yun et al. (Eds.): ADHIP 2023, LNICST 547, pp. 64–74, 2024.
https://doi.org/10.1007/978-3-031-50543-0_5

group counseling with classmates with similar psychological problems to help college students alleviate psychological stress, prevent mental illness, adapt to the environment well, handle interpersonal relationships well, help them solve problems in personality development and emotional regulation, and fully tap their psychological potential. With the modernization of national education, the external environment and internal needs of teachers' teaching and research have undergone changes [1]. In order to solve the above psychological education problems in colleges and universities, the networked and information-based working mode has gradually entered colleges and universities. However, students' mental health is a comprehensive reflection of many factors, and the common psychological test data is relatively single, leading to a certain degree of subjectivity in the test results. Moreover, during the use of the system, most teachers, counselors, and mental health educators are limited to data addition, deletion, and modification, without systematically integrating the data in the system. As a result, important data attributes have not been effectively applied, and active warning mechanisms cannot be established, failing to play a proactive warning role. Therefore, it is necessary to intelligently integrate the psychological education performance data of college students.

The purpose of data fusion technology is to obtain simpler and more accurate judgments through multi-source information, that is, by integrating multiple types of data with information from relevant databases to obtain more accurate data. Reference [2] proposed a data fusion algorithm based on distributed Compressed sensing and hash function. Firstly, the distributed Compressed sensing method is used for sparse observation of sensing data to remove redundant data; Secondly, a one-way hash function is used to obtain the hash value of the observed values of the perception data, and it is filled with unrestricted disguised data into the observation values of the perception data to achieve the purpose of hiding the real perception data; Finally, after extracting the camouflage data from the aggregation node, the hash value of the perception data is obtained again and data fusion is completed. Reference [3] designed a deep learning network (DLN) model and its training method, and further proposed a multi-source data fusion algorithm based on DLN-DS.

When the above methods are applied to the data fusion of college students' psychological education achievements, there are problems such as slow Rate of convergence, low fusion efficiency, etc., and the effect of intelligent fusion is poor. To this end, an improved intelligent fusion method for college students' psychological education score data based on BP algorithm is proposed. Innovatively optimize the weights and thresholds of BP neural network through genetic algorithm to obtain a stable network structure and solve the problem of slow Rate of convergence. Standardize the hierarchical data using matrix transformation and preprocess the fused data using the Grobeis criterion; Extract multi-source heterogeneous level data features, establish a data intelligent fusion model based on improved BP neural network, and complete the intelligent fusion of college students' psychological education level data. The test results demonstrate that this method has the following contributions: it reduces fusion delay, improves packet switching rate and fusion efficiency, and has good fusion performance. The overall structure of the current study is shown in Fig. 1:

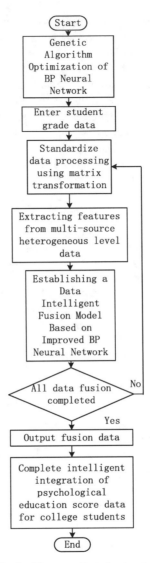

Fig. 1. The overall study structure

2 Improved BP Network Algorithm

Although BP neural network has great advantages in data prediction and image segmentation due to its simple operation and strong practicality, it inevitably has some problems, such as slow convergence speed, poor algorithm stability, and easy falling into local minima. Genetic algorithm performs well in global search. By optimizing the weights and thresholds of the BP neural network through genetic algorithm, a stable network structure can be obtained. It has good improvements and improvements in image segmentation processing, data calculation, and trend prediction, making it a worthwhile optimization solution. The structure of the BP neural network used for data fusion is shown in Fig. 2. Each layer has multiple neural network nodes with different thresholds, and the neural network node is the type of psychological performance data used for fusion.

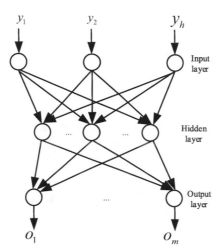

Fig. 2. The BP neural network structure

As shown in Fig. 2, the number of nodes in the output layer, hidden layer and input layer of BP neural network are set as m, n and h successively. The neuronal node connection weights of the input and hidden, hidden and output layers are ϖ_{ij}, ϖ_{jh}. Then the fusion result o_m of the output is:

$$o_m = \sum_{j=1}^{m} g\left(\sum_{i=1}^{h} \varpi_{ij}\varpi_{jh}y_i - \sigma_j\sigma_m\right) \tag{1}$$

In the formula, the input data is y_i; the hidden level threshold of the neural network node j and the output node threshold are σ_j and σ_m; the excitation function of the hidden level node is $g()$.

In order to fully leverage the advantages of genetic algorithm and BP neural network, this article improves and optimizes BP neural network through genetic algorithm, complementing each other's advantages. Genetic algorithm is a biological intelligence optimization algorithm. This solution set is the most ideal solution for the problem to be solved [4, 5]. The steps for solving practical problems using genetic algorithms are as follows:

(1) Initialize. Randomly select an initial population P (0) from N samples through genetic algorithm parameter encoding, forming a set of feasible solutions.
(2) Individual evaluation. The initial population P (0) is introduced into the objective function, and then the fitness of various populations in the sample is calculated.
(3) End condition judgment. Given an initial condition, determine the end condition of the algorithm, and if the condition is met, directly skip to step (8).

(4) Select an operation. Adopt the rule of survival of the fittest and survival of the fittest for the target group, and select a large number of excellent individuals.
(5) Mutation operation. Perform mutation operations on target individuals through mutation probability.
(6) Cross operation. Perform crossover operations on target individuals through crossover probability.
(7) After mutation and crossover random operations, if the next population P (t + 1) composed of N new individuals is obtained, proceed to step (2), and vice versa, proceed to step (4).
(8) Evolutionary individuals. Get the individual with the highest fitness in the function, which is the most ideal solution of the problem, and end the operation.

In order to fully leverage the advantages of genetic algorithm and BP neural network, this article improves and optimizes BP neural network through genetic algorithm, complementing each other's advantages. The specific steps are as follows:

(1) Initialization: Initialize the entire population, using parameter encoding to generate N individuals, which form a new initial population. Then, calculate the population size, determine the chromosome length and range, and provide the ideal error value.
(2) Calculation of fitness: fitness is calculated as follows:

$$G = \frac{1}{2} \sum_{h=1}^{M} (o_h - \hat{o}_h)^2 \tag{2}$$

In the formula, o_h and $\hat{o}_h$ are the basic fusion results and the actual fusion results of BP neural network. Because the value of the fitness function G must be non-zero and positive, the value of G is large in many states, indicating the high accuracy of the underlying result.

The individual is the feasible solution of each connection weight and threshold. If the size of the connection weight and threshold initial population is N, and the fitness of the i individual is G_i, then the possibility of the i individual being left is q_i:

$$q_i = \frac{G_i}{\sum_{i=1}^{N} G_i} \tag{3}$$

The cross-operator action is mating reconstruction, and in the study of this paper, it is mainly used to reconstruct the BP neural network structure. The cross-over operation method is:

$$\begin{cases} \varpi_{ij} = \beta \varpi_{jh} + \varpi_{ij} - \beta \varpi_{ij} \\ \varpi_{jh} = \beta \varpi_{ij} + \varpi_{jh} - \beta \varpi_{jh} \end{cases} \tag{4}$$

where, β is a random number, and its value interval is 0 to 1.

The mutation operator can fine-tune the gene of a coding string, which can ensure the population diversity of the genetic algorithm and realize the improvement and optimization of BP neural network.

3 Intelligent Pre-processing of College Students' Psychological Education Achievement Data

The matrix conversion is used to standardize the score data, and the score information domain is $X = \{x_1, x_2, \cdots, x_n\}$, where x_n is the object to be classified, and each object is measured by n indicators, where x_{i1} is the score nickname information, x_{i2} is the score password information; x_{in} is the score feature. The raw data matrix of the achievement information is:

$$x_{in} = \begin{Bmatrix} x_{11} \ x_{12} \ \cdots \ x_{1n} \\ x_{21} \ x_{22} \ \cdots \ x_{2n} \\ \vdots \quad \vdots \quad \cdots \quad \vdots \\ x_{i1} \ x_{i2} \ \cdots \ x_{in} \end{Bmatrix} \tag{5}$$

After obtaining the specific original data, it is standardized, and the expression formula is:

$$x_{ij} = \frac{x_{ij} - \overline{x}_j}{s_j} \tag{6}$$

In the formula, $\overline{x}_j$ represents the mean of the standardized score data; s_j Represents the unit dimensions after processing, and all variable units have been effectively removed. The data mean is 0, the standard deviation is 1, and the variable mean for grades is between [0, 1].

Due to certain deviations in the collection process of multi-source psychological education score data, the Grobeis criterion should be used to preprocess the merged data. The detailed process is as follows:

Assuming that $x_1, x_2, \cdots, x_n$ is subject to normal distribution, based on this, the data can be preprocessed according to Grobeis criteria. According to the processing results, the two-dimensional information entropy within the cluster is calculated. When the nodes in the cluster belong to the same attribute, the node data is fused in order, so that all the data are converged to the fusion pool [6–8].

In order to realize uneven clustering, it is necessary to introduce different competition radii at the first node of the cluster to reduce the occurrence times of the "hot zone" problem [9]. Candidate nodes use the existing data to construct a new node and use it as the initial cluster head [10, 11]. After the initial cluster first is determined, broadcast the cluster first message, the unselected cluster first will no longer be dormant, and the cluster first with the lowest communication cost is selected to complete the cluster creation. During the fusion process of intra-cluster and intercluster data, the generating tree with the smallest weight was determined to construct the multi-channel fusion path. The weight calculation formula is:

$$\omega_{ij} = \varepsilon \frac{d_{ij}}{Q_i} + \varphi \frac{d_{ij}}{Q_j} \tag{7}$$

In formula (8), ε and φ respectively represent the adjustment coefficients of two nodes i, j Q_i and Q_j Represent the remaining energy of two nodes i, j separately; d_{ij} Represents the distance between two nodes.

Under the premise of determining the multi-channel fusion path, set the size of the multi-channel spatial window and fuse the central subband data of the two nodes within the window.

In response to the changes in two spatial windows, it is necessary to convert them into fuzzy values in order to obtain local decision results in the data center. When two data centers cannot support each other, they will not be able to merge [12, 13]. In order to distinguish between trustworthy and untrustworthy data, a multi-source data regularization method was adopted to determine the relative importance of fusion order and eliminate untrustworthy data. At the multi-channel fusion level, first fuse the sub band data to be fused at the center of two nodes, and then fuse each sub band data accordingly. Due to the good storage characteristics of the fused central data, it can not only integrate the original data but also effectively improve the storage effect of the fused space.

4   Intelligent Fusion of Performance Data Based on the Improved BP Algorithm

4.1   Extracting the Features of Multi-source Heterogeneous Achievement Data

Based on the above data pre-processing results, the multi-source heterogeneous data set is updated, the sample data size is marked M, the data status is expressed in t, and the mean standardization method is used to standardize the data. The results are as follows:

$$Y_{ij}^*(t) = \frac{Y_{ij}(t)}{\overline{Y}_j} \tag{8}$$

In the equation, the standardized results of the data are marked as $Y_{ij}^*(t)$, the data mean is marked $\overline{Y}_j$, and the constant is marked i and j. Based on the above data, the standardized processing results are obtained, and the feature vectors of multi-source heterogeneous data are extracted for the first attribute. The calculation results are as follows:

$$\phi(Y_{ij}) = \sum_{t=1}^{T} Y_{ij}^*(t)/T \tag{9}$$

In this equation, the feature vector of multi-source heterogeneous data is expressed in $\phi(Y_{ij})$, and the data attribute recording period is expressed in T.

4.2   Intelligent Fusion Model of Score Data

The flowchart of intelligent fusion of college students after optimizing BP neural network is shown in Fig. 3.

As shown in Fig. 3, the steps of fusion are specific as follows: (1) Create a BP neural network, including the determination of its activation function and the number of nodes. (2) Population initialization. The goal of GA optimization is all network parameters, which must be encoded as genetic individuals, thus constructing a mapping between neural network parameters and the space of genetic dimensions.

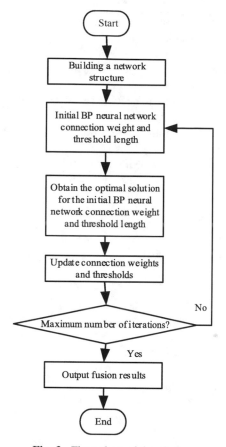

Fig. 3. Flow chart of data fusion

(3) q_i represents the characteristics of each data, and the fitness value of q_i is calculated as follows:

$$\alpha(q_i) = S_{SE} = \sum_{i=1}^{\beta} (\chi_i - \delta_i)^2 \tag{10}$$

In Eq. (9), S_{SE} refers to the sum of squared errors of the neural network model; δ_i Refers to the output value of the i network; χ_i Refers to the expected output of the i network; β refers to the actual number of neurons in the output layer.

(4) Implement genetic population optimization. For q_i, calculate the feature clustering range to represent ε_i; And calculate the number of features to represent ϕ_i.

(5) Implement offset and mutation operations.

(6) For all the population of this generation formed by heredity, calculate the fitness value of all the population, and select the best heredity as the heredity of the next

generation. For other individuals, select them through selection strategies, and combine them with the selected best heredity as the genetic population of the next generation.

(7) Judge the termination conditions: repeat steps (3) to (4). When the termination conditions are met, the optimization is directly ended to obtain the heredity of the optimal fitness.

(8) Implement optimal individual decoding and assign values to the BP neural network to achieve coarse accuracy updates of network parameters;

(9) Implement high-precision training optimization of thresholds and weights using the Levenberg Marquardt algorithm;

(10) After the maximum number of iterations or reaching the target error, end the training and complete the establishment of the optimal model; Otherwise, return to the previous step to continue training. Complete the intelligent integration of college students' psychological education score data according to the above steps.

5   Experiments and Analysis

In order to verify the performance of the intelligent fusion method for college students' psychological education score data, the fusion delay time and packet fusion exchange rate were selected as evaluation indicators to measure the ability of the fusion method. Select the performance data of students majoring in mental health education from a certain university as the test sample, with a total of 3000 data volumes. The above grade data was fused using the methods presented in this article, traditional method 1 (Method proposed in reference [2]), and traditional method 2 (Method proposed in reference [3]), and the fusion effect of the method was calculated.

5.1   Fusion Delay Comparison Test

During the data fusion process, there may be delays. When using the proposed method, traditional method 1, and traditional method 2 for data fusion, the delays generated by the above three methods during data fusion were tested, and the results are shown in Fig. 4.

During the data fusion process, there will be delays. The longer the delay time, the worse the fusion performance of the data fusion method, and vice versa.

Analyzing Fig. 3, it can be seen that as the amount of data to be fused increases, the data fusion delays detected by the three methods all show varying degrees of increase. This is because the method in this paper innovatively optimizes the weights and thresholds of BP neural network with a stable network structure, improves the convergence rate and reduces the latency. Among them, the fusion delay generated by the proposed method during data fusion is lower than the test results of traditional method 1 and traditional method 2.

5.2   Fusion Effect Comparison Testing

Based on the above test results, the packet exchange rate was selected as the data fusion performance detection indicator to test the fusion performance of the data under the

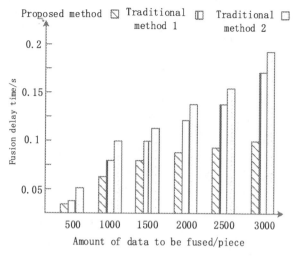

Fig. 4. Results of fusion delay tests for different methods

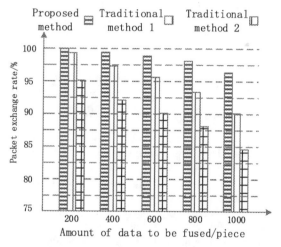

Fig. 5. Test results of packet exchange rate using different methods

proposed method, traditional method 1, and traditional method 2. The results are shown in Fig. 5.

During the data fusion process, the higher the packet exchange rate detected, the better the data fusion effect. The lower the packet exchange rate tested, the worse the data fusion effect.

Analyzing the experimental data in Fig. 5, it can be seen that as the amount of data to be fused increases, the packet exchange rates detected by the three data fusion methods all show varying degrees of decline. However, the packet exchange rate detected by the proposed method during data fusion is the highest among the three methods. From this, it can be seen that the fusion effect of the proposed method in data fusion is higher than that of traditional method 1 and traditional method 2. In summary, the proposed method has low energy consumption, low latency, and high packet exchange rate in data fusion, indicating good fusion performance in data fusion.

6   Conclusion

As the scope of computing technology increases, data fusion algorithms become particularly important. This method innovatively utilizes genetic algorithm to improve the BP neural network, extracts data features based on data preprocessing results, establishes a data fusion model, and completes data fusion through model output. The experimental results show that the data fusion effect of this method is good. However, due to limited conditions, this article did not significantly improve the accuracy of fusion. Future research will improve fusion accuracy while ensuring fusion efficiency.

References

1. Zhang, L., Zuo, X.: Credit of small and medium sized scientific and technological enterprises based on BP neural network evaluation research. MATEC Web Conf. **336**(5), 901–905 (2021)
2. Kou, L., Liu, N., Huang, H., et al.: Data aggregation privacy protection algorithm based on distributed compressive sensing and hash function. Appl. Res. Comput. **6**(1), 239–244 (2020)
3. Li, B., Li, K., Zhong, S., et al.: Research and application of deep learning algorithm based on multi-source data fusion of power grid. Electron. Des. Eng. **29**(10), 116–119, 124 (2021)
4. Stipanovic, I., Bukhsh, Z.A., Reale, C., Gavin, K.: A multiobjective decision-making model for risk-based maintenance scheduling of railway earthworks. Appl. Sci. **11**(3), 965–971 (2021)
5. Baldan, M., Nikanorov, A., Nacke, B.: A novel multi-surrogate multi- objective decision-making optimization algorithm in induction heating. COMPEL **39**(1), 144–157 (2020)
6. Kitaeva, A.V., Kolupaev, M.V., Stepanova, N.V., Zhukovskiy, O.I.: Convergence rate of mean-square error of kernel estimators of non-homogeneous Poisson process intensity function. J. Phys: Conf. Ser. **1680**(1), 012021 (2020)
7. Martín-Hernández, E., Martín, M., Mohammadi, M., Harjunkoski, I.: Modeling and analysis of organic waste management systems in centralized and decentralized supply chains using generalized disjunctive programming. Ind. Eng. Chem. Res. **60**(4), 1719–1745 (2021)
8. Lahmiri, S., Tadj, C., Gargour, C.: Biomedical diagnosis of infant cry signal based on analysis of cepstrum by deep feedforward artificial neural networks. IEEE Instrum. Meas. Mag. **24**(2), 24–29 (2021)
9. Singh, Y.P., Gautam, P.: Development of data mining algorithm for giving the loan in banks introduction. High Technol. Lett. **27**(1), 284–295 (2021)
10. Hasheminejad, S., Khorrami, M.: Clustering of bank customers based on lifetime value using data mining methods. Intell. Decis. Technol. **14**(4), 1–9 (2020)
11. Negara, Y., Doni, A.F.: Comparison of data mining algorithm performance on student savings dataset. J. Phys: Conf. Ser. **1569**(2), 022081 (2020)
12. Liu, S., Li, Y., Fu, W.: Human-centered attention-aware networks for action recognition. Int. J. Intell. Syst. **37**(12), 10968–10987 (2022)
13. Kanaan, L., Haydar, J., Samaha, M., Mokdad, A., Fahs, W.: Intelligent bus application for smart city based on LoRa technology and RBF neural network. WSEAS Trans. Syst. Control **15**(1), 725–732 (2020)

Optimal Planning Method of Rural Tourism Route Based on Multi Constraint and Multi Objective

Qingqing Geng and Yi Liu[✉]

Chongqing College of Architectural and Technology, Chongqing 401331, China
liuyi13808301860@163.com

Abstract. In order to recommend rural tourism routes that best meet users' tourism needs, a multi constraint and multi-objective optimal planning method for rural tourism routes is proposed. Firstly, collect and mine tourist generated data, obtain personalized preferences of tourists, construct a comprehensive evaluation model for tourist attractions, and obtain comprehensive evaluation results for tourist attractions. Secondly, from the perspective of multiple constraints and planning objectives, establish an optimal planning model for rural tourism routes. Finally, by applying the evaluation function method to solve the above construction model, the optimal planning results of rural tourism routes can be obtained. The experimental data shows that after the application of the proposed method, the optimal planning results of rural tourism routes are consistent with the optimal routes, and the minimum number of iterations for solving the optimal planning results of rural tourism routes is 8, fully confirming the better application performance of the proposed method.

Keywords: Multiple Objectives · Rural Tourism · Route Planning · Route Optimization · Tourism Routes · Multi Constraint

1 Introduction

In recent years, with the increasing prosperity of the national economy, people's demand for spiritual aspects has also been increasing. In this trend, tourism has gradually become a popular way of leisure, and as an important driving force for rural development, rural tourism has become an important topic of concern for all sectors of society in recent years. At the same time, a large number of rural tourism areas across the country also showed explosive growth, which to some extent promoted the development of rural tourism industry. Currently, the tourism industry is developing rapidly and occupies an important position in the national economy. The rapid development of the Internet has brought new vitality to the tourism industry, changing the way residents travel [1].

If a person wants to travel at a certain time, they have some basic preferences and constraints, such as planned time constraints, money expenses, types of favorite attractions, peers, accommodation requirements, etc. Therefore, when planning tourist routes for

L. Yun et al. (Eds.): ADHIP 2023, LNICST 547, pp. 75–90, 2024.
https://doi.org/10.1007/978-3-031-50543-0_6

tourists, it is necessary to consider the constraints of tourists from multiple aspects and achieve as many tourist goals as possible. Among the existing research results, the time-based route planning problem model and the tourism route recommendation model based on pattern and preference perception are more widely applied [2]. The former focuses on time and proposes a corresponding route planning problem model. By solving it, a travel route planning result that includes time can be obtained and recommended to users; The latter constructed a system architecture based on the proposed travel route recommendation, and then modeled the mobility patterns of each user. Finally, a tourism route recommendation plan was proposed to develop tourism routes for target users and recommend personalized services. The above two methods consider fewer influencing factors and cannot obtain the optimal rural tourism route, which reduces tourists' interest and restricts the subsequent development of rural tourism. Therefore, a multi constraint and multi-objective optimal planning method for rural tourism routes is proposed.

2 Research on the Optimal Planning Method of Rural Tourism Route

2.1 Collection and Mining of Visitor Generated Data

Each tourist has different preferences, and their corresponding rural tourism route planning models also have some differences. In order to meet the personalized needs of tourists, the data generated by tourists are collected and mined to obtain tourists' preferences, which lays a solid foundation for the establishment of the optimal planning model of the follow-up rural tourism routes.

The data generated by tourists are mainly image information and text information. Among them, image features are used to express the collected image information, as shown below:

First, global image features

Color feature: color feature is a feature based on pixels, that is, all pixels in the image contribute to the statistical results of the image feature. Therefore, color feature cannot describe the local changes of the image well.

Texture feature: The advantage of texture feature is that it is calculated statistically in the area containing multiple pixels, rather than based on pixel features. In pattern matching, it will not fail to match due to local deviation.

Shape features: The extraction of shape features is very important for object recognition and retrieval [3]. In order to retrieve or recognize objects or objects in images, it is very effective and necessary to extract the shape information of objects or objects. However, the extraction and application of shape features still face great challenges. The most common method, Hough transform, has a large amount of computation, and cannot effectively detect the line segments in the image, as well as the lines including the line segments, and cannot complete the detection of line segments with fewer pixels.

Second, local image features

SIFT feature: SIFT feature is a local image feature based on key points, first proposed by David Lowe, and is considered to be one of the most robust local image features. SIFT features first find the key points in the image and express them in vector form one by one. These feature vectors have the characteristics of rotation invariance, scale invariance, and illumination invariance.

Color SIFT feature: Color SIFT feature was proposed by Alaa E. Abdel Hakim and Aly A. Farag. This feature considers both color and geometric shape information when describing objects or objects in the image. The color invariance is guaranteed by the model of Geusebroek et al., while the geometric invariance is guaranteed by the SIFT feature, so that the feature makes good use of the discriminability of color information, and guarantees the color invariance (the stable feature that the color of the object surface in the scene is independent of the object shape is called color invariance).

The collected text information is described by text feature units. The representation unit of text information can be words or phrases from the perspective of statistical machine learning. The perspective of natural language processing is to define various syntactic rules. This research mainly introduces the expression of text information from the perspective of statistical machine learning. In addition, Chinese text expression involves word segmentation. Compared with phrases, word segmentation is much easier than phrase segmentation. Therefore, when processing Chinese data, words are often used as text features to express the content of the text. Text features are mainly as follows:

First, word frequency

Word frequency is the number of times a word appears in a document. Text expression based on word frequency is proposed based on the assumption that words with high frequency have a relatively important impact on the expression of document content. But information retrieval theory believes that words with low frequency may also contain more information. Therefore, simply using word frequency to describe the content of a document is sometimes inaccurate.

Second, document frequency

Document frequency refers to the number of documents containing a word in the entire dataset. Calculate the document frequency for each feature in the document set. If the value is large, it indicates that many documents contain this feature, and then it is considered that this feature has no discrimination; If the value is very small, it indicates that very few documents contain this word, and this feature is not representative for this document set. The features in these two extreme cases cannot well express the content of the document [4].

Third, TF-IDF

Compared with text expression based on word frequency and document frequency, TF-IDF is a very effective text expression method, proposed by Salton. TF is the word frequency, which is used to calculate the ability of each word in the document to describe the document content; IDF is called anti document frequency, which is used to calculate the ability of each word in a document to distinguish content differences between documents, and is inversely proportional to document frequency DF. TF-IDF is based on the assumption that words that appear many times in a document will also appear many times in documents expressing similar meanings, and vice versa. Therefore, the larger the TF and the smaller the IDF, the better the distinguishability of this feature, and the better the content of such documents can be expressed. The TF-IDF calculation formula is:

$$\Omega = \#(t_k, d_j) * \log_2 \frac{|T_r|}{\#T_r(t_k)} \tag{1}$$

In Eq. (1), Ω represents the TF-IDF value; $\#(t_k, d_j)$ represents the number of occurrences of the word t_k in article d_j; $|T_r|$ represents the number of documents; $\#T_r(t_k)$ represents the total number of documents where the word t_k appears.

Based on the collected tourist generated data mentioned above, apply the Location Topic model to conduct in-depth mining on it. The Location Topic model is a statistical model used for text analysis and topic modeling. It combines the geographic location information and thematic information of documents, aiming to reveal the correlation between space and themes in text data. This model assumes that the document generation process includes two main steps: topic selection and word generation. In the topic selection step, the document selects topics from a topic distribution that is related to the geographic location of the document. In the word generation step, the document generates specific words based on the word distribution of the selected topic. In this way, the Location Topic model can combine themes with geographic locations to reveal spatial thematic patterns present in text data. This is very useful for many application scenarios, such as social media analysis, regional feature analysis of news reporting, etc.

Due to the inevitable connection between tourist generated data and location, location information was taken into account when establishing the model. It is not difficult to imagine that the following information usually appears in tourist generated data: travel expenses, local characteristics, travel time, transportation, accommodation, playing mood, and so on. Among them, travel expenses, transportation, accommodation, and entertainment mood are all irrelevant information to the location, while local characteristics are closely related to the location. Considering the above domain knowledge, the Location Topicl model first divides the possible semantics in tourist generated data into two categories when modeling, one is location related and the other is location independent. Based on this, semantic mining of tourist generated data is carried out. The process of generating a document in the LT model is as follows: firstly, the generation of the document is simplified to the generation of paragraphs. When several paragraphs are all generated, a set of tourist generated data is also completed. The advantage of this is that each segment can be constructed according to different locations. For each paragraph, first determine the type of theme in the paragraph, which semantic themes are related to the location, and which semantic themes are not related to the location. For each topic, generate words based on their distribution on the topic, thus forming a discourse.

In addition, with the spatio-temporal data becoming more and more accessible, the analysis and mining of spatio-temporal data has gradually become a new research field and hotspot. The partition group framework extracts common parts from scattered route sets by dividing long routes into small segments for clustering. The concept of route pattern uses route mining to analyze the user's behavior in time and space, such as the geographical region the user passes, the time the user walks, etc.

The above process completes the collection and mining of visitor generated data, determines user preferences, and provides support for subsequent research.

2.2 Construction of Comprehensive Evaluation Model for Scenic Spots

The quality of scenic spots is one of the main influencing factors of rural tourism route planning, so before the establishment of the optimal planning model of rural tourism routes, it is necessary to build a comprehensive evaluation model of scenic spots to obtain the comprehensive evaluation results of scenic spots.

The construction of a comprehensive evaluation model for scenic spots is mainly divided into three steps. Firstly, it is necessary to analyze and summarize the influencing factors of the comprehensive rating of scenic spots. A comprehensive analysis and induction of the influencing factors of the comprehensive evaluation of scenic spots is an important step in the construction of a comprehensive evaluation model for scenic spots. Otherwise, it will greatly affect the objectivity and reliability of the final comprehensive rating of the tourist attraction, and thus greatly affect the tourist experience of the final planned route. Secondly, use random forest calculation to get the importance ranking of all the influencing factors summarized in the first step, select some influencing factors according to the importance ranking, and build a more objective comprehensive evaluation model of scenic spots than the simple tourist rating [5]. Finally, the weight coefficients of several influencing factors selected in the second step are calculated by using the information entropy assignment method, and the comprehensive evaluation model of scenic spots is obtained.

Drawing on existing research and further analysis of practical problems, this paper analyzes the influencing factors related to scenic spot rating from the perspectives of scenic spots, transportation and tourists. In fact, when analyzing the rating of scenic spots, one of the factors can be selected to measure the rating of scenic spots, or multiple factors can be selected to comprehensively measure the rating of scenic spots through previous experience. However, such measurement method is relatively not objective and comprehensive. Therefore, this section selects random forests to obtain the importance ranking of these factors, and constructs a comprehensive evaluation model of scenic spots according to the importance ranking. Thus, relatively objective comprehensive scores of scenic spots can be obtained.

The influencing factors of comprehensive evaluation of scenic spots are as follows:

(1) Scenic spot factors

For travel, scenic spots are closely related to tourists and routes. From the perspective of scenic spots, the relevant influencing factors are: scenic spot resources, tourist seasons, scenic spot tickets, specific categories of scenic spots, levels of scenic spots and recommended travel duration of scenic spots. These factors will affect the planning of tourism routes to some extent. This paper mainly collected the relevant data of scenic spot resource indicators, travel duration and scenic spot level. The specific data of scenic spot resources were obtained through the tourism census conducted by the National Bureau of Statistics, and the scenic spot resources were evaluated from various aspects to obtain the final data.

(2) Traffic factors

Traffic accessibility is the main consideration of traffic factors. According to the existing research, tourists mainly gather in a certain area centered around the city. If this range is exceeded, the greater the cost of transportation time is, the fewer tourists will be. A large number of relevant studies have proved that the more densely distributed the scenic spots are, the more likely they will be selected and planned by tourists. That is to say, if a scenic spot has better traffic accessibility than other scenic spots, then the scenic spot is more likely to be selected, and the degree of accessibility is measured by time, expressed as

$$T_i = \frac{\sum_{j=1}^{n} L_{ij}}{V} \tag{2}$$

In Eq. (2), T_i represents the total transportation time from the i-th scenic spot to the other scenic spots; n represents the total number of rural tourist attractions; L_{ij} represents the distance between attraction i and attraction j; V represents the speed of the mode of transportation.

(3) Tourist factors

The purpose of planning tourist routes is to serve tourists. Therefore, the evaluation results of tourists on tourist attractions have great reference significance in planning tourist routes. Here we mainly collect the data of comprehensive environmental score. The comprehensive environmental score mainly refers to the tourists' evaluation of the infrastructure, transportation, health and environment of the scenic spot through the indicators of very good, good, average, poor and very poor. The corresponding scores are 5, 4, 3, 2 and 1 respectively. The average score is the comprehensive environmental score, and the calculation formula is:

$$H_i = \frac{\sum_{j=1}^{4} \Upsilon_{ij}}{4} \tag{3}$$

In Eq. (3), H_i represents the comprehensive environmental rating of the i-th scenic spot; Υ_{ij} represents the corresponding scores for infrastructure, transportation, health, and scenic environment.

After analyzing the influencing factors, usually several important factors can be selected based on previous experience to construct a model and calculate the comprehensive rating of the scenic spot. However, such an approach may not be objective and comprehensive enough. Therefore, it is necessary to obtain the importance ranking of multiple factors through specific calculations before constructing a comprehensive evaluation model for scenic spots. The random forest method has two main advantages: the first is that it can measure and analyze the utility of multiple variables in a large amount, mainly because it can improve the accuracy of analysis and measurement, but will not increase the amount of calculation significantly; The second point is that the processing results of missing and unbalanced data are relatively stable, mainly because of its low sensitivity to multivariate collinearity. The calculation of variable importance is to first change the value of independent variable of out of pocket data in random forest, and then measure the importance of the factor by the mean reduction of classification accuracy of the observed variable before and after the change [6–8].

The specific analysis process of the importance of influencing factors is as follows:

(1) First, analyze and measure each decision tree through out of pocket data, and then record the deviation of each out of pocket data analysis and measurement. Use $\varepsilon_1, \varepsilon_2, \cdots, \varepsilon_b$ to represent the deviation value of each decision tree;ourist factors

(2) The second step is to generate new out of pocket data. The specific operation method is to perform unrestricted noise interference on the sample feature P on the data outside the original bag. Then use the old data to test the new data to get the new error, which is expressed as $\varepsilon_{11}, \varepsilon_{12}, \cdots, \varepsilon_{1b}$;

(3) Finally, calculate the importance of variables. The specific operation method is to subtract the deviation measured by analyzing the data outside the new bag from the deviation measured by analyzing the data outside the old bag, and then calculate the average value of the sum. Set the number of decision trees to, and the formula for calculating the importance of variables is:

$$\omega_i = \frac{(\varepsilon_{11} - \varepsilon_1) + (\varepsilon_{12} - \varepsilon_2) + \cdots + (\varepsilon_{1b} - \varepsilon_b)}{M} \tag{4}$$

In Eq. (4), ω_i represents the importance value of the influencing factors for the comprehensive evaluation of the i-th scenic spot.

Formula (4) is used to calculate the importance ranking of multiple influencing factors, and select several relatively important influencing factors to build a comprehensive evaluation model of scenic spots. If A/B/C is used to represent the selected factors, the expression of the comprehensive evaluation model of scenic spots is:

$$Y_i = (\alpha * a_i + \beta * b_i + \chi * c_i) * 100\% \tag{5}$$

In Eq. (5), Y_i represents the comprehensive evaluation result of the i-th scenic spot; a_i, b_i and c_i represent the relative scores of the i-th scenic spot regarding the influencing factors A/B/C; α, β and χ represent the weights of the influencing factors A/B/C for the i-th scenic spot.

The above process completes the construction of the comprehensive evaluation model of scenic spots and the acquisition of the comprehensive evaluation results of scenic spots, which facilitates the establishment of the optimal planning model of the follow-up rural tourism routes.

2.3 Construction of Optimal Planning Model for Rural Tourism Routes Based on Multi-objective and Multi Constraint

In this section, we mainly solve the problem of helping tourists to determine the tour sequence of scenic spots and generate a tourism route suitable for tourists after they determine their hotel and the scenic spots to travel, on the premise of meeting the tourists' multi constraint and multi tourism goals. This section establishes a multi constraint and multi-objective tourism route planning model.To facilitate our calculation, we first set up reasonable model assumptions according to the actual problems, and then define the relevant constraints to establish the model [9]. When tourists need to plan a tour, they need to input their own personalized preferences, such as the start time and end time of the tour, the type of scenic spots they want to visit, the estimated travel costs, peer groups and other information. After receiving the request from tourists, we recommend hotels and scenic spots with higher scores to users according to the hotel scenic spot scoring model shown in the previous section, and then use the tourism route planning model to plan the travel for tourists. Under the condition of meeting the constraints of tourists, we recommend appropriate tourism routes for tourists.

For the convenience of calculation, this article needs to set reasonable assumptions based on the actual travel environment, so that unimportant factors do not affect the planning results of the route. On the basis of collecting a large amount of tourism industry growth data, tourist data volume, and other information, combined with recent tourism trends, under the guidance of experts, the following assumptions are made:

(1) Tourists always start from their hotel. Because the luggage and articles taken by tourists are stored in the hotel, tourists will always return to the hotel after the tour, regardless of the possibility of changing the hotel.
(2) By default, tourists use private cars for self-help travel, regardless of fuel fees, tolls, parking fees and other possible accidental charges during driving. That is, the toll is not considered in the model.
(3) The starting price of the hotel is taken as the price for staying in the hotel, and the hotel is not full, that is, tourists can always book the hotel they want to stay in.
(4) There is no change in ticket prices for tourist attractions in off season and peak season, and there is no preferential price for students, teachers and other different groups.
(5) The food cost of tourists during the travel is not considered.
(6) The travel time of tourists only includes travel time and travel time, without considering other time that may be spent, tourists will not visit the same scenic spot repeatedly, and the time of each tourist destination is always fixed.
(7) Tourists always enter the scenic spot during its opening hours, regardless of the closure of the scenic spot.
(8) The influence of weather, traffic jam and other conditions on the day of travel is not considered.

This section recommends hotels and scenic spots to tourists according to the hotel scenic spot rating model to help tourists plan the tourist routes that meet their preferences. The goal of route planning is to maximize the total score of routes and the number of scenic spot types, and time, cost, etc. are constraints of the model [10–12]. To facilitate the subsequent description of the article, based on the above assumptions, the constraints of the construction model are defined as follows:

(1) Tourist information: Each tourist has their own preference constraints, including time constraints UT, cost constraints UC, and other information.
(2) Attraction Information: Each recommended attraction set $S = \{s_1, s_2, \cdots, s_n\}$ should include information such as recommended travel time ST, attraction rating SS, attraction ticket price SP, attraction type SG, suitable season SD, suitable audience SC, longitude SL, latitude SA. For different attractions, tourists will prioritize the attraction with the highest total constraint score.
(3) Play route: If the edge set of the path is represented by R, then $R = \{r_1, r_2, \cdots, r_K\}$. The weight of any edge is the travel time T between two scenic spots. According

to the assumption in the previous text, tourists arrive between two different scenic spots by car. The shortest driving time obtained from Baidu Maps is used as the distance time between the two places, and the driving time is represented as the weight of each edge.

(4) Route map: The route map refers to the network information of tourist attractions during their travels. If Q is used, then $Q = (S, R)$, S represents the set of tourist attractions, and R represents the edge set between attractions. For example, the route composed of tourists departing from Hotel A, visiting attractions B, C, D, E, and F, and returning to Hotel A is the tourist's itinerary.

(5) Total travel path: If TR represents the total travel path for tourists, then $TR = \{h_0, s_k, s_{k+1}, \cdots, s_{k+n}\}$, which includes the hotels where tourists stay and the tourist attractions. s_k represents the scenic spots that tourists have passed by, and the time of arrival at the scenic spot s_k is represented by $TRT(s_k)$.

(6) Total cost of play TRF: Based on the above assumptions, in this article, except for the hotel fees and ticket prices of tourists, other expenses of tourists are not considered, and only individual expenses are considered. The total cost of a tourist's total travel path $TR = \{h_0, s_k, s_{k+1}, \cdots, s_{k+n}\}$ is the ticket price of all attractions plus the price of the tourist's hotel stay.

(7) Time spent in a single scenic spot: the time spent in a scenic spot is the average time recommended on the website.

(8) Total travel time: the total travel time of tourists includes the driving time of the user from the hotel to the target scenic spot plus the travel time of the target scenic spot plus the time of the last return to the hotel.

(9) Attraction Total Constraint Score: For each recommended attraction, there is a total constraint score for the attraction, which is calculated by the hotel attraction rating model mentioned above as the total constraint score SRS for that attraction. If the total score of a tourist attraction is higher, it indicates that it can better meet the personalized preferences of tourists.

(10) Total score of route TRS: After a tourist chooses a tourist route, $TR = \{h_0, s_k, s_{k+1}, \cdots, s_{k+n}\}$ will have a total score of the route. The calculation method for the total score of the route is the sum of the scores of the various scenic spots that the tourist has passed through.

(11) Effective route: The effective route for tourists is the route that meets the total travel time constraints of tourists, and the cost does not exceed the maximum cost of tourists' budget. If one item does not meet the needs of tourists, the route cannot be said to be effective.

According to the above constraints, the optimal planning model of rural tourism routes is established, and the expression is

$$\begin{cases} \max TRS(TR) \\ \max SG(TR) \\ s.t. \ TRT(TR) \leq UT \\ \quad TRF(TR) \leq UC \end{cases} \tag{6}$$

The above process completed the construction of the optimal planning model of rural tourism routes, laying the foundation for the realization of the final research goal.

The construction process of the optimal planning model for rural tourism routes is shown in Fig. 1.

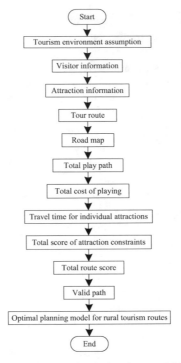

Fig. 1. Construction process of the optimal planning model for rural tourism routes

2.4   Solution to the Optimal Planning Model of Rural Tourism Routes

There are many methods to solve the multi constrained multi-objective programming problem, and the evaluation function method is one of them. Its purpose is to transform the multi-objective optimization into the single objective optimization and then solve it [13]. By using the evaluation function $u[f(x)]$, a multi constraint multi-objective planning problem - the optimal planning model for rural tourism routes - is transformed into the following problem:

$$(Pu)\begin{cases} \min\ u[f(x)] \\ s.t.\ \ x \in X \end{cases} \qquad (7)$$

In Eq. (7), (Pu) represents the auxiliary calculation function; $f(x)$ represents the optimal planning objective for rural tourism routes; X represents the optimal planning target set for rural tourism routes.

The key of the geometric weighting method is to create a new evaluation function to simplify the multi-objective optimization problem and then solve it [14, 15]. The specific method of linear weighting method is to define different weight coefficients v_i for each optimization objective $f_i(x)$ based on the different importance of each optimization objective in the minds of decision-makers, and then convert multi-objective optimization into single objective optimization, with the expression:

$$u[f(x)] = \sum_{i=1}^{m} v_i f_i(x) \qquad (8)$$

Then the solution steps of the optimal planning model of rural tourism routes are as follows:

(1) Unify dimensions and standardize the objective functions $f_i(x)$. Then calculate the two maximum values of each optimization objective function: $f_{i-\max}(x)$ and $f_{i-\min}(x)$. Specific application functions:

$$\psi_i(x) = \frac{f_i(x) - f_{i-\min}(x)}{f_{i-\max}(x) - f_{i-\min}(x)} \qquad (9)$$

In Eq. (9), $\psi_i(x)$ represents the normalization result of the objective function.
(2) Based on the different importance levels of different optimization objectives in the minds of decision-makers, the weight coefficient v_i that matches the optimization objective function $\psi_i(x)$ is given, and the evaluation function is represented as

$$u[f(x)] = \prod_{i=1}^{m} \psi_i(x)^{v_i} \qquad (10)$$

Solve single objective optimization problems. By solving the optimal solution x^* of the single objective optimization problem, the non inferior solution of the multi-objective optimization problem before transformation is obtained. The two objectives of the multi constraint multi-objective programming problem are: the maximum number of tourist attractions and the highest overall score of tourist attractions. The weight coefficients representing their importance will also change with the individual needs and preferences of tourists. Therefore, choosing the geometric weighting method to solve the multi-objective optimization problem of scenic spot screening has to some extent improved the reliability of the final route in the practical application of the model.

Through the above process, the multi constraint and multi-objective planning problem is solved, and the optimal planning results of rural tourism routes are obtained, providing better rural tourism experience for tourists.

3 Experiment and Result Analysis

3.1 Selection of Experimental Objects

This paper selects a town and village as the research area of short-term tourism route planning. Its good geographical location and unique cultural history have created a wealth of tourism resources. Natural ecology, rural ancient towns, characteristic blocks, exhibition halls and other tourism resources are concentrated, with good geographical combination conditions, which enables tourists to experience more types and combinations of scenic spots in a short time.

The distribution of scenic spots of experimental subjects is shown in Fig. 2.

Fig. 2. Distribution of scenic spots of experimental subjects

As shown in Fig. 2, the number of scenic spots in the rural area, the experimental object, is large, but the distribution is not obvious, which meets the application performance test requirements of the proposed method.

3.2   Experimental Data Acquisition and Pre-processing

For the collection of tourist attraction data, tourist route data and travel time data, use Python's urllib library to obtain relevant information from the corresponding platform. Urllib is Python's built-in HTTP request library, which provides a way to obtain web page data from a specified URL address, and then parse it to obtain the desired data.It includes the request module, the exception handling module error, the URL parsing module parse, and the robots.txt parsing module robotparser. The methods of parsing data include direct processing, JSON parsing, regular expression processing, Beautiful Soup parsing, PyQuery parsing, and XPath parsing. In this study, we mainly select direct processing, JSON parsing and regular expression processing for the acquired data, and finally select relational data Oracle to store the collected data in a unified way. In addition to the tourist route data, the construction of the scenic spot association map also needs the taxi track data. In order to make the data meet the requirements of the subsequent algorithm, the original taxi GPS data needs to be processed, including data cleaning, starting and ending point extraction, and grid processing.

Due to the limitation of research space, only the traffic time data processing process is described in detail, as shown below:

(1) Request Web Service API key.

Create a new application in the Gaode Map Console, click the "Add New Key" button on the created application, enter the application name in the pop-up dialog box, select the binding service as "Web Service API", and click the Submit button to obtain the key.

(2) Splice HTTP request URL

Fill in the required request parameters according to the URLs of different APIs. The key applied in the first step should be sent together as a required parameter.

(3) Receive the data returned from the HTTP request (in JSON format), and parse the JSON data using Python

The Walking Path Planning API can obtain data for walking commuting options up to 100 km. The URL is https://restapi.amap.com/v3/direction/walking?parame ters. Parameters indicates the request parameter to be entered.

Through the above process, the collection and processing of experimental data are completed, providing support for the smooth progress of subsequent experiments.

3.3 Analysis of Experimental Results

Select the time-based route planning problem model as the comparison method 1, and the tourism route recommendation model based on pattern and preference perception as the comparison method 2, and design the comparative experiment of rural tourism route optimal planning. The application effect of the proposed method is shown by the results of the optimal planning of rural tourism routes and the number of solving iterations.

In order to clearly show the optimal planning results of rural tourism routes, 10 rural tourist attractions are selected as tourist destinations, and the optimal planning results of rural tourism routes are shown in Fig. 3.

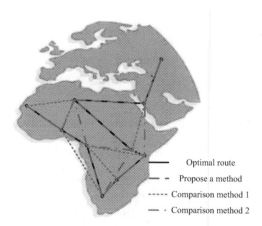

Optimal route

Propose a method

Comparison method 1

Comparison method 2

Fig. 3. Schematic diagram of optimal planning results of rural tourism routes

As shown in Fig. 3, the proposed Location Topic model exhibits higher accuracy in the optimal planning of rural tourism routes. Compared with comparison methods 1 and 2, the optimal planning results of rural tourism routes obtained by this method are consistent with the actual optimal routes. There is a significant difference between the optimal planning results of rural tourism routes obtained by comparing methods 1 and 2 and the actual optimal routes.

This indicates that the method proposed in this paper can more accurately capture the correlation between geographical location and themes, thereby providing more reliable and accurate results in rural tourism route planning. By combining geographic location information and thematic information, our model can better understand the characteristics and thematic needs of different regions, and consider these factors into the planning process of the optimal route.

This high-precision rural tourism route planning result is of great significance to tourists. It can help them better arrange their itinerary, choose the best travel route, and obtain a richer and more satisfying travel experience. At the same time, it also has a positive impact on the development and promotion of rural tourism. Accurate route planning can attract more tourists, enhance the attractiveness and competitiveness of rural tourism, and promote the sustainable development of rural tourism.

The number of iterations for solving the optimal planning results of rural tourism routes obtained through experiments is shown in Table 1.

Table 1. Iterations of Solving the Optimal Planning Results of Rural Tourism Routes

Working condition	Propose method	Comparison method 1	Comparison method 2
1	12	20	26
2	10	25	35
3	9	27	34
4	8	30	36
5	15	19	28
6	14	20	27
7	13	24	25
8	9	21	21
9	10	23	20
10	11	18	24

As shown in the data in Table 1, under different experimental conditions, the number of iterations for solving the optimal planning results of rural tourism routes obtained after the application of the proposed method is less than that of the comparison methods 1 and 2.

Specifically, the comparison methods 1 and 2 generally have higher iteration times for solving under each experimental condition, while the proposed method has relatively fewer iteration times for solving. This indicates that the proposed method has higher efficiency in solving the optimal planning of rural tourism routes.

Reducing the number of solving iterations means that our method can find the optimal solution more quickly, thereby saving computational resources and time costs. This is very important for practical applications, especially in large-scale rural tourism route planning.

Therefore, based on the data analysis results in Table 1, we can conclude that the proposed method has significant advantages in the efficiency of optimal planning of rural tourism routes, and can solve the optimal solution more quickly, providing users with more efficient and convenient rural tourism planning services.

4 Conclusion

The vigorous development of the economy has brought progress to the tourism industry. The rapid popularization of internet technology has led to the emergence of massive amounts of tourism information on the internet, and the methods for obtaining tourism information have also become diverse. Currently, the tourism information on the internet has reached a point where people cannot browse it one by one. In this situation, when tourists want to travel to a strange Urban tourism, it is difficult to determine their own travel routes according to the travel information on the Internet. Therefore, helping tourists plan a travel route that meets their basic preferences and improving their decision-making efficiency is of great significance.

Based on this situation, this article studies the tourism route planning problem based on multiple constraints and objectives, establishes a comprehensive rating model for scenic spots and a tourism route planning model, and plans tourism routes for tourists. The experimental data shows that the proposed method greatly improves the accuracy and efficiency of optimal planning for rural tourism routes, providing better route planning services for tourists.

In terms of future prospects, with the vigorous development of the economy and the progress of the tourism industry, the following trends can be foreseen:

Firstly, with the continuous development of intelligent technology, future tourism planning will become more intelligent and personalized. By combining artificial intelligence and Big data technology, the intelligent system can provide each user with customized travel route planning according to their preferences, interests and needs to meet their personalized travel needs.

Secondly, combining real-time data and user feedback will become an important direction for future tourism planning. By obtaining real-time information on transportation, weather, and scenic spot congestion, combined with user feedback and evaluations, tourism routes can be adjusted and optimized in a timely manner, providing more accurate and practical planning services.

In addition, cross-border cooperation and resource integration are also one of the trends in future tourism planning. Collaborating with transportation departments, hotel and scenic area management departments, and sharing data and resources can provide more comprehensive and efficient tourism planning services.

Finally, sustainable development and ecological protection will become important considerations for future tourism planning. Pay attention to protecting the natural environment and cultural heritage, promote the development of Sustainable tourism, and ensure the sustainable impact of tourism on society, economy and environment.

In summary, future tourism planning will be more intelligent and personalized, combining real-time data and user feedback, cross-border cooperation and resource integration, focusing on sustainable development and ecological protection. The method proposed in this article provides useful exploration and foundation for the development of future tourism planning.

References

1. Su, J., Li, J.: Decentralized intelligent search of tourist routes based on check-in data. Int. J. Mobile Comput. Multimedia Commun. **12**(3), 16–21 (2021)
2. Castillo, S.A., Hornillos, N.B., Val, P.A., et al.: Machine learning to predict recommendation by tourists in a Spanish province. Int. J. Inf. Technol. Decis. Mak. **21**(4), 1297–1320 (2022)
3. Zyrianov, A.I., Zyrianova, I.S.: Planning of the interregional tourist route in the Urals. Quaestiones Geographicae **40**(2), 109–118 (2021)
4. Chen, C., Zhang, S., Yu, Q., et al.: Personalized travel route recommendation algorithm based on improved genetic algorithm. J. Intell. Fuzzy Syst. Appl. Eng. Technol. **40**(3), 4407–4423 (2021)
5. Han, S., Liu, C., Chen, K., et al.: A tourist attraction recommendation model fusing spatial, temporal, and visual embeddings for flickr-geotagged photos. Int. J. Geo-Inf. **10**(1), 20–26 (2021)
6. Ishizaki, Y., Koyama, Y., Takayama, T., et al.: A route recommendation method considering individual user's preferences by Monte-Carlo tree search and its evaluations. J. Inf. Process. **29**(11), 81–92 (2021)
7. Zhang, Q., Liu, Y., Liu, L., et al.: Location identification and personalized recommendation of tourist attractions based on image processing. Traitement du Signal: Signal Image Parole **38**(1), 197–205 (2021)
8. Liu, X., Jan, N.: Personalized recommendation algorithm of tourist attractions based on transfer learning. Math. Probl. Eng. **25**(15), 1–7 (2022)
9. Guowei, L.: Optimization method of tourist route in peak period based on ant colony algorithm. Comput. Simul. **37**(12), 353–357 (2020)
10. Zhang, Y., Tang, Z.: Cross-modal travel route recommendation algorithm based on internet of things awareness. J. Sens. **56**(10), 981–989 (2021)
11. Li, X.F.: Personalized planning method of rural tourism route based on landscape gene. J. Hebei North Univ.: Nat. Sci. Ed. **37**(07), 47–51 (2021)
12. Li, C.J.: Design of ecotourism personalized route planning system based on weather forecast data. Mod. Electron. Tech. **44**(04), 103–106 (2021)
13. Luo, Y.: SWOT analysis of new trend of rural tourism development in Xinjiang. Asian Agric. Res. **14**(6), 51–54 (2022)
14. Couto, G., Pimentel, P.M., Batista, M.D.G., et al.: The potential of rural tourism development in the Azores islands from the perspective of public administration and decision-makers. WSEAS Trans. Environ. Dev. **17**(5), 713–721 (2021)
15. Uri, N., Svitlica, A.M., Brankov, J., et al.: The role of rural tourism in strengthening the sustainability of rural areas: the case of Zlakusa village. Sustainability **13**(12), 1–23 (2021)

Vertical Search Method of Tourism Information Based on Mixed Semantic Similarity

Honghong Chen[✉] and Hongshen Liu

Heilongjiang Polytechnic, Haerbin 150080, China
gyg210422@163.com

Abstract. With the rapid development of the tourism industry, the volume of tourism information has increased exponentially, making it difficult for tourists to obtain the tourism information they need, which has become the main factor restricting the development of the tourism industry. In order to solve the above problems, the research on vertical search method of tourism information based on mixed semantic similarity is proposed. The Heritrix web crawler is used to collect tourism information and de duplicate it. On this basis, the Nutch structure is used to process tourism information, calculate the mixed semantic similarity between tourism information and known topics, and determine the corresponding topics of tourism information based on this, and develop an adaptive vertical search algorithm for tourism information. The vertical search results of tourism information can be obtained by executing the formulation algorithm. The experimental data shows that after the application of the proposed method, the maximum recall rate of vertical search of tourism information is 96%, the maximum precision rate of vertical search of tourism information is 98%, and the minimum response time of vertical search of tourism information is 0.56s, which fully proves that the proposed method has better application performance.

Keywords: Mixed Semantic Similarity · Vertical Search · Similarity Calculation · Tourism Information

1 Introduction

The rapid development of the national economy, the continuous improvement of people's living standards and the government's policy support for the domestic tourism industry have led to the rapid development of the domestic tourism industry, which plays an increasingly important role in the national economy [1]. China is rich in tourism resources and has a large consumer group. Tourism has developed into a wide range of industrial groups, including people's basic consumption needs of clothing, food, housing and transportation, as well as higher level needs of play, entertainment and shopping. The development of Internet technology has fundamentally changed the way people obtain information. For the tourism industry and its related fields, there are a large number of websites in China, including tourist attraction information websites, professional

L. Yun et al. (Eds.): ADHIP 2023, LNICST 547, pp. 91–107, 2024.
https://doi.org/10.1007/978-3-031-50543-0_7

websites such as hotels and airlines, comprehensive vertical search websites, and tourism channels of large portal websites. However, these websites have more or less problems, which are mainly reflected in: less effective information, more garbage information, poor user experience, and lack of features in content. On the one hand, users hope to quickly and accurately understand tourism related information through the Internet [2], on the other hand, they are confused and at a loss when facing the huge amount of mixed information. The general search engine is usually used as the entrance of information retrieval, but it is ineffective in accurately querying the target information. Although a large number of results can be retrieved according to keywords, it is difficult to really meet the needs of users. Most of the time, users also need to manually analyze and filter these information accurate location of target information is a time-consuming and laborious task for ordinary users. In this context, tourism vertical search engine came into being.

The tourism vertical search engine is aimed at the tourism industry and its related fields, automatically collecting relevant data, sorting out, filtering and artificially optimizing the data, and providing users with accurate tourism information retrieval services. Compared with general search engines, tourism vertical search engines have structurally extracted and integrated information related to the tourism industry, such as hotels, scenic spots, air tickets, catering, transportation and other information, and provided special services according to users' personalized needs. The tourism vertical search engine automatically extracts the relevant data of the tourism industry, saving a lot of time in searching for information. At the same time, it carries out structural processing on these data and integrates the data in combination with the different needs and interests of users, which can not only meet the current needs of users, but also explore the potential needs of users [3]. The tourism vertical search engine has completed a lot of work for users, such as information search, sorting, filtering, etc., making it easier and faster for users to obtain the information they want, and making the process of obtaining information more intelligent and humane.

The main core function of tourism vertical search engine is to vertically search tourism information and provide users with relevant tourism information services. The emergence and development of search engines have greatly facilitated people's access to information and played a positive role in promoting social progress. However, with the advent of the era of big data, the amount of data has exploded, and the amount of data generated every day in the world has reached 1EB. Although the general search engine continues to innovate in technology and improve its processing speed, it is still unable to meet people's diverse and personalized information retrieval needs [4]. According to the existing research results, the information search methods that are frequently used are the information search method based on correlation coefficient and the tourism vertical search method based on MongoDB. There is a time delay (mainly because the search engine cannot respond to information updates in a timely manner, the time required for classification and indexing has increased significantly due to the dramatic increase in data volume, and many effective information cannot be added to the index in a timely manner, with a certain time delay) The search accuracy is low (users have increasingly high requirements for the efficiency and accuracy of information retrieval, while general search engines have no advantages in the efficiency and accuracy of search) and other defects, which cannot meet the needs of tourists. Therefore, the research on vertical search method of tourism information based on mixed semantic similarity is proposed.

2   Research on Vertical Search Method of Tourism Information

2.1   Tourism Information Collection

Tourism information collection is the primary link of tourism information vertical search, which plays a vital role in the follow-up research. This research uses the Heritrix web crawler to collect tourism information, which lays a solid foundation for subsequent tourism information de duplication.

Heritrix is an open source web crawler tool developed by Java and used to crawl and archive web pages on the Internet. Herritrix provides interface management tools to control Herritrix, and also provides command line tools for users to select and call. Its structure is shown in Fig. 1.

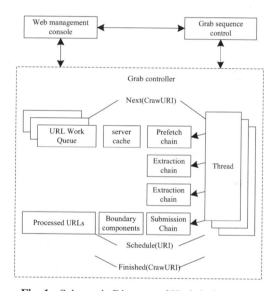

Fig. 1. Schematic Diagram of Heritrix Structure

As shown in Fig. 1, Heritrix is mainly divided into the following components: the Web management console, the crawl sequence controller, and the crawl controller [4]. The crawl controller is composed of Frontier, multithreaded processor, server cache and crawl range controller.

The functions of each module of Heritrix are as follows:

The web management console is implemented by jetty to provide users with a visual web-based management page. Through this page, you can set the modules used by the crawler at runtime. A good and convenient management interface promotes the wide application of Heritrix.In addition, Heritrix also provides a set of command line management tools to manage crawlers.

The crawl order controller is a configuration file based on XML format, which configures the crawl task. When the crawler runs, it calls the corresponding module according to the information in the configuration file to implement the crawl task [5]. In addition to the configuration of key modules, the configuration file also includes other important information, such as the definition of the captured URL range and the captured entry URL.

The grabbing controller is the core of the Heritrix web crawler, the general controller of the grabbing process, and the coordinator of each functional module to ensure that each module cooperates to complete the grabbing task together during the grabbing process. The crawl controller consists of three core modules, namely: Frontier, server cache and multithreaded processor.

Among them, Frontier is the URL scheduler, which is used to balance the fetching task, allocate the task to multiple different fetching threads to complete, and reduce the access pressure on a single Web server.

The server cache stores the persistent information of the captured server, including the server's IP address, DNS and other information, which can be called by various components of the crawler.

The multithreaded processor is the thread that actually executes the fetching work. In the running process, each thread will execute the URL processing chain once to complete the relevant operations of page fetching [6]. Herritrix generally runs in a multi-threaded way, that is, multiple URL processing chains are performing page fetching operations at the same time, and the fetching of each URL is independent of each other, and the fetching threads of different pages do not interfere with each other.

The advantages and disadvantages of Heritrix are shown in Table 1.

According to the content analysis in Table 1, Heritrix is suitable for vertical search of tourism information, so using it to collect tourism information on the Internet mainly includes two stages, namely web page collection and information extraction, as shown below:

(1) Web page collection.

Web page collection includes the collection of information on domestic and foreign tourism websites. Taking domestic tourism websites as an example, provide the specific collection process. Collect hotel data from Tongcheng, Ctrip and Xinxin websites, including HTML pages of hotel information and photos related to hotels, and store them locally in the form of mirror files. Collect scenic spot data from Tongcheng, Xinxin and Tuniu websites, including HTML pages of scenic spot information and photos related to scenic spots, and save them locally in the form of mirror files [7]. This function is mainly used to retrieve specific theme data from the target website using a web crawler and save it locally for subsequent operations.

(2) Information extraction.

Extract data in a specific format from the captured hotel data, generate Json format files, and store them in the collection. The root node of hotel data is hotel, and the specific data extraction format is shown in Table 2.

Table 1. Analysis of Advantages and Disadvantages of Heritrix

S/N	advantage	inferiority
1	Heritrix is fully functional.Heritrix itself only provides a framework for web crawlers, and there are many options in the same mode.Using the default provider module can record the captured pages well and avoid repeated work	Heritrix is only suitable for capturing Internet pages, that is, storing the image of the page. It cannot parse the content of the page itself. This part of parsing needs to be completed separately
2	Heritrix has excellent performance.Because Herritrix controls I/O and operations very well, it uses very little system resources when it runs. Even if it runs for a long time, it only generates very little system garbage, and its performance will not be reduced	Heritrix itself has many imperfections, which need to be expanded and improved to give full play to its potential.For example, Heritrix can't handle Chinese characters very well
3	Heritrix configurability.The Heritrix framework is flexible. Users can configure key modules to control the capture time, strategy, speed and scale of different websites. In addition, they can configure file archiving methods and capture target file formats	The configuration of Heritrix is very complicated, and there are many places that need customization.When changing a feature, you need to change many things. Some modules need to be changed to take effect
4	–	The fault tolerance and recovery mechanism of Heritrix are not perfect, and it needs to be expanded to achieve

Table 2. Hotel Information Extraction Format Table

content	Json	remarks
Original URL	OriginUrl	Original page link address
theme	Title	Original page subject information
Hotel name	HotelName	Name of the hotel
level	Level	Hotel star
Description	Description	Hotel related information
city	City	Hotel City
District/County	Zone	District/County where the hotel is located
address	Address	Address of the hotel
Picture Path	PicUrl	Hotel image storage path
floor price	LowestPrice	Hotel minimum price

Extract specific format data from the captured scenic spot data and integrate it into the sites collection. The root node of scenic spot data is site. Json format files are generated and stored in the specific data extraction format as shown in Table 3.

Table 3. Format table of scenic spot information extraction

content	Json	remarks
Original URL	OriginUrl	Original page link address
theme	Title	Original page subject information
Name of scenic spot	SiteName	Name of the scenic spot
Description	Description	Information about tourist attractions
city	City	City where the scenic spot is located
District/County	Zone	District/County where the scenic spot is located
Types of tourist attractions	SiteType	Types of tourist attractions
Picture Path	PicUrl	Storage path of scenic spot pictures

Collect tourism information through Heritrix according to the above process to provide basis for subsequent research.

2.2   De Duplication of Tourism Information

Most of the collected tourism information belongs to structured information, which has duplication and affects the effect of subsequent information search. Therefore, it is necessary to de duplicate the tourism information.

The de duplication processing of structured information in vertical search engines is one of the important structured information processing technologies in the improved vertical search engine model, which directly affects the accuracy of vertical search engines' search results for web information [8, 9]. The de duplication function of vertical search engines is mainly used in four parts, as shown in Table 4.

In the section of freeing storage space: By deduplicating data, duplicate data can be identified and deleted, thereby reducing the occupancy of storage space. This is particularly important in large-scale data storage and processing, as it can save costs and improve the overall performance of the system.

In the section of improving the efficiency of web page information collection: In web crawling and information collection tasks, data deduplication can effectively reduce the downloading and processing of duplicate web pages or data, and improve the efficiency of collection [10]. By removing duplicates, it is possible to avoid duplicate crawling of the same content, saving network bandwidth and computing resources.

Table 4. Application table of de duplication function

Application part	Application description
Free storage space	In the explosive development trend of the Internet, the repetition rate of information is constantly improving. De duplication of information with high repetition rate can significantly reduce the information storage space. Structured data is the main part of the data used by vertical search engines, and its storage requirements are high, so the problem of de duplication of structured information is crucial
Improve the efficiency of web information collection	In the process of crawling web information resources, search engines collect the content related to the search request sent by users. A large number of duplicate data resources in professional fields will lead to low efficiency of information collection of vertical search engines. After the de duplication of structured information of information resources, the collection efficiency of structured information resources can be better improved
Enhance user experience and improve user utilization	In the improved search engine model, the de reprocessing design of structured information effectively improves the sorting results of the index sorting module. Index sorting improves the accuracy of data retrieval results for the user interface. It can not only enhance users' experience of vertical search engines, but also improve users' utilization of vertical search engines
Improve the quality of retrieval data and the accuracy of index sorting	Vertical search engines have a preliminary problem of page duplication removal in the process of page crawling and page extraction. However, due to the diversity of the format of massive web information resources, the repetition rate of search results can only be optimized in a small range. Therefore, further data de reprocessing is needed in the vertical search engine to improve the quality of data retrieval results and the accuracy of index sorting

There is no distinction between enhancing user experience and improving user usage: for user generated content (such as social media, comments, etc.), data deduplication can avoid duplicate and redundant information being displayed to users, improving their reading and browsing experience. At the same time, reducing duplicate content can also increase user stickiness to the platform, improve user usage and satisfaction.

In improving the quality of retrieved data and the accuracy of index sorting, data deduplication can improve the quality of retrieved data in search engines and database systems. By removing duplicate data, the impact of redundant information on search results can be reduced, and search accuracy and efficiency can be improved. In addition, when building indexes and Sorting algorithm, data de duplication can avoid the interference of duplicate data on the sorting results, and improve the accuracy and effect of sorting.

The information processing technology carries out the pattern separation, data adjustment and relevant link analysis of the body content of the web page information stored in the database, and carries out a structural analysis process. Based on the structured data, the information is further processed such as de duplication and classification [11, 12]. An efficient de recalculation method should be used to process structured data, which improves the security performance of structured information. Through the study of the above several commonly used de duplication algorithms and in-depth analysis of their advantages and disadvantages, this paper proposes an improved algorithm with high efficiency and high security [13].

The basic idea of the improved de duplication method is that $T_i = \{t_1, t_2, \cdots, t_n\}$ represents the set of feature items of the top n tourism information with the highest weight, $W_i = \{w_1, w_2, \cdots, w_n\}$ represents the corresponding feature vector of the feature item set, $A(P_i)$ represents the webpage summary, $C(T_i)$ represents the string concatenated from the top n tourism information, $C[S(T_i)]$ represents the string concatenated from the top n tourism information after alphabetical sorting, $MD5(X)$ represents the N hash value of string X, $M(P_i, P_j)$ represents P_i and P_j are mutually repeated webpage. Use $A \Rightarrow B$ to represent 'A holds, then B holds'. Then the de duplication algorithm expression is.

$$\left.\begin{array}{c}(MD5(C(T_i))) = \left(MD5\big(C\big(T_j\big)\big)\right) \\ \left(\frac{|w_i - w_j|^2}{|w_i|^2 - |w_j|^2}\right) < \alpha\end{array}\right\} \Rightarrow M\left(P_i, P_j\right) \tag{1}$$

In formula (1), α represents an auxiliary parameter with a value range of [0, 1]; $MD5$ processes input information in 512 bit packets. Each packet is divided into 16 32-bit sub packets. After a series of processing, the output of the algorithm consists of four 32-bit packets. Cascading the four 32-bit packets will generate a 128 bit hash value. According to formula (1), the tourism information is de duplicated to obtain more indirect tourism information, which facilitates the follow-up research.

2.3   Theme Identification of Tourism Information

The reason for choosing the Nutch framework for tourism information segmentation is its open source, configurability, distributed architecture, Data cleansing and de duplication functions, as well as scalability and active community support. These advantages can quickly establish a tourism information segmentation system and achieve efficient and accurate data collection and processing. Among them, (1) Open source framework: Nutch is an open source web crawler framework that can be freely accessed and used. (2) High configurability: The Nutch framework provides rich configuration options that can be flexibly configured and adjusted according to specific needs. (3) Distributed architecture: The Nutch framework supports distributed crawling and processing, which can execute tasks in parallel on multiple machines, improving crawling speed and efficiency. (4) Data cleansing and de duplication: The Data cleansing and de duplication functions are built into the Nutch framework, which can clean and de duplicate crawled data through configuration and plug-ins. (5) Scalability and flexibility: The Nutch framework adopts a modular design, allowing you to expand functionality and customize development based on requirements and scenarios. (6) Community Support and Activity: Nutch is an active open source project with a large user and developer community. Therefore, based on the above de duplicated tourism information, the Nutch architecture is used to segment tourism information. On this basis, the mixed semantic similarity between tourism information and known topics is calculated, and the corresponding topics of tourism information are determined, providing support for the subsequent launch of adaptive vertical search algorithm.

The processing of Chinese information is mostly based on word processing. The Chinese information stored in the computer does not have obvious segmentation marks between words. Therefore, we must use the segmentation specification of Chinese words to convert Chinese information into words, which is the so-called Chinese word segmentation problem [14]. Chinese word segmentation module is a part of the preprocessing module, which is mainly used by the indexer. The indexer transmits the original text information to be processed to the word segmentation module for processing, while the Chinese word segmentation module returns the processed word segmentation results to the indexer with the corresponding data structure for subsequent processing, so that the indexer can use the word segmentation results to index documents.

Nutch is developed for English environment, so its Chinese word segmentation needs to be improved. The NutchAnalyzer class is an extension point for extending analysis text in Nutch. Writing your own Chinese plug-in must extend from this extension point. The specific word segmentation method is determined by the user according to the application. GB2312–80 divides the Chinese character code table into 94 areas, corresponding to the first byte; Each area has 94 bits, corresponding to the second byte. The value of two bytes is the area code value and the tag number value plus 32 (20H) respectively. The collected Chinese characters are placed in the 16–87 area, and the 0 bit of each area does not store Chinese characters. The offset calculation formula of Chinese characters in the code table is.

$$O = (C_1 - 0xB0) * 94 + (C_2 - 0xA1) \tag{2}$$

In formula (2), O represents the position of a Chinese character in the code table; C_1 and C_2 represents the internal code of Chinese characters.

The logical structure of the word segmentation dictionary is the data structure form after the dictionary is added to the memory. The logical structure of the dictionary used in this paper is shown in Fig. 2.

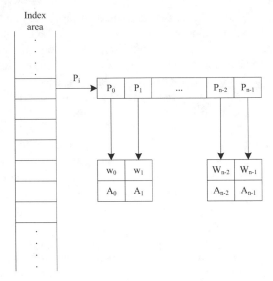

Fig. 2. Structure diagram of Chinese word list in memory

Among them, P_i is a pointer to all entries with the first character being the i-th Chinese character C_i; w_i is the k-th word with the initial character C_i (entries are arranged in internal code order from smallest to largest), excluding the initial character; A_i is the attribute of w_i (including value, ambiguity, position, part of speech, etc.); I_i is the index entry of the word C_i, occupying 5 bytes.

The tourism information is segmented according to the rules shown in Fig. 2. Based on this, the mixed semantic similarity between the tourism information and the known topics is calculated [15]. The semantic computing method proposed in this section emphasizes the importance of syntactic structure in computing sentence semantic similarity. The similarity between two sentences can be obtained by weighted summation of similarity between terms of syntactic components shared between sentences [16]. Unlike SyMSS, which only selects core words in dependency pairs and compares them, this paper calculates all words in the same dependency relationship. The syntactic structure and semantic hybrid algorithm in this paper uses the following formula to measure the semantic similarity between sentences. The expression is.

$$sim_1(S_1, S_2) = \frac{1}{n} \sum_{i=1}^{n} sim(E_{1i}, E_{2i}) - (N_1 + N_2 - N) \cdot PF \tag{3}$$

In Eq. (3), S_1 and S_2 are composed of N_1 and N_2 syntactic components, respectively, with N common syntactic components. Each syntactic component in S_1 is represented by E_1. Sentence S_2 is the same, where E_{1i} and E_{2i} have the same syntactic function; $(N_1 + N_2 - N)$ represents indicates the number of redundant syntactic components between two sentences, because if there are different grammatical components between two sentences, this method will ignore the syntactic information of these different components, and the syntactic information of these different components may also cause semantic differences. Due to different syntactic components, the similarity between these sentences cannot be calculated directly. Here, a balance parameter is used to indicate that there is additional information between these sentences; PF represents auxiliary calculation parameters.

Semantic similarity recognition will give different weight values to different syntactic components, so this paper also proposes a weighted syntactic combination method, which can measure the similarity between sentences through formula (4), which is.

$$sim_2(S_1, S_2) = \frac{w_0 A_0 + \cdots + w_n A_n + \sum_{i=1}^{n} w^o \cdot sim(E_{1i}, E_{2i})}{w_0 + \cdots + w_n} - (N_1 + N_2 - N) \cdot PF \tag{4}$$

In Eq. (4), $w_n A_n$ represents the product of the weight assignment of the subject and the semantic similarity value at the lexical level between the subjects. The product of the direct object and the verb is omitted. E_i is no defined common syntactic component. According to the formula, the calculation result of the above example is 0.7261. Since the subject and verb of the two sentences are consistent, their similarity increases after weighting.

The syntactic information is used to improve the semantic similarity at the sentence level, which largely avoids the neglect of sentence meaning in previous sentence level semantic similarity calculation, and improves the effect of sentence level semantic similarity. However, depending on the syntactic information of the sentence, there are also shortcomings. For example, in the two sentences "the train from Nanjing to Beijing leaves at ten o'clock" and "the train from Beijing to Nanjing leaves at ten o'clock", "the train from Nanjing to Beijing" and "the train from Beijing to Nanjing" are both attributes of the subject "train", so they are used as a dependency unit to calculate the lexical similarity between them. However, the algorithm at the lexical level is insensitive to the order of words, so the final similarity result of the two sentences is 1, which means they are identical. However, it is obvious that the two sentences are not identical semantically. Therefore, the order of words in the sentence is also related to semantics. This paper uses formula (5) to introduce editing distance as a feature to correct the effect of similarity calculation. The expression is.

$$sim^*(S_1, S_2) = sim_1(S_1, S_2) + \beta \frac{\min dist(w_{i1}, w_{i2})}{|S_1| + |S_2|} \tag{5}$$

In formula (5), $\min dist(w_{i1}, w_{i2})$ refers to the editing distance of the new sentence formed by extracting the same syntactic information vocabulary in order between two sentences. For example, the editing distance of the new sentence extracted in the above example is 1. $|S_1| + |S_2|$ refers to the sum of the number of words in the sentence, so that the influence of the editing distance of long and short sentences is consistent, and the value of this item is less than 1; β is a feature parameter configuration, and its value range is $(-\infty, 0]$.

According to the calculation result of formula (5), the tourism information theme recognition rules are formulated, as shown in the following formula:

$$\begin{cases} sim^*(S_1, S_2) \geq \delta & \text{Subject matching} \\ sim^*(S_1, S_2) < \delta & \text{Contradictory themes} \end{cases} \tag{6}$$

In formula (6), δ refers to the recognition threshold of tourism information subject, which needs to be set according to the actual situation.

Through the above process, the identification of tourism information topics is completed, which facilitates the subsequent adaptive vertical search.

2.4   Introduction of Adaptive Vertical Search Algorithm

Based on the above recognition results of tourism information topics, an adaptive vertical search algorithm for tourism information is developed, and the vertical search results of tourism information can be obtained by implementing the developed algorithm, so as to provide better services for tourists.

In this paper, the vector space model is used A_P and H_P. Set an initial threshold value to judge the relevance between the content of the web page corresponding to the network node and the query subject, and only the web pages with high relevance can be included in the root set or the corresponding base set. The algorithm is implemented as table 5:

Table 5. Implementation process of adaptive vertical search algorithm

Step	Content
1	Construct the root collection
2	Expand the base set
3	Loop through steps 1 and 2 and determine if the required number of extension pages that meet the pre-set threshold has been reached
4	Calculate the *Hub* and *Authority* values and normalize them. Until A_P and H_P converge, otherwise return to step 3
5	Output vertical search results for tourism information

Step 1: Construct the root collection [17]. The construction of the root set is very critical. It first requires users to focus on keywords in the field or industry; Secondly, set the weight according to the role of these keywords; Finally, we began to use these keywords to construct the root set. When constructing the root set, the first thing to be calculated is the relationship between query q and page t in the root set, expressed as.

$$\begin{cases} \zeta(q, t) < \gamma \quad \text{delete} \\ \zeta(q, t) \geq \gamma \quad \text{reserve} \end{cases} \tag{7}$$

In Eq. (7), $\zeta(q, t)$ represents the calculated value of the relationship between query q and page t; γ represents the limit value.

According to formula (7), if their relationship is not greater than the predetermined limit, t will be removed from the root set; If the calculated value exceeds the pre-set limit, then t becomes a node in the root set.

The number of root sets can be set as an upper limit according to the actual situation. When the upper limit is reached, the expansion of the number of root sets will stop. Then the root set expansion rules are as follows:

$$\begin{cases} K < \psi & \text{continue} \\ K \geq \psi & \text{cease} \end{cases} \tag{8}$$

In Eq. (8), K represents the number of root set nodes; ψ represents the upper limit of the root set node.

Step 2: Expand the base set. The root set is used to extend the base set. The extension method is the same as that of constructing the root set;

Step 3: Step 1 and Step 2 can operate circularly until the number of expansion pages that meet the preset threshold reaches the required number;

Step 4: Calculation *Hub* and *Authority* Value and normalize it. Until A_P and H_P convergence, otherwise return to step 3;

Step 5: Select the top n with the highest A_P and H_P values as the return result, which is the vertical search result for tourism information.

The above process completes the vertical search of tourism information, provides more abundant and accurate tourism information support for tourists, and promotes the development of the tourism industry.

3 Experiment and Result Analysis

3.1 Experiment Preparation Stage

The main task of the experiment preparation stage is to select evaluation indicators, and the selection results are as follows:

First, recall: refers to the calculation formula of the results returned most by the search engine according to the user's query criteria:

$$L = \frac{Q_1}{Q_{total}} \times 100\% \tag{9}$$

In Eq. (9), L represents the recall rate; Q_1 represents the number of pages related to the subject in the search results; Q_{total} refers to the number of pages related to all topics.

The second is the precision ratio: refers to the ratio of the number of theme related pages to the total number of returned pages in the results returned to users by the search engine. The calculation formula is.

$$G = \frac{Q_1}{Q_2} \times 100\% \tag{10}$$

In Eq. (10), G is the precision; Q_2 indicates the number of all pages in the search results.

Third, response time: refers to the time spent by the search engine from the user submitting the query criteria to returning the results, which directly reflects the efficiency of the search engine query mechanism. Due to the limitation of research space, its calculation formula will not be repeated.

3.2 Analysis of Experimental Results

Based on the above selected evaluation indicators, the information search method based on correlation coefficient and the tourism vertical search method based on MongoDB are used as comparison methods 1 and 2 to carry out the comparative experiment of tourism information vertical search. The specific analysis process of the experimental results is as follows:

3.2.1 Analysis of Recall Rate

The recall rate obtained through experiments is shown in Fig. 3.

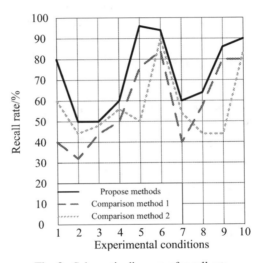

Fig. 3. Schematic diagram of recall rate

As shown in the data in Fig. 3, under the background conditions of different experimental conditions, the vertical search recall rate of tourism information obtained after the application of the proposed method is far higher than that of comparison method 1 and comparison method 2, and the maximum vertical search recall rate of tourism information obtained under the background of the fifth experimental condition is 96%.

3.2.2 Precision Analysis

The precision obtained through experiments is shown in Fig. 4.

As shown in the data in Fig. 4, under the background conditions of different experimental conditions, the vertical search precision of tourism information obtained after the application of the proposed method is far higher than that of comparison method 1 and comparison method 2, and the maximum vertical search precision of tourism information obtained under the background of the fifth experimental condition is 98%.

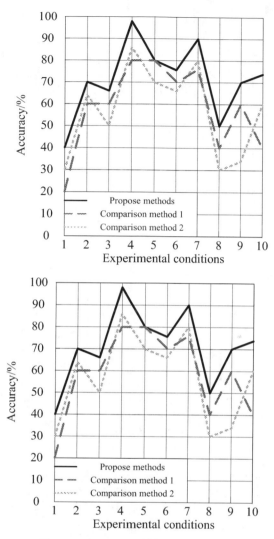

Fig. 4. Schematic diagram of precision

3.2.3 Response Time Analysis

The response time obtained through experiments is shown in Table 6.

As shown in the data in Table 6, under the background conditions of different experimental conditions, the vertical search response time of tourism information obtained after the application of the proposed method is far lower than that of comparison method 1 and comparison method 2, and the minimum vertical search response time of tourism information obtained under the background of the first experimental condition is 0.56s.

Table 6. Response Time Data Table/s

Test conditions	Propose method	Comparison method 1	Comparison method 2
1	0.56	2.56	4.26
2	1.23	3.02	5.78
3	0.80	4.56	6.35
4	0.75	5.01	4.15
5	1.11	5.89	7.25
6	1.25	6.12	6.59
7	1.04	4.12	8.01
8	0.89	4.52	6.89
9	1.01	6.35	7.45
10	1.20	7.01	8.10

4   Conclusion

With the development of the tourism industry, self-help travel has also developed rapidly. Tourism strategy information search has become a new demand. The development model of tourism strategy information search engine website, which takes tourism strategy information search as a development opportunity, and integrates a series of profit points such as hotels, air tickets, tourism products, is quietly emerging. Under this development trend, this paper proposes a new vertical search method for tourism information, introduces vertical search engine technology, provides users with more efficient and faster search tools, and creatively adds mobile elements. Users can download tourism information to mobile clients in the form of maps, upload photos anytime and anywhere, and combine GPS functions. Realize mobile phone search for surrounding living facilities and send scenic spots introduction to users. Experimental data show that the proposed method greatly improves the recall and precision of vertical search of tourism information, shortens the response time of vertical search of tourism information, and can provide users with more effective tourism information services.

Acknowledgement. 2022 Project of Heilongjiang Vocational College: "Preliminary Exploration of the Training Strategies for Higher Vocational Tourism Talents in the Background of Digital Economy" (No. XJYB2022097).

References

1. Jabalameli, M., Nematbakhsh, M., Ramezani, R.: Denoising distant supervision for ontology lexicalization using semantic similarity measures. Expert Syst. Appl. **177**(2), 114922 (2021)
2. Li, X., Li, H., Pan, B., et al.: Machine learning in internet search query selection for tourism forecasting. J. Travel Res. **60**(6), 1213–1231 (2021)

3. Chandrasekaran, D., Mago, V.: Evolution of semantic similarity—a survey. ACM Comput. Surv.Comput. Surv. **54**(2), 1–37 (2021)
4. Jiang, S., Cao, L.: Research on the secret homomorphism retrieval method of multiple keywords in privacy database. Comput. Simul. **39**(4), 408–412 (2022)
5. Liu, S., Xiyu, X., Zhang, Y., Muhammad, K., Weina, F.: A reliable sample selection strategy for weakly-supervised visual tracking. IEEE Trans. Reliab.Reliab. **72**(1), 15–26 (2023)
6. Kamran, A.B., Naveed, H.: GOntoSim: a semantic similarity measure based on LCA and common descendants. Sci. Rep. **12**(1), 1–10 (2022)
7. Lou, B., Zhao, W., Liu, X., Li, L., Ma, H.: Teaching design of tourism management based on information-based teaching method: a case study of selection of hotel construction site. Asian Agric. Res. **13**(12), 55–61 (2021)
8. Zhao, F., Zhu, Z., Han, P.: A novel model for semantic similarity measurement based on wordnet and word embedding. J. Intell. Fuzzy Syst. **40**(5), 1–12 (2021)
9. Yang, Z., Yang, L., Huang, W., et al.: Enhanced deep discrete hashing with semantic-visual similarity for image retrieval. Inf. Process. Manage. **58**(5), 102648 (2021)
10. Orlando, L., Ortega, L., Defeo, O.: Perspectives for sandy beach management in the Anthropocene: satellite information, tourism seasonality, and expert recommendations. Estuarine Coastal Shelf Sci. **262**, 107597 (2021). https://doi.org/10.1016/j.ecss.2021.107597
11. van der Vegt, A., Zuccon, G., Koopman, B.: Do better search engines really equate to better clinical decisions? If not, why not? J. Assoc. Inf. Sci. Technol. **72**(2), 141–155 (2021). https://doi.org/10.1002/asi.24398
12. Dias, L., Aldana, I., Pereira, L., et al.: A measure of tourist responsibility. Sustainability **13**(6), 3351 (2021)
13. Sánchez-Cervantes, J.L., Alor-Hernández, G., Paredes-Valverde, M.A., Rodríguez-Mazahua, L., Valencia-García, R.: NaLa-Search: a multimodal, interaction-based architecture for faceted search on linked open data. J. Inf. Sci. **47**(6), 753–769 (2021). https://doi.org/10.1177/0165551520930918
14. Varthis, E., Poulos, M., Giarenis, I., Papavlasopoulos, S.: A novel framework for delivering static search capabilities to large textual corpora directly on the Web domain: an implementation for Migne's Patrologia Graeca. Int. J. Web Inf. Syst. **17**(3), 153–186 (2021). https://doi.org/10.1108/IJWIS-10-2020-0062
15. Wu, C., Zhuo, L., Chen, Z., et al.: Spatial spillover effect and influencing factors of information flow in urban agglomerations—case study of china based on Baidu search index. Sustainability **13**(14), 8032 (2021)
16. Ting, X.: Chemistry course network teaching based on key information search and big data cloud platform. J. Intell. Fuzzy Syst. Appl. Eng. Technol. **40**(4), 7347–7358 (2021)
17. Wicaksono, A.F., Moffat, A.: Modeling search and session effectiveness. Inf. Process. Manage. **58**(4), 102601 (2021)

Retrieval Algorithm of Digital Information Resources for Legal Theory Teaching Based on Multi-scale Dense Network

Zefeng Li[(✉)], Lu Zhao, and Peihua Zhang

School of Marxism, Xi'an Eurasia University, Xi'an 710000, China
zhangyulan526@yeah.net

Abstract. To solve the problem that the existing retrieval algorithms have poor retrieval ability when retrieving digital information resources for legal theory teaching, the loss value in the iterative process of the retrieval algorithm is too large, which affects the retrieval performance and cannot meet the high-precision requirements of users for resource retrieval, the multi-scale dense network theory is introduced, Research on the design of digital information resources retrieval algorithm for legal theory teaching based on multi-scale dense network. Based on web crawler technology, obtain legal theory teaching document data sources, merge and transform different documents, introduce multi-scale dense networks, combine K-means algorithm and Canopy algorithm to cluster resources, allocate legal theory teaching digital information resources, and extract resource retrieval features based on word matching features, central sampling index library features, and topic related features, efficient retrieval of resources through information fusion. The experimental results show that the new retrieval algorithm has stronger retrieval ability in practical applications, and the loss value can rapidly decline in the algorithm iteration process, thus providing users with higher quality retrieval services.

Keywords: Multi-Scale Dense Network · Teaching · Information Resource Retrieval · Digitization · Legal Theory

1 Introduction

Since the beginning of the 21st century, the rapid increase of network bandwidth, the rapid development of network technology and information technology and the wide range of applications have greatly promoted the development of digital and networked education resources. The education resource network has also developed from the original campus LAN with single function and limited users to the Internet [1] that can connect to the Internet anytime and anywhere to provide teaching services for teachers and students. Educational resource network also plays an increasingly important role in modern scientific research and teaching. As more and more people access and share educational resources through the Internet and research institutions in colleges

L. Yun et al. (Eds.): ADHIP 2023, LNICST 547, pp. 108–121, 2024.
https://doi.org/10.1007/978-3-031-50543-0_8

and universities digitize the original educational resources, the educational resources on the Internet are growing exponentially. Therefore, the management and effective use of online educational resources has become a research topic for many researchers [2].The users of network education resources are network users. Therefore, the management and effective utilization of network education resources should also be discussed from the perspective of users in many aspects and from a new perspective.

Using the Internet to find resources has become one of the important ways to learn scientific research in today's Internet developed era. There are a lot of legal resources in the network resources, including shared resources of law schools and scientific research institutions, online legal resources provided by the government and official institutions, legal resources in the comprehensive network database, and so on. Different ways of legal theory digital information resources retrieval should use different search methods and retrieval skills.At present, law schools and scientific research institutions in many universities have established special websites as their own office aids, external publicity positions and teaching and research platforms. Among them, most law schools have relevant shared resources on their websites, Such as the recent teaching situation of the law school, the scientific research achievements of teachers, the latest development of the discipline and various legal resources. This kind of legal website is more academic and has a high value of use. In addition to professional law schools and scientific research institutions, there are many special websites established by governments or law related institutions at home and abroad to publish some bulletins, legal news, official documents, legal interpretations, case explanations and other legal resources. The resources provided by these websites are generally authoritative and timely, and also have high reference value.

Digital information resources refer to hardware resources, software resources and various information resources [3] connected by the Internet and distributed in different places. At present, the management and utilization of digital information resources in China mainly focus on the collection and construction of resources, as well as the construction of resource networks and corresponding management systems. However, due to the heterogeneous, dynamic and autonomous nature of the digital information resources of legal theory teaching, there are many problems in the management and utilization of resources, such as each school and scientific research unit has its own resource database, independent development, repeated construction of resources, etc. The result of these problems is that teachers and students need to spend a lot of time to find appropriate resources, It has greatly hindered the rapid development of current education [4, 5]. Therefore, based on enabling teachers and students to quickly retrieve the required resources, combined with the current research results of multi-scale dense network, it has a positive and practical significance for the research of digital information resources retrieval algorithm of legal theory teaching. In response to the problem of low resource retrieval efficiency, web crawler technology is used to obtain data sources for legal theory teaching documents. Combined with multi-scale dense networks, digital information resources are allocated to improve resource retrieval accuracy. According to different user retrieval needs, the extraction of digital information resource retrieval features for legal theory teaching is completed, and the retrieval results are fused, realize efficient retrieval of digital information resources for legal theory teaching.

2 Retrieval Algorithm of Digital Information Resources for Legal Theory Teaching Based on Multi-scale Dense Network

2.1 Data Source Acquisition of Legal Theory Teaching Documents

To establish the index of digital information resources for legal theory teaching, documents can be obtained from different places. The general method is the data obtained by web crawlers [6]. However, this article is a teaching resource retrieval, which can be directly linked to the corresponding legal theory teaching database, so it is mainly aimed at obtaining and saving MySQL data. The core processing class is abstracted as the Csource class. In order to have better data extensibility, it is an abstract class. This class mainly includes the method link data source. The CSource class has established the database link, and can flexibly obtain the target source data according to the configuration file. The core idea of the wordhit, that is, the wordhit encapsulated pos encapsulation, is to save the field in the first 24 bits of the byte, and the bit pos to save it to the second half of the byte. The byte operation is used to complete the encapsulation. CSource Document inherits CSource as the final document format of target source document conversion, including the construction process of Hit and the construction of document field. [7] CSource SQL is the inheritance of CSource Docment. Every record in the database is converted to the target Doc format.

Class Source Document inherits the base class Csource, creates a Hit according to the domain and index, and saves it. It is mainly used to obtain data from the database and save it to a temporary file in the specified format. The temporary file format is designed as shown in Fig. 1.It mainly stores the document ID of the word, the ID of the word in the word segmentation dictionary, and the specific location of the word in this document.

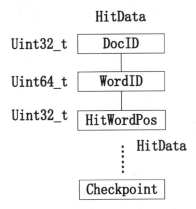

Fig. 1. Format Description of Temporary File

To specifically process each document retrieved from a Source, it is mainly to initialize the corresponding attribute field pattern through IteratieHits Start, that is, to build the mtSchema. The core of the pattern is the parsing of sql. The core of the pattern is vector < CcolumnInfo >, so in the initialization process, it is mainly to build the parsing container for each column.

In addition, it is necessary to establish a data link to the target database, obtain the text in the corresponding database according to the configuration, and use it as the data source of enterprise document index. The configuration file will include: first, the required parameters of the database link, such as user name, password, IP, port, etc. You can switch between different databases (https://www.westlaw.com/) quickly and effectively. Second, the query statement of the monthly data source, that is, the select statement query of the corresponding fields in the database to be full-text indexed, supports the set of select language. Flexible configuration of different target data. Third, establish the field configuration related to sorting of full-text index. Due to business requirements, there will be sorting methods different from classic relevance sorting, which will be explained in the search sorting module. Fourth, the tag attribute of the incremental index. In this part, the core domain of the incremental index is set in the configuration file for better incremental indexing, and incremental changes are made according to this domain.

Flexible expansion modules can be used in different systems through configuration files to quickly build education resource retrieval parts in different fields [8]. And it is easy to merge different documents. Parse the configured file, obtain the monthly standard source document, and then read and convert it. According to the above operations, complete the acquisition of the data source of the required legal theory teaching documents.

2.2 Allocation of Digital Information Resources Based on Multi-scale Dense Network

After obtaining the required data source of legal theory teaching documents, multi-scale dense network [9] is introduced to allocate the digital information resources of legal theory teaching. Consider a two-layer UDN, with MBS (macrobase stations) distributed in the center, multiple FBS, and multiple users (UE, user equipment) [10]. Figure 2 shows the random distribution of FBS and UE in MBS.

Set for FBS F Means:

$$F = \{1, \cdots, f, \cdots, F\} \tag{1}$$

UE set U indicates that the macro user set is U_m represented by the collection of micro base station users U_f means:

$$U_f = \{1, \cdots, u_f, \cdots, U_f\} \tag{2}$$

Because all FBS in the downlink of multi-scale dense network reuse the same spectrum, each spectrum is divided into L set for orthogonal subchannels L means:

$$L = \{1, \cdots, l, \cdots, L\} \tag{3}$$

user n from macro base station in sub channel m the signal to interference and noise ratio (SINR) can be expressed as:

$$\gamma_{m,n}^l = \frac{P_m^l g_{m,n}}{\sum P_m^l g_{m,n} + \delta^2} \tag{4}$$

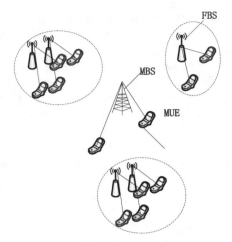

Fig. 2. Heterogeneous model of multi-scale dense network

In the formula, $\gamma_{m,n}^{l}$ indicates the user n from macro base station in sub channel m received letter dryness ratio; p_{m}^{l} represent macro base station m through subchannels l power allocated to users; $g_{m,n}$ indicates a subchannel l shanghong BTS m to its service users n channel gain; δ^{2} represents the variance of additive white Gaussian noise.

Similarly, users n on subchannels l from micro base station i the signal to interference and noise ratio (SINR) can be expressed as:

$$\gamma_{f,i,n}^{l} = \frac{p_{f,i}^{l} g_{f,i,n}}{\sum p_{f,i}^{l} g_{f,i,n} + \delta^{2}} \tag{5}$$

In the formula, $\gamma_{f,i,n}^{l}$ indicates the user n on subchannels l from micro base station i received letter dryness ratio; $p_{f,i}^{l}$ represents a micro base station i through subchannels l power allocated to users; $g_{f,i,n}$ indicates a subchannel l micro base station i to its service users n channel gain of. Further, the transmission rate of the base station user on the subchannel can be obtained:

$$R_{m,n}^{l} = W \log_{2}(1 + \gamma_{m,n}^{l}) \tag{6}$$

$$R_{f,i,n}^{l} = W \log_{2}(1 + \gamma_{f,i,n}^{l}) \tag{7}$$

In the formula, $R_{m,n}^{l}$ indicates from macro base station m to base station user n interchannel l transmission rate on; $R_{f,i,n}^{l}$ represent slave micro base station i to base station user n interchannel l transmission rate on; W indicates the bandwidth of each subchannel. Define binary variables $y_{m,n}^{l}$, $y_{f,i,n}^{l}$, this variable indicates whether the subchannel is multiplexed. When $y_{m,n}^{l}$, $y_{f,i,n}^{l}$ when the value of is 1, the user n multiplex subchannel l; When $y_{m,n}^{l}$, $y_{f,i,n}^{l}$ when the value of is 0, the user n no multiplexed subchannels l ◦ Through further calculation, the total throughput can be obtained.

On this basis, consider FBS resource allocation. From the analysis of multi-scale dense networks, it can be seen that finding a solution to the resource allocation of densely deployed FBS is the key to reduce interference and improve throughput. Therefore, this paper mainly studies the resource allocation algorithm of micro base stations in multi-scale dense networks. The problem solving in the above operation is divided into two sub problems, which become the deterministic optimization problem [11]. However, the densely deployed micro base stations will cause serious cross layer interference and same layer interference, resulting in the insignificant performance of the throughput improvement of the micro base stations.

The traditional K-means algorithm is simple and fast. In general, the closer the distance is, the greater the interference between the base stations will be. Therefore, the base stations are divided into the same cluster according to the principle of close distance. In order to reduce the interference between clusters, the clusters are allocated with orthogonal subchannels. However, the number of clusters and the location of the cluster center point need to be manually set when the traditional K-means algorithm clusters the micro base stations, External factors will lead to poor clustering results and cannot effectively reduce interference [12]. Based on this, in order to cluster the micro base stations more accurately, this paper introduces the Canopy algorithm to first "coarse cluster" the FBS and then "fine cluster" the K-means algorithm.

Canopy algorithm "coarse clustering" FBS distance threshold T_1 and T_2 selection is important, T_1 determine the number of clusters, T_2 determine the coverage of the cluster, T_1 slightly greater than T_2 although if some points are less than T_1 then this point will not be deleted from the micro collection. The condition of loop termination is that the element in the micro base station collection is empty, so a T, if the distance is less than T the point is directly divided into this cluster and deleted from the collection of micro NodeBs. If the point is larger than T then directly take this point from the collection of micro NodeBs as the cluster center to create a new cluster.

$$T = \lambda \frac{\sum_{i=1}^{N} d_{i,j} R_{m,n}^l R_{f,i,n}^l}{F(F-1)} \tag{8}$$

In the formula, $d_{i,j}$ represents the distance between two FBS; λ indicates the coefficient for adjusting the cluster size, which can be changed according to the real scene as the number of FBS changes dynamically. Canopy algorithm calculates the distance between all points to get the approximate number of clusters and the location of the cluster center. According to the distance between the initial cluster center and the FBS, K-means algorithm is used to "fine cluster" the FBS to finally get the clustering result of the FBS. The detailed clustering implementation process is as follows:

The first step is a hybrid clustering algorithm based on Canopy algorithm and K-means algorithm;

Step 2: initialize, and calculate the distance threshold of FBS according to the above formula T, FBS candidate set F, select a point from F at random f_i as the first Canopy, cluster center candidate set C;

Step 3: Judgment $d_{fi,fj}$ whether the value is less than $T \circ$ if $d_{fi,fj}$ less than T, FBS j from candidate set F deleted in;if $d_{fi,fj}$ greater than or equal to T, set f_i as a new Canopy;

Step 4, repeat the above operation until the cluster center does not change its position, and return the calculus clustering result.

In this section, Lagrange duality method is used to solve the sub channel allocation sub problem. Different sub channels involve different signal drying ratios, which affects the system throughput. While the traditional greedy algorithm based on signal drying ratio or channel gain is simple and easy, it cannot accurately allocate sub channels for users. Combining the above clustering results, select one cluster to allocate subchannels. This allocation method is also applicable to other clusters.Each micro base station allocates the average power on each orthogonal subchannel, which can transform the optimization problem in the above formula into the subchannel optimization problem of solving variables. The variable is extended to the interval [0,1], that is, the variable can take any value in the interval, and the above formula is transformed into a convex optimization problem to realize the allocation of digital information resources. The specific process is shown in Table 1.

Table 1. The Implementation Process of Hybrid Clustering

Step	Operate
Initialization	Calculate the distance threshold of FBS according to the formula, randomly select a point from F as the first Canopy, and construct the cluster center candidate set C
Canopy clustering	Perform Canopy clustering operation on each FBS: determine if the distance is less than the threshold, and if it is, delete the FBS from the selection set; If greater than or equal to, treat it as a brand new Canopy
K-means fine clustering	Select the center obtained from Canopy clustering as the initial cluster center, and use the K-means algorithm to fine cluster the FBS
Recurrent iteration	Repeat Canopy clustering and K-means fine clustering until the cluster center no longer changes

2.3 Retrieval Feature Extraction of Digital Information Resources for Legal Theory Teaching

After the distribution of digital information resources for legal theory teaching is completed, in order to ensure the subsequent retrieval accuracy, it is necessary to extract features of the allocated resources according to the different retrieval needs of different users [13]. The features that affect the retrieval of resource database are extracted from three aspects: term matching features, central sampling index database features, and topic related features. Among them, term matching is the most basic retrieval technique, which compares the keywords in the user query with the term items in the resource library to determine the correlation between the document and the query; Central sampling index library is a technology that aggregates large-scale data into smaller samples. By centrally sampling allocated resources, representative samples can be obtained, reducing the burden of computation and storage, and improving retrieval efficiency; Theme related features can help provide more personalized and accurate search results based on users' interests and needs. By analyzing users' query history, click behavior, and interest preferences, topic related features can be extracted to further optimize the ranking and recommendation of search results. Query likelihood: used to calculate the probability of query occurrence in the language model of the resource library:

$$\log P(q|C_i) = \sum \log(\lambda P(w|C_i) + P(w|G)) \tag{9}$$

In the formula, $\log P(q|C_i)$ represents a resource pool C_i query appears in q logarithmic probability of; $P(w|C_i)$ represented in the resource library C_i term appears in w probability of, $P(w|C_i)$ the value of is calculated by C_i the sum of the ratios of word frequency values and document length of all documents in; $P(w|G)$ it indicates the probability value of words appearing in all resource databases $P(w|C_i)$ the mean value of, which indicates that there are words in the global scope w the probability value of. By using global sub terms w probability value of $P(w|G)$ the term item appears in the value pair resource library w probability value of $P(w|C_i)$ smoothing to prevent $P(w|C_i)$ the value of is 0; λ represents the smoothing coefficient, and sets the λ the value is set to 0.8.Where the term w unigram and Bigram were used to calculate the title and body content of the document, and a total of four features were obtained.

Query term statistics: mainly calculate the frequency of query words in the resource database.

Interest. It includes the maximum and minimum value of the word frequency value of the query word in the resource database, and the maximum and minimum value of the TFIDF value. The calculation formula is:

$$tf_{\max}(q, r_i) = \max_{t \in q} tf(t, r_i) \tag{10}$$

In the formula, $tf(t, r_i)$ represents a resource pool r_i middle term t similarly, the word item t is calculated by Unigram and Bigram, and a total of 16 features are obtained by calculating the title and body content of the document respectively.

The characteristics of the central sampling index database are calculated by establishing the central sampling index database. The retrieval agent samples a certain proportion of documents from each resource database as the sampling document set, and uses the sampling document set to establish a central sampling index library. Given a specific query, the relevance score of the resource database is calculated by the resource database of the documents retrieved from the central sampling index database.

ReDDE and ReDDE.top features: ReDDE and ReDDE.top features are calculated using ReDDE and ReDDE.top resource database selection algorithm. In addition, the reciprocal of resource database ranking obtained by calculating ReDDE.top is used as Reverse_Rank feature.

Centroid_distance between the retrieval result and the center of the resource database: the distance between the top-k document in the retrieval result obtained from the central sampling index database and the center of the resource database. The assumption of this type of feature is that the closer the top-k document in the retrieval result in CSI is to the center of the resource library, the more relevant the resource library is to the query. By calculating the KL distance and cosine similarity between the vector mean of the top-k document and the center vector of the resource library, select k The value of is {10,50100}, and a total of 6 features are obtained. Considering that KL distance is a measure of distance rather than similarity, the reciprocal of KL distance is used as the corresponding eigenvalue.

Through the above operations, use the corresponding feature extraction methods to complete the extraction of features in the digital information resources of legal theory teaching.

2.4 Digital Information Resources Retrieval and Retrieval Results Integration

In information resource retrieval, the task of result fusion is to combine the retrieval results of multiple resource databases or subsets of data sets to obtain the final retrieval results. Generally, according to the access of the retrieval agent to the resource library, it can be divided into cooperative and non cooperative environments. In collaborative application scenarios, such as dividing large document data sets into small data subsets, when retrieving documents, select relevant data subsets for document retrieval. In this scenario, the efficiency of retrieval can be greatly improved by dividing a large data set into multiple document subsets, and then selecting the relevant document subset retrieval instead of the full set retrieval. In addition, by merging the result lists of multiple resource databases, the relevance of the final document sorting list results can be improved. In a non cooperative environment, if multiple search engine results are integrated and a single list is returned to the user's application scenario, we hope to obtain diversified relevant results through result fusion and improve the retrieval effect.

According to the characteristics of the retrieval environment of digital information resources for legal theory teaching, a ranking learning framework for result fusion is proposed, which combines the characteristics of multiple factors such as documents, result lists, resource databases and vertical fields. Based on this ranking learning framework, the accuracy of the task of result fusion is improved.

Retrieval result fusion is the fusion of retrieval results in the result list from different resource databases. The purpose of multi-source data fusion is to improve the efficiency of retrieval. In application scenarios in a collaborative environment, for example, by splitting a large document set into multiple document subsets, the cost of retrieval in a small subset of related documents is lower than that in full set retrieval. In application scenarios such as federated retrieval, it is often the result of merging multiple search engines, However, the condition of the document set in the search engine can only be obtained by distributing queries to the search engine. Figure 3 shows the frame structure of search result fusion.

When the retrieval results from different search engines need to be merged, because different search engines use different retrieval algorithms, the results in different result lists cannot be directly compared. Traditional result fusion algorithms, such as CORI, use CORI resource library selection algorithm to select resource library, and then weight the document score according to the resource library selection score to get the global score of the document. In this algorithm, the score formula of resource database or document needs to be fitted in both the resource database selection stage and the result fusion stage.

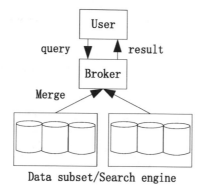

Fig. 3. Framework of search result fusion

The parameters in the formula are difficult to determine and there are many parameters that need to be debugged. Combining the framework of retrieval results fusion and the selected fusion algorithm, the fusion of digital information resources retrieval results is realized.

3 Experimental Analysis

3.1 Experiment Preparation

The program running environment of the experiment is developed and trained using Pycharm tool and Tensorflow deep learning framework.The Inter Xeon CPU E5–2683 v3 @ 2.00 GHZ dual core processor machine is equipped with the Ubuntu 14.04 operating system as the hardware experimental environment. The memory size is 256GB, and the GPU processor is NVIDIA K40.1000 cases of digital information resource data of law theory teaching were used in the experiment. The main purpose of this experiment is to verify whether the retrieval algorithm based on multi-scale dense network proposed in this paper has practical application feasibility, and to verify whether the new retrieval algorithm effectively overcomes the problems existing in the application of existing retrieval algorithms. The experiment uses Hamming distance to compare the application results of all retrieval algorithms. To facilitate the comparison, the retrieval algorithm based on multi-scale dense network in this paper is set as the experimental group, the retrieval algorithm based on ant colony algorithm is set as the control group I, and the retrieval algorithm based on TrieTree algorithm is set as the control group II. The application results of three retrieval algorithms are compared, So as to verify its application performance. Hamming distance is used in data transmission error control coding. Hamming distance is a concept, which represents the number of different characters at the corresponding position of two (same length) strings $d(x, y)$ two words x, y hanming distance between. Perform XOR operation on two strings, and count the number of 1, then this number is the Hamming distance.

After the corresponding experimental results are obtained, the performance of the three retrieval algorithms is compared and evaluated. The loss function can reflect the retrieval performance of the algorithm. The smaller the loss, the better the retrieval performance of the algorithm. Compare multi-scale loss with single scale loss. The loss value of the three retrieval algorithms in the iterative process is calculated, and the quantitative evaluation of the three retrieval algorithms is realized according to the calculation results. In this experiment, the following formula was selected for the loss function:

$$L_o = \arg\min \frac{1}{N}(Y - f(x)) \tag{11}$$

In the formula, L_o represents the loss value; N represents the number of iterations; Y represents the true value; $f(x)$ represents the search results obtained by the search algorithm. The results calculated by the above formula can realize the quantitative evaluation of retrieval ability, L_o the smaller the value, the better the retrieval performance of the retrieval algorithm; L_o the larger the value, the worse the retrieval performance of the retrieval algorithm.

3.2 Result Analysis

According to the above discussion, the results obtained under the application of three search algorithms are compared with the standard results of digital information resources for legal theory teaching required by users, and their corresponding Hamming distances are obtained respectively, and the results are drawn as shown in Fig. 4.

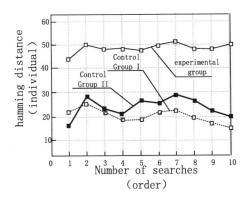

Fig. 4. Comparison of Hamming Distance of Three Retrieval Algorithms

Through further analysis of the number of Hamming distance 1 obtained from the above statistics, the Hamming distance can reflect the retrieval ability of the algorithm. The larger the Hamming distance, the stronger the retrieval ability. Similarly, the more Hamming distance 1 obtained from the statistics, the stronger the retrieval ability of the corresponding retrieval algorithm. According to this theory and the curve shown in the figure, the retrieval algorithm of legal theory teaching digital information resources based on multi-scale dense network in this paper has stronger retrieval ability, and can retrieve the legal theory teaching digital information resources needed by users.

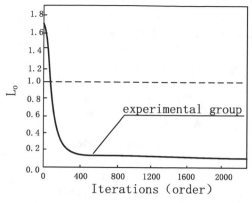

Fig. 5. Change of loss value during iteration of experimental group retrieval algorithm

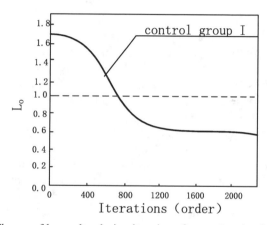

Fig. 6. Change of loss value during iteration of group I retrieval algorithm

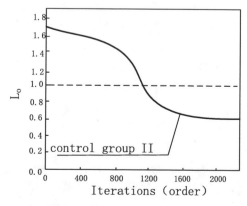

Fig. 7. Change of loss value during iteration of group II retrieval algorithm

The loss values of the three retrieval algorithms in the iterative process are calculated by the above formula, and the experimental results are respectively plotted as shown in Figs. 5, 6 and 7.

By analyzing the changes of the loss values of the three retrieval algorithms in the figure during the iteration process, it can be concluded that the experimental group retrieval algorithm can reduce its loss values to the ideal range with fewer iterations, and the loss values are always at the lowest level compared with the other two retrieval algorithms. The loss value of the retrieval algorithms of control group I and control group II declined slowly during the iteration process, and could not quickly reach the ideal retrieval state, and the loss value was still high at the end of the iteration. Therefore, the above results can prove that the experimental group's retrieval algorithm based on multi-scale dense network has higher retrieval performance.

4   Conclusion

As a supplement to law printing resources, closed databases, CD ROMs and other resources, Digital information resources of legal theory teaching play an important role in the learning and scientific research of relevant personnel. Therefore, only by effectively searching such resources can we regard the digital information resources of legal theory teaching as professional encyclopedias and make full use of resources. The retrieval of digital information resources is of great significance to the teaching of legal theory. In view of the shortcomings of existing retrieval algorithms in practical applications, propose a digital information resource retrieval algorithm for legal theory teaching based on multi-scale dense networks. Obtaining legal theory teaching document data sources through web crawler technology, using multi-scale dense networks to allocate digital information resources for legal theory teaching, and conducting information fusion processing to achieve efficient resource retrieval according to different user retrieval needs.

Considering the rapid changes and updating speed in the field of law, in future research, it is necessary to improve the updating speed of resources and adjust algorithms in a timely manner to provide more accurate and user-friendly resource recommendations.

Acknowledgement. Campus level project of Xi'an Eurasia University: Copyright protection of artificial intelligence creations (Project number: 2020XJSK23).

References

1. Huiping, L., Wenjuan, Z.: A generalized classification and coding system of human disease animal model resource data with a case study to show improving database retrieval efficiency. PloS one **18**(2) (2023)
2. Peng, J., Li, X.: Comprehensive retrieval method of MOOC teaching resources based on eigenvalue extraction. Int. J. Continuing Eng. Educ. Life-Long Learn. **33**(2–3) (2023)
3. Li, Q.: Information retrieval method of natural resources data based on hash algorithm. Int. J. Adv. Comput. Sci. Appl. (IJACSA) **14**(3) (2023)

4. Huan, W.: Fast retrieval method of massive library literature resources based on an online hash algorithm. Library, Harbin University of Science and Technology, Harbin, Heilongjiang, China **51**(3) (2023)
5. Fei, X., Saravanan, V.: An LDA based model for semantic annotation of web English educational resources. J. Intell. Fuzzy Syst. **40**(2), 3445–3454 (2021)
6. Huang, C., Chen, R., Lin, L.: Design and implementation of english teaching resources retrieval algorithm model based on deep learning. Adv. Comput., Signals Syst. **6**(3) (2022)
7. Xiaofeng, S., Tianjing, Z.: Research on intelligent retrieval method of teaching resources on large-scale network platform. Math. Probl. Eng. 2022 (2022)
8. Li, C., Zhen, Z., Jianjun, W., et al.: Dynamic retrieval model of quantitative data of power grid resources based on 3D geographic information systems (GIS). J. Nanoelectr. Optoelectr. **17**(2) (2022)
9. Wang, P., Zhu, H.: Single-image de-raining using joint filter and multi-scale deep alternate-connection dense network. Neurocomputing **457**, 306–321 (2021)
10. Soboleva, M.S.: The frequency and efficiency of the application of online sources for searching drug information and self-treatment. Pharmacophore an Int. Res. J. **13**(6) (2022)
11. Ning, W., Hui, Q., Yong, D., et al.: Transmission and drug resistance characteristics of human immunodeficiency virus-1 strain using medical information data retrieval system. Comput. Math. Methods Med. 2022 (2022)
12. Liqing, Z., Xiaowen, Y.: Intelligent retrieval method of mobile learning resources in the intelligent higher education system. Int. J. Syst. Assur. Eng. Manag. **13**(6) (2021)
13. Liu, S., He, T., Dai, J.: A survey of CRF algorithm based knowledge extraction of elementary mathematics in Chinese. Mobile Netw. Appl. **26**(5), 1891–1903 (2021)

A Storage Method of Online Educational Resources for College Courses Based on Artificial Intelligence Technology

Mingjie Zheng[1]($\boxtimes$) and Xu Wang[2]

[1] College of Humanities and Information, Changchun University of Technology, Changchun 130122, China
Zhengmingj541@163.com

[2] Tianjin Maritime College, Tianjin 300150, China

Abstract. In the Internet era, the scale of online education resources small files of college courses in online learning is becoming larger and larger, and the workload is large. The traditional storage method meets the storage requirements of massive education resource small files, and designs a storage method of online education resource of college courses based on artificial intelligence technology. Through artificial intelligence technology to recognize the characters and voice in online education resources of college courses, complete the recognition of online education resources of college courses in the platform. Gzip compression algorithm is used to compress recognition results. Design the distributed cloud storage architecture of online education resources to realize the distributed storage of compressed data resources. The test results showed that this method took less memory and took less time to read small files randomly, and has high reliability.

Keywords: Artificial Intelligence Technology · YOLOv3 · Online Education Resources · Gzip Compression Algorithm · Resource Storage

1 Introduction

With the rapid development of information technology, people's production, life and working methods are changing. The world we live in is becoming increasingly connected, interconnected and intelligent. The educational objectives and teaching models in the context of the traditional industrial era are no longer suitable for the needs of today's information and big data era. The concept and model of "smart education" came into being [1]. Various institutions and researchers have put forward many new innovations for the realization of intelligent education, such as the application practice form of online classroom and flipped classroom, changing the previous teaching mode, so that learners can obtain more independent learning time and corresponding learning resources, and make the learning process more efficient.

In recent years, with the introduction and gradual popularization of new education models, e-Learning and MOOC (Massive Open Online Courses), more and more learners

L. Yun et al. (Eds.): ADHIP 2023, LNICST 547, pp. 122–138, 2024.
https://doi.org/10.1007/978-3-031-50543-0_9

have the opportunity to learn on online platform. Accordingly, the platform has accumulated a large amount of teaching behavior data and knowledge resources, providing a good foundation for the update and improvement of the platform itself. However, the continuous growth of the data volume of online education resources for college courses has brought difficulties with its storage. At present, there are the following problems in the storage of online education resources for college courses:

(1) Capacity demand: the platform has established a file server to store digital data uniformly, but it still cannot meet the storage capacity demand of rapid resource growth;
(2) Access performance problem: the file server read and write operations are limited to the IO speed of a single server. With the increase of file data, the performance of file location and reading also increases and decreases significantly;
(3) The security of a single file server is relatively poor. It is very likely that the server storage device will be damaged and cannot be recovered. There is a lack of good backup mechanism. There are hidden dangers in data security, and it does not have the performance requirements of high reliability and high security;
(4) Sharing problem: the traditional silo storage mode and application development mode have caused the lack of unified planning and management of the current education system, and the scattered information resources have caused information islands, which cannot be effectively shared;
(5) Application requirements: The file server can only simply provide storage space, which cannot meet the actual needs of users. Its functions do not match the actual needs, and it is inconvenient for teachers and students to use. It is facing difficulties in resource retrieval, grouping, sharing, and so on.

Based on this background, this paper studies the storage of online educational resources for college courses. For the research on the storage of educational resources, many research results have been achieved. Reference [2] proposed an online learning resource compression storage method based on edge computing. In order to improve the compressed storage efficiency of online learning resources and reduce the occupation of cloud storage space, this method uses distributed compression sensing as the edge computing method. It collects online learning resources by constructing the first joint sparse model JSM-1 compression, reconstructs resources by combining the joint reconstruction algorithm composed of synchronous orthogonal matching pursuit algorithm and dictionary learning algorithm, obtains sparse dictionary atoms and measured values, and transmits them to the cloud server to establish a complete cloud sparse dictionary, realize compressed storage of online learning resources. Experiments show that the larger the sparse dictionary and the more training and learning samples are, the higher the SNR of resource compressed storage is, and the better the resource processing performance is. Using this method to compress online learning resources, the picture has a higher definition, and the compression time is shorter, which can save the compression time and reduce the storage space. Reference [3] building a digital resource library for higher education English through cloud platforms. The cloud platform enables the integration, efficiency, scalability, and interactivity of English resource libraries, while vocational colleges achieve the co construction and sharing of multi platform resources.

The construction of the English resource database in vocational colleges cannot be separated from the collaboration between institutions and enterprises. The requirements for various aspects of the vocational English digital resource database system are provided from three aspects: functional requirements, performance requirements, and operational requirements. The overall technical architecture and network topology of the education cloud platform where the vocational English digital resource database system is located are established to ensure the security, scalability, and operability of the platform, to ensure the collection and storage of resources. Cloud computing technology can integrate and manage various resources through technologies such as networking, virtualization, and distributed storage. Based on the proposed optimization strategy, the core module of the vocational English digital resource library system was designed, and some important interfaces and source code were displayed. However, the storage efficiency of this method's digital resource library is relatively low.

The problem is to design a storage method of online educational resources for college courses based on artificial intelligence technology.

2   Design of Online Educational Resources Storage Method for College Courses

The storage method for online education resources of university courses is mainly divided into three parts: identifying characters and speech in the online education resources of university courses, compressing the recognized characters and speech using Gzip compression algorithm, and distributed storage of compressed data resources. The storage structure diagram of online education resources for university courses is shown in Fig. 1.

2.1   Character Recognition and Speech Recognition

Through artificial intelligence technology to recognize the characters and voices in online education resources of college courses, complete the recognition of online education resources of college courses on the platform. Character recognition is divided into two steps: first, recognize the text part in the picture content and implement it into horizontal and vertical coordinates, which is called target detection; Then character recognition and extraction are carried out for the image content within the coordinate range. These two parts need different neural network models to complete.

The target detection algorithm for character recognition uses open-source YOLO. In the prediction process of YOLO, the target detection problem is first transformed into a linear regression problem in a high-dimensional space. First, the size of the graph to be detected is proportionally scaled and adjusted, and then the adjusted graph is sent into the model as the input of the convolutional neural network for recognition, finally, the output of the model is processed to get the detected target. The advantage of this is that the object center and object type will be judged only once for each grid area in the image, and the prediction analysis will not be repeated for a certain area when judging multiple possible results in the prediction process, which greatly improves the prediction efficiency.

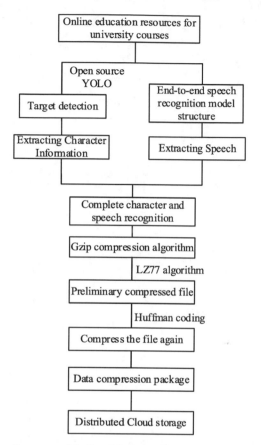

Fig. 1. Storage Structure of Online Education Resources for University Courses

YOLO's basic network model is GoogLeNet, and it has made corresponding improvements on its basis, giving up the inception. Instead, it alternately uses square convolution cores with compensation ranging from 1 to 3. A total of 24 convolutions are required, including fixed pooling steps and normalization processing, followed by two layers of fully connected neural networks. In the whole process mentioned above, the function of convolutional neural network is to extract features from the image, convert the features into one-dimensional data through normalization, input the full connection neural network, and then predict the coordinates and types of objects. This project uses an upgraded version of YOLO YOLOv3. Table 1 shows the neural network structure of YOLOv3.

Table 1. YOLOv3 Neural Network Structure

type	filters	size	output
convolutional	32	3X3	256 X 256
convolutional	64	3X3/2	128X 128
convolutional	32	1X1	64X 64
convolutional	128	3X3	64X 64
residual	64	3X3/2	32X32
convolutional	128	1X 1	32X 32
convolutional	256	3X3	16X 16
convolutional	128	3X3/2	16X 16
residual	256	1X1	8X8
convolutional	512	3X3	8X8
convolutional	256	3X3/2	64X 64
convolutional	512	3X3	256 X 256
residual	1 024	3X3/2	128X 128
convolutional	512	1X 1	64X 64
convolutional	I 024	3X3	64X 64
convolutional	I 024	global	32X32
residual	32	T000	32X 32
convolutional	64	1X 1	16X 16
convolutional	32	3X3	16X 16
convolutional	128	3X3/2	8X8
residual	64	1X1	8X8
avgpool	128	3X3	16X 16
connected	256	3X3/2	256 X 256
softmax	128	1X 1	128X 128

Here YOLO uses a convolutional neural network with a depth of 53 layers, in which the adjacent convolutional layers will form a convolutional residual network to resist the problem of gradient disappearance that may occur with the gradual deepening of the network layer. YOLOv3 also absorbs the idea of FPN121, and considers the characteristics of different sizes when performing target detection, which improves the accuracy of target detection in images, it is more sensitive to the content with smaller size on the image.

After the target detection algorithm extracts the text box in the image, the character recognition algorithm is required to extract the text information from the image content [4]. CRNN, a neural network structure, is used to extract the features of the text content in the image and make it one-dimensional, and then CTC is used to predict the text content sequence from the one-dimensional results output by CRNN. Since the task of character recognition is to extract the character information from the two-dimensional image, this task is suitable for both feature extraction using the CNN convolutional neural network model commonly used in computer vision and semantic association analysis using the RNN recurrent neural network model commonly used in natural language processing, so a CRNN neural network structure combining the two is designed.

The CNN structure of CRNN adopts the VGGI27 structure, and makes some fine adjustments to the VGG network. Because the final training results of the CNN part need to be imported to the RNN part, the CNN part must be output in one dimension, so the pool size of the last two largest pool layers of VGG is changed from 2 * 2 to 2 * 1 to increase the width of the feature map. The adjusted CNN structure neural network structure is shown in Table 2.

Table 2. CNN Structure of CRNN

Serial Number	Type	Configurations
1	Transcription	–
2	Bidirectional-LSTM	#hidden units:256
3	Bidirectional-LSTM	#hidden units:256
4	Map-to Sequence	–
5	Convolution	#maps:512, k:2 X 2, s:1, p:0
6	MaxPooling	Window:1 X 2, s:2
7	BatchNormalization	–
8	Convolution	#maps:512, k:3 X 3, s:1, p:1
9	BatchNormalization	–
10	Convolution	#maps:512, k:3 X 3,s:1, p:1
11	MaxPooling	Window:1 X 2, s:2
12	Convolution	#maps:256, k:3x 3, s:1, p:1
13	Convolution	#maps:256, k:3 X 3, s:1, p:1
14	MaxPooling	Window:2 X 1, s:2
15	Convolution	#maps:128, k:3 X 3, s:1,p:1
16	MaxPooling	Window:2 X 1, s:2
17	Convolution	#maps:64, k:3 X 3, s:1, p:1
18	Input	W X 32 Gy-scale image

After the CNN part is completed, the feature map is one-dimensional and transmitted to the RNN part. In order to prevent gradient dispersion in the RNN part during training, CRNN uses LSTM, a deformed RNN structure.

A training set containing 34W pieces of training data was made using all the vocabulary sets in the open source word segmentation library on Github. Each training data is a picture, and the content on each picture is a sentence with a length of 2–14 characters, including Chinese characters, Arabic numerals and Chinese punctuation marks. Draw all the words and sentences contained in the stutter on the complex background, and change the size and angle of each character, so as to approach the artistic font recognition in the real use scene; On this basis, about 15W labeled real scene data sets have been added, forming a total of 50W training data sets. Finally, CRNN model is used to train on this data set, and the final model obtained has an accuracy rate of 99.8% on the training set and 88% on the reserved real scene test set, which meets the expected requirements [5].

A TCN Transformer CTC end-to-end speech recognition model is designed, and the model structure is shown in Fig. 2. It consists of pre-processing module (acoustic pre-processing module, text pre-processing module), decoder, and mixed CTC/attention loss. It regards ASR as a sequence to sequence task. The encoder maps the input frame level acoustic characteristics (Formula (1)) to a sequence high-level representation (Formula (2)). The decoder decodes the generated text (Formula (3)) and the attention adjusted hidden state (Formula (2))β_l and finally generate the target transcriptional sequence (Formula (4)).

$$\alpha = (\alpha_1, \alpha_2, \ldots, \alpha_T) \tag{1}$$

In formula (1)α_T refers to the no T frame level acoustic characteristics for input.

$$\chi = (\chi_1, \chi_2, \ldots, \chi_N) \tag{2}$$

In formula (2)χ_N refers to the no N advanced representation of series.

$$\beta = (\beta_1, \beta_2, \ldots, \beta_{l-1}) \tag{3}$$

In formula (3)β_{l-1} refers to the no $l-1$ generated text.

$$\beta = (\beta_1, \beta_2, \ldots, \beta_L) \tag{4}$$

In formula (4)β_L refers to the no L target transcripts.

The pre-processing module is divided into acoustic pre-processing module and text pre-processing module. K 2-D convolution modules are used in the acoustic front-end module. Each convolution module contains a 2-D convolution layer and a ReLU activation layer [6]. Finally, we use Positional Encoding to obtain the absolute position information of acoustic features. In the text pre module, J TCN modules are used to learn the implicit positional relationship.

Speech recognition can be regarded as a timing problem. TCN is a new model structure based on convolution, which can deal with timing problems. Compared with using RNN structure, TCN has reached the level of RNN model in a variety of scene tasks. TCN can be divided into causal convolution and dilation convolution, of which causal convolution has strong temporal constraints, and dilation convolution can obtain a large receptive field through the setting of dilation factors. Using the TCN structure in the text front model can maintain the original good parallelism of the Transformer. In addition, it has the advantages of making the gradient more stable and occupying less memory than other convolution or RNN based structures.

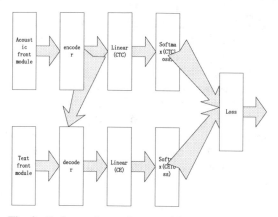

Fig. 2. End to end speech recognition model structure

The encoder and decoder are composed of several identical module stacks. Each module has two main sub layer structures, namely, the Multi Head Attention layer and the Feed Forward layer. After each sub layer, residual connection and layer normalization are used. The difference between the decoder and the encoder is that it uses a multi head attention mechanism to mask future information, so that future tag information cannot be seen during decoding, and the second Multi Head Attention layer uses cross attention. Different from the Transformer model in Linhao Dong et al., firstly, this paper adjusts the structure of the decoder. In the encoder part, the parallel TCN structure is used, which is used to fuse with the features processed by the Multi Head Attention layer, extract more features and slow down the disappearance of position information; Secondly, the encoder output part will also be input into the CTC structure to speed up the model training convergence speed and improve robustness.

In many mainstream model work, CTC target function is used as an auxiliary task, which can further improve the model effect. Because different from the attention model, CTC's forward backward algorithm can force monotonic alignment between voice and tag sequences, make up for the lack of attention alignment mechanism, and make the model more robust in noisy external environments. TCN Transformer CTC model combines the advantages of CTC and attention, so the total loss function is defined as the weighted sum of CTC loss and attention loss:

$$H_{\text{loss}} = \sigma CTC_{\text{loss}} + (1-\sigma)ATT_{\text{loss}} \tag{5}$$

Parameters in formula (5)σ is used to measure the importance of CTC loss and attention loss.

2.2 Compression of Recognition Results

Gzip compression algorithm is used to compress the recognition results. Gzip takes the Deflate algorithm as its core work. The Deflate algorithm mainly combines two algorithms to work together to complete compression and decompression. One is LZ77 algorithm, and the other is Hoffman algorithm. When Gzip uses the Deflate algorithm to work, it first reads the external incoming file stream. LZ77 algorithm first performs a preliminary compression of the file, and then transmits the compression result to the Hoffman algorithm. The result is compressed again through Hoffman coding to obtain the final data compression package.

LZ77 algorithm is an efficient lossless compression algorithm based on a sliding window mechanism similar to TCP network to solve network congestion and ensure reliable transmission. Its core principle is to use this sliding window mechanism to buffer encoded data in a sliding window of a set size. Then, when processing unencoded data, new characters will be used to match the encoded characters in the buffer, and the longest string will be found and recorded. The method of recording the matched maximum string is a triple, and its format is:

$$\omega = (\xi \ \psi \ \zeta) \tag{6}$$

In formula (6)ξ indicates the string offset; ψ indicates the length of repeated characters; ζ represents a new character that is not encoded.

In a triple, the string offset is the distance between the previous repeated character and the character being processed, and its size cannot exceed the size of the window. The length of the repeating character is the maximum matching length between the character in the window and the character being processed. An uncoded new character is the first character that is not repeated. If new characters cannot be matched from the window, output (0,0, new characters).

For example, if the string abcdbcddadcde is compressed, the compression process of LZ77 is as follows: when encoding the first character in the string, the character a is used to match the string in the sliding window. Since there is no character in the window at this time, the first character is encoded as (0,0, a). Then the second character is encoded. We find that the first four characters are all like this, so the offset and length are both 0. When the fifth character is encoded, the repeated string in the window is bcd, so the encoding offset starts from b, is 1, the repeated character is 3, and the next character is a, so the encoding triplet is (1,3, a). And so on until the last character. The detailed coding of each step is shown in Table 3.

Table 3. Detailed coding of each step

Compressed characters	Character to be processed	code
-	abcdbcdadcde	(0,0,a)
a	bcdbcdadcde	(0,0,b)
ab	cdbcdadcde	(0,0,c)
abc	dbcdadcde	(0,0,d)
abcd	bcdadcde	(1,3,a)
abcdbcda	dcde	(3,3,e)

It can be seen from the compression process in Table 3 that when the position of the same string in the whole string to be compressed is far away, that is, when the same string is sparse, each character must be encoded with a triplet. At this time, the compression efficiency of the algorithm will be relatively low, and even in extreme cases, the compressed data may be larger than before compression [7].

The decompression of each step is shown in Table 4.

Table 4. Decompression of each step

Read in code	Decoding processing	Decoded Data
(0,0,a)	From position 0, insert 0 decoded characters, and insert a new character	a
(0,0,b)	Starting from position 0, insert 0 decoded characters, and insert new characters as b	ab
(0,0,c)	From position 0, insert 0 decoded characters, and insert new characters as c	abc
(0,0,d)	Starting from position 0, insert 0 decoded characters, and insert new characters as d	abcd
(1,3,a)	From position 1, insert 3 decoded characters, and insert a new character	abcdbcda
(3,3,e)	Starting from position 3, insert 3 decoded characters, and insert new characters as e	abcdbcdadbce

There is a stack of data that has not yet been compressed. The data items in the data have different repetition frequencies. To distinguish different data items, a single code is allocated for each data item. The principle of allocation is that the code corresponding to the higher frequency is shorter, and on the contrary, it is longer. The above overview is the key implementation logic of Huffman coding. It can achieve that the average value of all data items is always in the shortest state. It is worth noting that the Huffman code can be created by scanning the data that has not yet been encoded twice. The two functions are respectively: sorting out the reproduction frequency of each data that has not yet been encoded and the actual coding work.

The implementation process of Hoffman algorithm is divided into two steps: building Hoffman tree and generating Hoffman code. First, we need to count the frequency of each character in the data worm, and use them as leaf nodes to build a binary tree. Then, by combining the two nodes with the lowest frequency, a complete Hoffman tree is continuously constructed. When generating Hoffman code, start from the root node to traverse the Hoffman tree. When encountering the left subtree, add 0 to the code, and when encountering the right subtree, add 1 to the code, until the leaf node is traversed, the Hoffman code of the character can be obtained.

2.3 Compressed Data Resource Distributed Storage

Design a distributed cloud storage architecture for online education resources to realize distributed storage of compressed data resources. The architecture design includes logical architecture design and physical architecture design. Logical architecture focuses on describing the relationship between different levels, while physical architecture focuses on describing the layout and composition of hardware devices. In the design of this architecture, we mainly focus on the design of distributed storage, that is, the design and use of distributed file architecture. This architecture is implemented in the resource layer of the logical architecture, which is mainly to organize and uniformly manage different storage resources (servers) under the physical architecture.

From the perspective of logical architecture, the distributed cloud storage architecture can be divided into four layers. The lowest layer is the resource layer, which corresponds to the physical resources to provide the most basic underlying distributed storage architecture. Then the upper layer is the service layer, which provides the user with a file read/write function interface. Above this layer is the application layer. Developers can use the lower layer service interface to improve management functions, here are the functional modules of the distributed cloud storage architecture. The top layer is corresponding to different access systems, including support for PCs, tablets, mobile phones, etc. The design of the logical architecture is shown in Fig. 3.

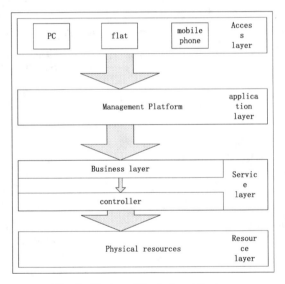

Fig. 3. Design of logical architecture

(1) Resource layer: it is the hardware foundation of the whole platform, including many servers/storage and other infrastructures. These resources are distributed in different places, including the existing data network data center room/data center hosted by a third party, and the computer room for disaster recovery. These resources are organized and managed using distributed file system and database cluster, and become the basic resource provider of distributed storage [8]. We use MFS open source architecture at this level.

(2) Service layer: carry out permission control, and provide standardized encapsulation of resources, application interfaces, etc., persistent storage of data, format conversion between different form data and background database, unified operation of

unstructured data and structured data, and users' online multi module asynchronous operation, etc., for upper layer applications, as well as third-party developers [9].

(3) Application layer: It is a cloud storage business system, mainly for teachers and students and other users to achieve resource management business operations. It is mainly responsible for the realization of business logic, modular division (business and general modules), and the realization of simultaneous online operation of the system by different users, that is, the functional requirements in the architecture requirements.

(4) Access layer: allows users to access through computer web pages, tablets, mobile phones, etc.

In terms of physical architecture, the distributed cloud storage architecture consists of multiple types of physical servers with different roles, and the composition relationship between them needs to be described. The distributed file system working with multiple server clusters is used as the underlying support, as shown in the following formula:

$$\psi = \begin{pmatrix} \text{MasterServer, Chunkserver,} \\ \text{Client, \ldots, Metalogger} \end{pmatrix} \tag{7}$$

In Formula (7)MasterServer refers to the metadata server; Chunkserver refers to data storage services; Client means the Customer; Metalogger refers to the metadata log server.

The upper layer sets up an application server to carry the business system and accept user requirements.

Each server has installed Centos 5.8bit, where:

(1) Application server: The application server uses open source Tomcat, which is the most popular free J2EE container at present; Deploy JBoss and system support services, including message queue, bus and other middleware services and JVM environment;

(2) Control node master: the cloud storage system adopts distributed storage architecture at the bottom, and uses MFS architecture for secondary development. The control node is the master node of the metadata server of the distributed file system;

(3) Data node chunkserver: MFS distributes data on different machines, called data nodes, and establishes multiple replicas to ensure reliability;

(4) Database MySQL node: MySQL database is used as the database of the business system, and the main language is the structured query language. Access the database through JDBC to realize the interaction between the database and the application. Read write separation configuration is used to improve the application access performance.

The data distribution of the distributed cloud storage architecture is implemented as follows: first, the file data is divided into 64 MB chunks, as shown in the following formula:

$$\upsilon = \left(CHUNK_1^{64}, CHUNK_2^{64}, \ldots, CHUNK_t^{64} \right) \tag{8}$$

Then they are stored on multiple chunkserver nodes, and there is a certain algorithm to select chunkserver during storage, so as to prevent uneven load distribution of data blocks between chunkservers of the system[10].

3   Experimental Test

3.1   Server Configuration

For the designed storage method of online education resources for college courses based on artificial intelligence technology, in order to test its performance, configure the server as shown in Table 5.

Table 5. Configured Servers

S/N	role	system	IP
1	MFS Master、keepalived Master	CentOS6.2 64bit	192.182.10.101
2	MySQL	CentOS6.2 64bit	192.182.10.102
3	MySQL	CentOS6.2 64bit	192.182.10.103
4	MFS Metalogger、keepalived Backu	CentOS6.2 64bit	192.182.10.104
5	MFS Chunkserver	CentOS6.2 64bit	192.182.10.105
6	MFS Chunkserve	CentOS6.2 64bit	192.182.10.106
7	APP Server、MFS Client	CentOS6.2 64bit	192.182.10.107
8	Virtual IP, configured by 192.182.10.101 and 192.182.10.104 respectively to provide services	-	192.182.10.108

The master is the core of MFS and needs to be installed on a server with high stability and high configuration. Large memory, large hard disk, and high CPU requirements. A machine is used as both the metalogger and the backup master server, so that machine should use the same configuration as the master.The installation configuration keeps alive: 192.182.10.101 and 192.182.10.14. Where 101 is the master and 104 is the backup. Monitor the master server through keepalived. When the mfsmaster service on the master server 192.182.10.101 has a problem, it will automatically switch to the backup server 192.182.10.104.

3.2 Test Items

The purpose of performance testing is to check whether the performance requirements are met. The test items are as follows:

(1) Memory usage

To analyze the memory usage of NameNode by different numbers of files, upload 2000, 4000, and 10000 small files respectively, and observe and record the memory usage.

(2) File reading experiment

Upload 2000, 4000 and 10000 small files of educational resources respectively. In the test, the online learning resources compression storage method based on edge computing (Reference [2] method) and a digital resource library of english for higher education based on a cloud platform (Reference [3] method) are used as the comparison method, which is represented by Method 1 and Method 2 respectively.The random reading operation experiment and sequential reading experiment of small files were conducted respectively, and the corresponding average value was obtained as the experimental data after several experiments.

(3) Use three methods to upload five types of files: text files, image files, video files, audio files, and compressed files, and compare the situation of uploading files using different methods.

3.3 Test Results

The test results of memory usage are shown in Fig. 4.

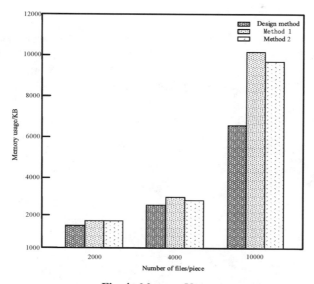

Fig. 4. Memory Usage

As can be seen from Fig. 4, the memory occupation of file name nodes of different orders of magnitude is different. When the number of files is small, the occupation of design methods is not much different from that of comparison methods. When the number of files is 10000, the memory usage of Method 1 and Method 2 is 10100KB and 9800KB, respectively. The memory usage of the design method is only 6700KB, indicating that the design method can reduce memory usage and has good stability.

In the file reading experiment, the random access experiment results of small files are shown in Fig. 5.

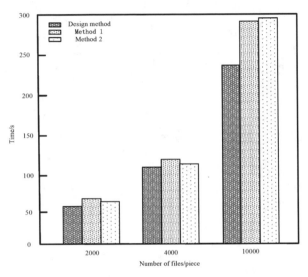

Fig. 5. Comparison of Random Reading Time of Small Files

From Fig. 5, it can be seen that as the number of small files increases, the random read time of small files using different methods also increases. When the number of small files is 10000, the random read times for Method 1 and Method 2 are 285s and 290s, respectively, while the design method only takes 240s, indicating that the design method is more efficient in randomly reading small files.

The file upload situation of different methods is shown in Table 6.

From Table 6, it can be seen that uploading image files using Method 1 failed, while uploading video files and compressed files using Method 2 failed. The design method can successfully perform upload operations for all file types, indicating its high reliability.

Table 6. File Upload Status by Different Methods

File type/type	Design method	Method 1	Method 2
Text file	Success	Success	Success
Image file	Success	Fail	Success
Video file	Success	Success	Fail
Audio files	Success	Success	Success
Compressed file	Success	Success	Fail

4   Conclusion

With the popularization and wide use of the Internet, online learning is becoming more and more popular, and online education resources of college courses in the network are becoming more and more abundant, especially the learning resources in the form of text. Design a storage method of online education resources of college courses based on artificial intelligence technology. Using artificial intelligence technology to recognize characters and speech in online educational resources, the text pre model in speech recognition uses TCN structure to maintain good parallelism of the Transformer and reduce memory usage. The Gzip compression algorithm is used to compress the identification results to design the distributed Cloud storage architecture of online education resources, realize the distributed storage of online education resources of college courses. The storage method of online educational resources for university courses based on artificial intelligence technology has great potential in the future, not only providing personalized, interactive, and efficient learning experiences for students, but also helping teachers better conduct teaching. With the development of artificial intelligence technology, the recognition process of characters and speech in online education resources will be improved in the future to achieve better storage effects.

References

1. Bowker, L.: Translating for Canada, eh?: Developing open educational resources to support localization into Canadian English and French. J. Int. Local. **8**(2), 156–164 (2021)
2. Yuan, L.: Research on compressed storage method of online learning resources based on edge computing. J. Ningxia Normal Univ. **43**(01), 76–83 (2022)
3. Wang, J., Li, W.: The construction of a digital resource library of English for higher education based on a cloud platform. Sci. Program. (Pt.10), 4591780.1–4591780.12 (2021)
4. Tomas, V., Solomon, P., Hamilton, J., et al.: Engaging clinicians and graduate students in the design and evaluation of educational resources about universal design for learning. Can. J. Speech-Lang. Pathol. Audiol. **45**(1), 59–75 (2021)
5. Barkhatova, D.A., Simonova, A.L., Lomasko, P.S., et al.: Features of "inverted" educational resources for distance learning of pupils. Open Educ. **25**(4), 4–12 (2021)
6. Hakim, A.: Study of open educational resources: a survey based approach. Acad. Int. Multidisc. Res. J. **11**(2), 1546–1553 (2021)

7. Hongmin, S., Gongjian, Z.: Research on a new algorithm of erasure correcting code for distributed storage data based on decision tree model. Comput. Integr. Manuf. Syst. **39**(6), 473–477 (2022)
8. Khan, N.A., Shafi, S.M.: Open educational resources repositories: current status and emerging trends. Int. J. Dig. Liter. Dig. Comp. **12**(1), 30–44 (2021)
9. Borzutzky, C.: Adolescent medicine and pediatric residency training: the value of collaboration and shared educational resources. The Journal of Adolescent Health: Official Publication of the Society for Adolescent Medicine **68**(5), 842–843 (2021)
10. Camel, V., Maillard, M.N., Descharles, N., et al.: Open digital educational resources for self-training chemistry lab safety rules. J. Chem. Educ. **98**(1), 208–217 (2021)

Allocation Method of Teaching Resources of Talent Training Course Based on BP Neural Network

Yan Liu[1]([envelope]), Senwei Wang[1], and Hua Sui[2]

[1] Dalian University of Science and Technology, Dalian 116038, China
15904962696@163.com
[2] Shenyang Aerospace University, Shenyang 110136, China

Abstract. The traditional allocation method of teaching resources is unbalanced, which leads to the low utilization value of teaching resources in talent training courses. In order to make the allocation of teaching resources of talent training course balanced and facilitate the management and application of teachers before, during and after teaching, a method of allocating teaching resources of talent training course based on BP neural network is proposed. Firstly, the BP neural network model is constructed according to the framework of teaching information resource allocation of talent training courses, and the information characteristics of teaching resources of talent training courses are extracted. According to the characteristics, the target object feature tag is managed to complete the information allocation of teaching resources of talent training courses. The experimental results show that the proposed method has good performance in practical application and can promote the utilization value of teaching resources in talent training courses.

Keywords: BP Neural Network · Personnel Training · Curriculum Teaching Resources · Configuration Method

1 Introduction

Talent cultivation is related to the future and development of a country, therefore, talent cultivation has always been a hot issue in education. With the rapid development of science and technology, artificial intelligence technology has been widely used in the field of education [1]. With the continuous development and improvement of information technology, artificial intelligence technology has also been widely used in the field of education. Among them, the teaching of talent training course based on BP neural network has become a hot research direction [2]. As a mature artificial intelligence technology, BP neural network is essentially a feedforward artificial neural network, which can be used to deal with the nonlinear relationship between input and output. In the teaching of personnel training courses, by using BP neural network, students' learning ability, personal needs, hobbies and other factors can be taken into account, so as to provide personalized curriculum resources and achieve higher teaching results [3].

© ICST Institute for Computer Sciences, Social Informatics and Telecommunications Engineering 2024
Published by Springer Nature Switzerland AG 2024. All Rights Reserved
L. Yun et al. (Eds.): ADHIP 2023, LNICST 547, pp. 139–151, 2024.
https://doi.org/10.1007/978-3-031-50543-0_10

In the teaching of personnel training course, BP neural network can be used to optimize the allocation of resources and improve the teaching effect. Specifically, through BP neural network, students' learning ability can be deeply analyzed, and course content and teaching material resources can be scientifically planned according to students' individual needs, so as to improve students' learning efficiency and effectiveness.

Reference [4] proposes a learning path recommendation system for programming education based on neural network. This system applies recursive neural network to learners' ability map, which shows learners' scores. In short, the learning path is constructed from the submission history of learners through the process of trial and error, and the ability chart of learners is used as an indicator of their current knowledge. Reference [5] puts forward the practical research on the construction of teaching resources of specialized courses under the background of educational informatization. The article points out the difficulties in the current construction of teaching resources of specialized courses in educational informatization, illustrates the construction ideas of teaching resources of specialized courses from the perspective of multi-party cooperation in Industry-University-Research, and puts forward the specific curriculum scheme and three-dimensional teaching content of Principles of Management, which is of great significance to enrich the construction of teaching resources of specialized courses under educational informatization. Reference [6] puts forward the configuration method of open education resource database oriented to graphic database. The progress of telecommunications stimulates personalized learning and collaborative learning, which enables the teaching and management personnel of educational institutions to carry out teaching innovation. These innovations include open educational resources, which have promoted teaching practice and students' free learning since its establishment. Therefore, the goal of this study is to design an open education resource library, configure a recommendation system and a graphics-oriented database.

On the basis of the above research, this paper puts forward a method of allocating teaching resources for talent training courses based on BP neural network. The innovation of this method is to construct a BP neural network model through the framework of teaching information resource allocation of talent training courses, extract the information characteristics of teaching resources of talent training courses, and set the feature label of feature management target object according to the extraction results to complete the information allocation of teaching resources of talent training courses. This method can help teachers better carry out personalized education, improve teaching effect and accelerate students' learning process. Tailor-made courses for students in an intelligent way, so as to realize personalized education and improve teaching effect and students' learning results.

2   Allocation of Teaching Resources of Talent Training Course Based on BP Neural Network

BP neural network can automatically learn and discover the rules and patterns of resource allocation by learning a large number of data samples, which is more data-driven, and can adjust itself according to the actual situation to improve the accuracy and effect of resource allocation. BP neural network can continuously optimize the allocation of resources through the back propagation algorithm, so that the teaching resources can more accurately match the needs of students and improve the teaching effect. The network can dynamically adjust according to students' learning feedback and performance, and correct and improve the resource allocation strategy in time.

The information allocation of teaching resources of talent training courses is based on the newly added information and historical information of teaching resources of talent training courses, and the teaching information resources of talent training courses are updated. Based on the information of teaching resources of talent training courses of BP neural network, this paper constructs a resource information allocation system framework with the best performance and the fastest efficiency, and its system framework is shown in Fig. 1.

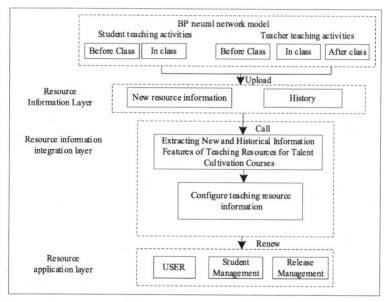

Fig. 1. Framework diagram of teaching information resources allocation for talent training courses

As can be seen from Fig. 1, students' activity information and teachers' teaching information generated by using BP neural network to teach talent cultivation courses are uploaded to the resource information layer, and the resource information allocation layer calls the newly added and historical teaching resource information of talent cultivation courses in the resource information layer, and realizes the allocation of teaching resource information of talent cultivation courses through BP neural network model and feature extraction of teaching resource information of talent cultivation courses, and finally updates the allocated information to the resource application layer for use by teachers, students and managers.

2.1 BP Neural Network Model

BP neural network is an artificial neural network based on back propagation algorithm, also known as multilayer feedforward neural network [7, 8]. It consists of an input layer, an output layer and several hidden layers, and each neuron is connected with the neurons in the previous layer and the next layer. BP neural network adjusts the connection weight between each neuron through multiple iterations and back propagation errors to realize the mapping of input and output. The specific structure is shown in Fig. 2.

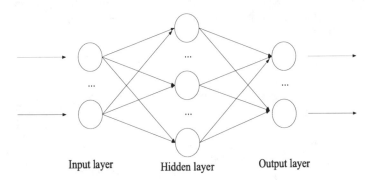

Input layer   Hidden layer   Output layer

Fig. 2. Schematic diagram of BP neural network model structure

In the training process of BP neural network, the parameters of the network are adjusted by making the input data undergo multiple back propagation errors until the error is minimized, so that the network can accurately map the input to the required output. In the allocation of teaching resources for talent training courses, BP neural network can be applied to many fields, such as talent selection, student evaluation, curriculum recommendation and so on [9]. Through the training and analysis of data, BP neural network can help teachers better understand students' learning patterns, learning behaviors and learning achievements, so as to design courses and allocate resources more scientifically and effectively. There are n neurons in the input layer, and these neurons simultaneously constitute a node on each subordinate line; There are n different groups in the same hidden layer, and the data value of n can be set according to different conditions, in which each group can be composed of k small neurons [10]; After reaching the second hidden layer, there will be n groups, which means that each group at this level is composed of k small neurons; And every neuron in the input level is connected with every other neuron in the hidden level; Each sub-network in the network model has only one output layer, and the final result is determined by the output layer of each sub-network.

2.2 Information Feature Extraction of Teaching Resources in Talent Training Course

The allocation layer of teaching resources of talent cultivation course extracts the information characteristics of teaching resources of talent cultivation course from the information layer of teaching resources, which needs to consider the new information and historical information of teaching resources of talent cultivation course [11–13], and extract the information characteristics of teaching resources of talent cultivation course from the overall point of view, so as to avoid missing hidden or repeated construction information when allocating teaching resources information of talent cultivation course [14, 15]. Using BP neural network model to process the new information and historical information of teaching resources, and complete the extraction of information characteristics of teaching resources [16, 17].

Suppose C_1 represents the information of the original window, C_2 represents the information of the incremental window, and the expressions of C_1 and C_2 are respectively:

$$C_1 = [c_1, c_2, \cdots, c_z] \tag{1}$$

$$C_2 = [c_{z+1}, c_{z+2}, \cdots, c_{z+m}] \tag{2}$$

In formula (1) and formula (2), z and m represent the information of the original and newly added windows respectively, and the teaching resource information C of the talent training course contains C_1 and C_2, and the mutual information matrices of C, C_1 and C_2 are represented by D, D_1 and D_2 respectively, and the D expression is expressed as:

$$D = (z + m)^{-1} \times (D_1 + D_2) \tag{3}$$

In formula (3), the characteristic decomposition of D_1 can be realized by diagonalization, and its decomposition can be expressed by unit matrix $\overline{E}_1$, and the expression of $\overline{E}_1$ is:

$$\overline{E}_1 = H_1^T \times D_1 \times H_1 \tag{4}$$

In formula (4), H_1 and H_1^T represent the overall features of the original window information source and the decomposed features.

The projection received by H_1 constitutes space, and the calculation formula is:

$$\overline{E}_2 = H_1^T \times D_2 \times H_1 \tag{5}$$

Using the above calculation process, the decomposition results of teaching resources characteristics of talent training courses are obtained [18–20]. Through the decomposition results, the characteristic values of teaching resources information of all talent training courses are obtained, and the expression is:

$$F_{GH} = (\alpha + \beta)^{-1} \times (1 + \alpha_k) \times (\overline{E}_1 + \overline{E}_2) \tag{6}$$

In formula (6), the feature vector of teaching resources of talent training courses is represented by P_Z, and the calculation formula of P_Z is:

$$P_Z = F_{GH} \times K_k \tag{7}$$

In Formula (7), K_k represents the k th feature vector, and according to the information feature P_Z of teaching resources of talent training courses, the allocation of teaching resources of talent training courses is developed.

2.3 Target Object Feature Tag Management

In order to ensure the accuracy of the allocation of teaching resources in talent training courses and meet the needs of different learners for personalized allocation of teaching resources, a learner portrait feature tag is established according to the feature vector P_Z of teaching resources information of talent training courses and learners' learning operation [21–23]. Based on the information such as learners' historical learning preferences, priority selection times for a specific type of resources, and targeted retrieval of resources in a short period of time, user portrait tags are established [24, 25]. When the user portrait label is used as the resource allocation, the type of allocation resources prefers the order, but the difficulty level of the specific allocation of teaching resources for talent training courses needs to be determined according to the learners' cognitive level.

When allocating resources, this method determines the cognitive level of learners according to their mastery of knowledge points, so as to allocate resources with less difficulty across the gradient. Suppose the learner is O, the user portrait tag sequence of the learner is $\{o_1, o_2, \cdots, o_m\}$, the knowledge set of the talent training course that the user has learned is $Q = \{q_1, q_2, \cdots, q_m\}$, and the test set of the corresponding knowledge point is $K = \{k_1, k_2, \cdots, k_m\}$. Represent learners' mastery of this knowledge point by the correct rate of answering questions in different time periods:

$$\zeta_{12} = \frac{Z_{CA} \times O \times Q \times K}{\eta_\omega^{\tau_{ij}} \times P_Z} \tag{8}$$

In formula (8), η_ω represents the judgment vector of learners' mastery of knowledge point ω; Learners have mastered the knowledge points of talent training courses, and the judgment value is determined to be 1, otherwise it is determined to be 0. Z_{CA} indicates the learners' failure rate in answering questions; τ_{ij} represents the time factor of forgetting effect; After obtaining the maximum likelihood of the error rate, the learners' cognitive ability level of the learned knowledge points is obtained, and the structure diagram is shown in Fig. 3.

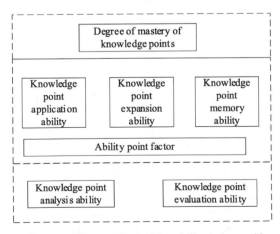

Fig. 3. Structure diagram of learners' cognitive ability to learned knowledge points

According to learners' cognitive level of knowledge points of talent training courses and user portrait labels, the teaching resource information of talent training courses is configured under the designed resource allocation mechanism.

2.4 Information Allocation of Teaching Resources for Talent Training Courses

According to the learners' mastery of this knowledge point, the information allocation of teaching resources of talent training courses is carried out, and the discrete sampling set of the distribution of the teaching resource information of the talent training course is obtained. The node link matrix is related to the teaching resource information of the talent training course in the node time slot, and the scheduling space vector matrix of the teaching resource information of the talent training course in the leading time slot is represented by F, and F is a two-dimensional matrix of $X \times X$. The teaching resource information of talent training course is allocated in a balanced way by using the probability allocation method. In the case of teaching resource allocation and storage, the teaching resource information allocation of talent training course is transformed into the maximum hop number of 2-hop neighboring nodes, and the conversion expression is as follows

$$X_n^2 = X_n^1 \times \left(\bigcup_{i=X_n^1} X_i^1 \right) \times P_Z \tag{9}$$

In formula (9), X represents the total number of transmission paths of teaching resources information of talent training courses, and n and i represent the multiple access protocols at the n time and the scheduled time slot allocation.

The link set of qualified teaching resources distribution of talent training courses can be obtained by correlation detection method, and the feature vector set S of teaching resources information allocation of talent training courses is defined as:

$$S = [F_A, F_B, F_C, F_D] \tag{10}$$

In formula (10), F_A represents time, F_B, represents weight, F_C represents weighted average ratio, and F_D represents overload ratio. In order to ensure the balanced allocation of teaching resources in personnel training courses, the resource allocation channel is designed by using BP neural network model, and the output channel X_D is obtained as follows:

$$X_D = \sum_{i=1}^{N} Q_i/N \times S \tag{11}$$

In formula (11), N represents the independent variable. Assuming that the evolution matrix is represented by $z(m)$, the eigenvector of teaching resource allocation of talent cultivation course is divided into $b(m)$ subvector Q_{mj} by singular value decomposition. In tangent space, the characteristic solution of teaching resource of talent cultivation course satisfies the condition of $Q_{mj} \geq N$, and the transportation module of teaching resource of talent cultivation course obtains the utilization rate L_{YL} after the allocation of teaching resource. The solution formula is as follows:

$$L_{YL} = \mu^{-1} \sum_{m=1}^{Q_{mj}} \iota_m = \left(\mu \times Q_{mj}\right)^{-1} \sum_{n=1}^{Q_{mj}} \sum_{m=1}^{\mu} z_{nm} \qquad (12)$$

In formula (12), μ represents the adaptive coefficient. By predicting and analyzing students' learning behavior and learning achievements through the utilization rate of resource allocation, we can get students' personalized learning mode, learning evaluation results and other information. According to this information, teachers can optimize the allocation of course teaching resources according to the different characteristics and needs of students, and the allocation results are shown in Fig. 4.

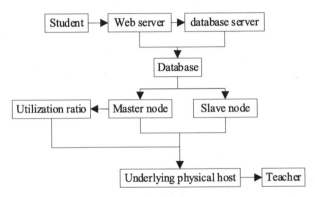

Fig. 4. The result diagram of teaching resources allocation of talent training courses

According to Fig. 3, through the coordinated operation of a Master node and a plurality of Slave nodes, while ensuring the interconnection of various devices, the overall allocation of teaching resources for talent training courses is more interactive. This resource allocation method can help teachers better understand students' learning needs and abilities, thus providing students with better learning experience and more personalized teaching resources. At the same time, it can also evaluate and improve the teaching quality of teachers. Therefore, the research on the allocation method of teaching resources for talent training courses based on BP neural network is completed.

3   Experimental Analysis

In order to verify the effect and feasibility of the teaching resource allocation method of talent training course based on BP neural network, the experiment takes a school in a certain area as the experimental object. The school covers an area of 530,000 m2, with 800 teachers and 16,000 students respectively, and randomly selects a class with 45 students. The experimental data is selected from the world's largest open enterprise database, which is the OpenCorporates database. The teaching resource information of the talent training course based on BP neural network in this class consists of the information of teaching activities before, during and after class. The original information and the newly added teaching information before, during and after class are shown in Table 1.

Table 1. Information Set of Teaching Resources for Personnel Training Courses

Name of teaching resource information set of talent training course	Original information quantity/GB	New information/GB
Information before class	30.6	5.2
Information in class	25.4	8.8
Information after class	40.5	3.6

Taking the original information of teaching resources before, during and after class in Table 1 as the experimental object, this paper tests the influence of different window sizes on the information extraction characteristics of teaching resources before, during and after class. The adaptive sliding window width is 50–300 bytes in the experiment. In order to ensure the objectivity of the experiment, it is introduced into X86 windows, hard disk 80G, Oracle 10g XE database, and the memory is more than 1GB. In the three configuration methods, the course teaching resource files shown in Table 2 are imported respectively, and the names and file formats of each resource file are recorded in Table 2.

Table 2. File types and corresponding formats of course teaching resources

Serial number	List files of type	File layout
1	sqoop-site.xml	Sqoop configuration file
2	Hbase-site.xml	Hbase configuration file
3	hadoop-policy.xml	Hadoop security profile
4	Hadoop-cnv.sh	Bash script
5	Spark-defaults.conf Spark	Default configuration file
6	Masters	Master node configuration
7	Slaves	Slave node configuration
8	Core-site.xml Hadoop	Core configuration

In view of the comparison of the effects of the three configuration methods, the response time of their respective servers is selected as the evaluation index, and five students are selected as experimental volunteers. The required teaching resources of talent training courses are obtained by using the three methods, and the historical retrieval resources of the above five students are analyzed. The configuration effects are compared by the user's preference for resources in the configuration results. The following formula is the calculation formula of students' preference for resources:

$$Z_{YP} = \sum_{U(u,s)} A_U \times G_h \tag{13}$$

In Formula (13), Z_{YP} represents students' preference for resources; $U(u, s)$ represents the data results produced by s retrieval behaviors of a student u in the recommendation system; A_U indicates the time-consuming factor for students to read the configuration results, and the longer it takes, the more interested students are in resources. The opposite is true; G_h indicates the similarity between the resources allocated to students and the resources retrieved by students in the past. According to the above formula, the calculation of students' preference for resources under the three methods is completed, and the calculated data is drawn as shown in Fig. 5.

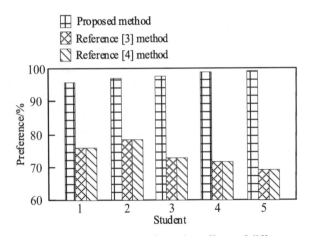

Fig. 5. Comparison diagram of configuration effects of different methods

From Fig. 5, it can be seen that the five students had a higher preference for the recommended resources configured by the proposed method, while the preference for the recommended resources configured by the methods in reference [3] and reference [4] was significantly lower. Specifically, students' preference for the resources allocated by the proposed method exceeded 96%, while the preference for the resources allocated by the methods in reference [3] and reference [4] did not exceed 80%. In summary, the BP neural network-based method for allocating teaching resources for talent cultivation courses proposed in this article not only has good operational performance in practical applications, but also can provide students with the online teaching resources they need for talent cultivation courses, bring higher quality resource recommendation services to students, and promote the utilization value improvement of teaching resources for talent cultivation courses.

Through the recall of teaching resource information management of talent training course, the configuration effect of this method is further verified. According to the output result of teaching resource information configuration of talent training course in Fig. 5, the recall rate of teaching resource information of talent training course before, during and after class is counted, and the statistical results are shown in Fig. 6.

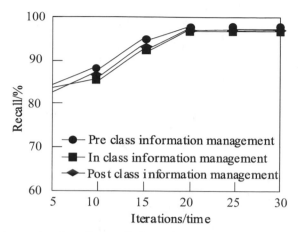

Fig. 6. Comparative results of recall rate of information allocation of teaching resources in talent training courses

From Fig. 6, it can be seen that the proposed method has a high recall of pre class, in class, and post class teaching resource information for talent cultivation courses, and the resource information allocation effect of this method is good. This means that using the method proposed in this article for resource information allocation can better meet the learning needs of students at different stages. The experimental results show that the application of this method in the allocation of teaching information resources for talent cultivation courses can achieve good results, and shows stability when allocating teaching resource information for different types of talent cultivation courses. This indicates that the method proposed in this article has applicability and reliability, and can effectively allocate resource information in different types of courses.

In summary, the methods proposed in this article can effectively allocate teaching resource information for talent cultivation courses, improve teaching effectiveness and students' learning experience. In addition, the application of this method in different types of courses also has stability, providing a feasible resource allocation plan for educational institutions and teachers.

4 Conclusion and Prospect

This paper puts forward the allocation method of teaching resources for talent training courses based on BP neural network, and draws the following conclusions through research:

(1) The proposed method has good performance in practical application, which can provide students with the online teaching resources they need for talent training courses, bring them higher quality resource recommendation services, and promote the utilization value of teaching resources for talent training courses.

(2) The application of teaching information of talent training courses has a good effect, and it is stable when configuring teaching resource information of different types of talent training courses.

The next step can be studied in depth in the following aspects:

(1) For the problem of insufficient and uneven allocation of teachers' resources, we should improve the incentive mechanism for teachers, fully consider the particularity of economy, culture and policy in ethnic areas, and innovate training methods according to the actual situation in Butuo County; It is necessary to strengthen the construction of teachers and improve the training mechanism to ensure the quality of physical education teachers. Relevant departments should define the management subject, formulate the management system, form a supervision mechanism for the implementation of the management system, and have a set of perfect mechanisms suitable for the development of schools, aiming at improving the teaching quality of schools, achieving educational balance and promoting the development of education in ethnic areas with students as the main body.

(2) To give full play to the linkage mechanism of departments to optimize the balanced allocation of resources and improve the supervision mechanism, relevant managers should pay attention to the construction of sports venues and the allocation of sports equipment in weak areas, establish the investment concept of "differential compensation", and deeply realize that the balanced allocation is not a simple quantitative balance, nor a formal balance; We should attach importance to the development of physical education class, aiming at promoting the all-round development of students; Offering ethnic project courses to make up for the lack of resources can be further studied in the future.

References

1. Dilworth, M.E.: Historically black colleges and universities in teacher education reform. J. Negro Educ. **81**(2), 121–135 (2022)
2. Shi, Y.A.N.G.: Research on the cultivation and development of creative talents in colleges and universities. Sci. J. Hum. Soc. Sci. **4**(1), 49–52 (2022)
3. Segura, J.: The teaching of usability in software development: case study in the computer engineering career at the university of matanzas. Int. Assoc. Online Eng. (IAOE) **12**(1), 1–13 (2021)
4. Saito, T., Watanobe, Y.: Learning path recommendation system for programming education based on neural networks. Int. J. Distance Educ. Technol. **18**(1), 36–64 (2020)
5. Aishan, Y., Li, G., Yangyang, D.: Practical research on the construction of professional curriculum teaching resources under the informationization of education. Jiangsu Bus. Rev. **02**, 124–126 (2023)
6. Morales, A., Gonzalez, J., Alquerque, K., et al.: Recommendation system with graph-oriented databases for repository of open educational resources. IOP Conf. Ser.: Mater. Sci. Eng. **1154**(1), 21–29 (2021)
7. Wang, X., Mu, C.: Reform of the classification and evaluation system for scientific and technological innovation talents in the intelligent age. E3S Web Conf. **251**, 02024 (2021). https://doi.org/10.1051/e3sconf/202125102024

8. Shrifan, N., Akbar, M.F., Isa, N.: An adaptive outlier removal aided K-means clustering algorithm. J King Saud Univ. – Comput. Info. Sci. **34**(4), 6365–6376 (2021)
9. Taheri, S.I., Salles, M., Nassif, A.B.: Distributed energy resource placement considering hosting capacity by combining teaching-learning-based and honey-bee-mating optimisation algorithms. Appl. Soft Comput. **113**(45), 107953–107966 (2021)
10. Liu, S., Li, Y., Fu, W.: Human-centered attention-aware networks for action recognition. Int. J. Intell. Syst. **37**(12), 10968–10987 (2022)
11. Wang, Y.Y., Luo, S., Wang, Z.J.: Photovoltaic power prediction combined with popular learning and improved BP Neural network. Comput. Simul. **39**(11), 153–157 (2022)
12. Tucker, B.V., Kelley, M.C., Redmon, C.: A place to share teaching resources: speech and language resource bank. J. Acoust. Soc. America **149**(4), 147 (2021)
13. Yao, S., Li, D., Yohannes, A., et al.: Exploration for network distance teaching and resource sharing system for higher education in epidemic situation of COVID-19 – Science Direct. Procedia Comput. Sci. **183**, 807–813 (2021)
14. Hu, S., Liu, Y., Wang, S.: Teaching exploration of case-based data modeling optimization for database system. Open J. Soc. Sci. **08**(3), 514–521 (2020)
15. Yan, S., Jingsheng, Z.: 2021 Design of informationbased teaching resources sharing system based on multimedia technology Mod. Electron. Tech. 44 20 32 36
16. Xi, L.: Design of MOOC idea based teaching resource sharing system for digital film and television production. Mod. Electron. Tech. **43**(16), 115–118 (2020)
17. Wang, Z., Muthu, B.A., Kadry, S.N.: Research on the design of analytical communication and information model for teaching resources with cloud: haring platform. Comput. Appl. Eng. Educ. **29**(2), 359–369 (2021)
18. Li, H., Zhong, Z., Shi, J., Li, H., Zhang, Y.: Multi-Objective optimization-based recommendation for massive online learning resources. IEEE Sens. J. **21**(22), 25274–25281 (2021)
19. Ying, F., Meng, L.: Design and research of online education course resource sharing based on cloud platform. Mod. Electron. Tech. **43**(1), 175–178 (2020)
20. Zou, F., Chen, D., Xu, Q., Jiang, Z., Kang, J.: A two-stage personalized recommendation based on multi-objective teaching–learning-based optimization with decomposition. Neurocomputing **452**(6), 716–727 (2021)
21. Liu, T., Yang, Z.: Personalized recommendation method of sports online video teaching resources based on multiuser characteristics. Math. Probl. Eng. **2022**(2), 1–8 (2022)
22. Chen, L., Bai, W., Yao, Z.: A secure and privacy-preserving watermark based medical image sharing method. Chin. J. Electron. **29**(5), 819–825 (2020). https://doi.org/10.1049/cje.2020.07.003
23. Tian, X., Ahmad, M T.: Method of Ideological and political teaching resources in universities based on school-enterprise cooperation mode. Math. Probl. Eng. 2022, 10(1):1098–1121 (2022)
24. Zhang, Z.: A Method of recommending physical education network course resources based on collaborative filtering technology. Sci. Program. **2021**(10), 1–9 (2021)
25. Liang, X., Yin, J.: Recommendation algorithm for equilibrium of teaching resources in physical education network based on trust relationship. J. Internet Technol. **23**(1), 133–141 (2022)

Anti Noise Speech Recognition Based on Deep Learning in Wireless Communication Networks

Yanning Zhang[✉], Lei Ma, Hui Du, and Jingyu Li

Beijing Polytechnic, Beijing 100016, China
witgirl316@126.com

Abstract. As a new high-tech industry, the application of speech recognition technology is becoming more and more competitive, with a wide range of application fields and application prospects, and has far-reaching significance for the development of science and technology. The communication environment of wireless communication network will bring various types of noise to speech, so an anti noise speech recognition method based on deep learning of wireless communication network is designed to achieve anti noise speech recognition in this environment. The voice signal of wireless communication network is preprocessed by anti aliasing filtering, analog-to-digital conversion, pre emphasis, framing and windowing, endpoint detection, etc. A series of denoising processes are implemented for the voice signal of wireless communication network, and different speech preprocessing methods are adopted for different characteristics of noise. A speech signal feature extraction method based on improved EMD is designed and implemented. The speech recognition model is designed based on the regression neural network in deep learning, and the anti noise speech recognition of wireless communication network is realized. Test results show that the lowest word error rate of this method is 0.156, and the word error rate is also low.

Keywords: Deep Learning · Improve EMD · Hidden Markov Model · Anti Noise Speech Recognition

1 Introduction

With the development of society, wireless communication has been used in many scenarios, and many new communication networks have emerged. The emergence and application of wireless ad hoc networks, wireless mesh networks, wireless sensor networks and other networks have improved people's living standards. In wireless communication networks, communication speech often produces noise, which affects people's auditory judgment. Speech is the most natural, effective and convenient means for human information exchange. Therefore, it is of great significance to study speech recognition in wireless communication networks [1].

The research work of speech recognition began in the 1950s. AT&T Bell Laboratories invented the first speech recognition system - Audry system, which can recognize

L. Yun et al. (Eds.): ADHIP 2023, LNICST 547, pp. 152–168, 2024.
https://doi.org/10.1007/978-3-031-50543-0_11

10 English numbers. In the late 1960s and early 1970s, with the development of computer technology, speech recognition technology has also made substantial development and become an important subject in the research field [2]. Firstly, the development of computer technology provides hardware and software conditions for the realization of speech recognition. Secondly, "Dynamic programming method" has effectively solved the problem of difficult alignment of speech signals on the time axis. At the same time, speech recognition has proposed two technologies, dynamic time warping technology and linear prediction coding technology, which provide effective methods for unequal length matching of speech signals and feature extraction of speech signals. The use of these two technologies can basically achieve the speech recognition of individual isolated words. At the same time, the hidden Markov model I5 technology has also been preliminarily applied and the vector quantization L0 theory has been proposed. In the 1980s, the research on speech recognition gradually deepened, and the focus of research turned to non-specific, large vocabulary continuous word speech recognition, and many new speech recognition algorithms emerged. Another breakthrough development is the proposal of the model technology based on statistics. Compared with the template matching technology, this technology does not require the refinement of speech features, but constructs a speech recognition system from the perspective of the overall analysis of speech signals. In addition, the successful application of hidden Markov model and artificial neural network in speech recognition system has made a great breakthrough in continuous speech recognition. At present, hidden Markov model has become the mainstream technology of speech recognition. Among them, Sphinx system developed by CMU in 1988 is a typical speech recognition system. It is the first non-specific, large vocabulary continuous speech recognition system with high performance in the world. Its database has 99 words and 4200 consecutive sentences, and the system's recognition rate can reach 95.8% at the highest. Since the 1990s, many developed countries in Europe and the United States and some famous technology companies have invested huge funds in the field of speech recognition to realize the practicality of speech recognition, and have begun to provide relevant products for the market. The recognition rate of the speech recognition system has been greatly improved, and the speech recognition technology has further matured and started to enter the practical application from the laboratory. At the same time, the recognition of Chinese speech has been paid more and more attention.

At present, the pure speech recognition technology has developed more maturely. In a quiet environment, the speech acquisition and template training can match the speech to be recognized very well. Therefore, the existing speech recognition system can complete the recognition very well, and the recognition efficiency is also very high. However, in the wireless communication network environment, speech will contain noise. Noise interference affects the matching between the speech to be recognized and the training template, which further affects the recognition rate of the system. Therefore, anti noise becomes one of the key technologies in modern speech recognition research.

For the research of anti noise speech recognition, some scholars proposed an anti noise speech recognition technology based on RBF neural network: to solve the problem of poor performance of speech recognition system in noisy environment at present, RBF neural network has the best approximation performance, fast training speed and other characteristics. The anti noise speech recognition system based on RBF neural network is implemented by clustering and fully supervised training algorithms respectively. The hidden layer training of the clustering algorithm adopts the K-means clustering algorithm, and the output layer learning adopts the linear least square method; The adjustment of all parameters in the fully supervised algorithm is based on the gradient descent method, which is a supervised learning algorithm and can select parameters with good performance. Experiments show that the fully supervised algorithm has higher recognition rate than the clustering algorithm under different signal-to-noise ratios. Other scholars have proposed an anti noise speech recognition technology based on 3F speech enhancement distortion compensation: in order to improve the robustness of speech recognition system based on hidden Markov model in noisy environments, this paper studies an anti noise speech recognition algorithm based on speech enhancement distortion compensation. At the front end, speech enhancement effectively suppresses background noise. Thus, the signal to noise ratio of the input signal is improved, and the spectral distortion and residual noise caused by speech enhancement are adverse factors for speech recognition. The influence will be compensated by parallel model merging in the recognition phase or cepstrum mean normalization in the feature extraction phase. The experimental results show that this algorithm can significantly improve the recognition accuracy of the speech recognition system in the noisy environment in a very wide SNR range, and the improvement of the system performance is especially obvious in the low SNR situation, such as 5 dB white noise. Compared with the baseline recognizer, this algorithm can reduce the error rate by 67.4%. There are complex nonlinear relationships in speech recognition, such as time-varying speech and noise interference. When using the above methods for noise resistant speech recognition in wireless communication networks, it is difficult to capture these nonlinear relationships, resulting in poor recognition performance. And deep learning models have strong nonlinear modeling capabilities, which can better adapt to complex speech features and improve recognition performance. Therefore, a wireless communication network anti noise speech recognition method based on deep learning is designed.

2 Design of Anti Noise Speech Recognition Method for Wireless Communication Network

2.1 Speech Signal Pre-processing

The voice signal of wireless communication network is preprocessed by anti aliasing filtering, analog-to-digital conversion, pre emphasis, framing and windowing, endpoint detection, etc.

The anti aliasing filter is selected as the band-pass filter. The power frequency interference of 50 Hz power supply is suppressed through its high pass filtering part, and all the components in the frequency domain components of the input signal that exceed the following formula are suppressed through its low pass part to prevent aliasing interference.

$$h = \frac{k_s}{2} \tag{1}$$

In formula (1) k_s it refers to the sampling frequency.

Human pronunciation is a continuous analog signal, which cannot be processed by computer. Therefore, it is necessary to convert analog voice signals into digital signals, that is, analog-to-digital conversion. Since the average power spectrum of speech signal is affected by glottal excitation and nose and mouth radiation, the high-frequency end drops significantly above 800 Hz, so when calculating the spectrum of speech signal, the higher the frequency is, the smaller the corresponding component is, and the spectrum of high-frequency part is more difficult to find than that of low-frequency part. Therefore, pre emphasis processing should be carried out in preprocessing. The purpose of pre emphasis is to improve the high frequency part so that the spectrum of the signal becomes flat. The pre emphasis part is realized by a digital filter to improve the high-frequency characteristics, which is generally a first-order digital filter:

$$g(s) = 1 - vs^{-1} \qquad (2)$$

In formula (2) v is the pre weighting coefficient; s it is a voice signal.

Among v The value of is generally between 0.9 and 1, and the typical value is 0.9375.

Speech signal is a typical non-stationary signal, its characteristics change with time. However, the formation process of speech is closely related to the movement of the vocal organs. This physical movement is much slower than the speed of sound vibration. Therefore, speech signals can often be assumed to be stable in a short time, that is, in a period of 10–30 ms, its spectral characteristics can be seen as nearly unchanged. In this way, the analysis and processing method of stationary process can be used. From this assumption, various short-time processing methods based on frames are derived, and various speech processing methods discussed later are based on this assumption. In order to smooth the transition between frames and maintain their continuity, the method of overlapping segments [3] is adopted here. The overlapping part of the previous frame and the next frame is called frame shift, which is generally 0–1/2 of the frame length. The specific method is shown in Fig. 1.

In order to reduce the Gibbs effect caused by truncation after framing, it is usually necessary to windowing each frame signal. Each 10 ms–30 ms frame of the voice is analyzed in turn. This operation is called windowing. The window slides on the voice signal to frame the voice signal. When adding windows to voice signals, use Hamming windows.

The primary problem of speech signal processing is to determine the pure noise segment of a noisy speech, the noisy speech segment, and the start and end points of each speech segment, that is, endpoint detection in signal processing. Endpoint detection is the basis of speech signal processing. If the endpoint detection of a speech recognition system is done well, it can not only reduce the amount of calculation, but also improve the recognition rate of the system [4]. The calculation process of the short-term average zero crossing rate is shown in Fig. 2, that is, first process the speech signal sequence in pairs to check whether there is sign transformation, and if there is, it means zero crossing once; Then make a first-order difference and take the absolute value; Finally, low-pass filtering is performed to output short-term average zero crossing.

Since the short-term average zero crossing rate can reflect the frequency to a certain extent, while the energy of voiced voice is concentrated in the low frequency band and the energy of unvoiced voice is concentrated in the high frequency band, the zero crossing rate is generally low in the voiced voice band and high in the unvoiced voice band.

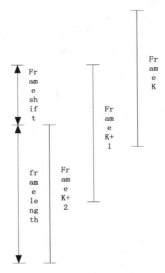

Fig. 1. Framing Diagram

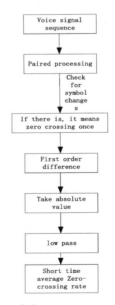

Fig. 2. Calculation of short-term average zero crossing rate

2.2 Speech Signal Denoising

A series of denoising processes are implemented for the voice signal of wireless communication network, and different speech preprocessing methods are adopted for different
characteristics of noise.

First, build a noisy speech model:

$$x(t) = q(t) + w(t) \tag{3}$$

In formula (3) $x(t)$ is noisy speech; $q(t)$ is pure voice; $w(t)$ it is a noise signal.

In this model, the pure speech signal can be regarded as a speech segment, and its physical characteristics and spectrum characteristics are regarded as invariable. The short-term spectrum analysis of speech is relatively stable, which is a time-varying, non-stationary random process. The noise signal is an additive noise, which is locally stationary, statistically independent or uncorrelated with the speech signal, and has no other reference signal.

Spectral subtraction is an effective method to deal with single frequency noise, and is the simplest method of speech enhancement. The spectral subtraction method is improved to overcome the shortcomings of spectral subtraction in broadband noise processing.

The improved method of spectral subtraction algorithm: after spectral subtraction of noisy speech frames and pure noise frames, the residual noise will produce music noise. How to eliminate music noise is the key to improve spectral subtraction algorithm. The analysis shows that if the estimated frame of music noise can be obtained, music noise can be eliminated.

The method of obtaining music noise frame: subtract the square of the amplitude spectrum of any two pure noise frames, which can be used as the estimation value of the square of the amplitude spectrum of music noise. Subtract the square of the amplitude spectrum of the same frame of pure noise (standard pure noise) from the square of the amplitude spectrum of multiple frames of pure noise to obtain the estimated square of the amplitude spectrum of different music noises. The specific methods are as follows:

First spectral subtraction: subtract the square of the amplitude spectrum of the noisy speech from the square of the amplitude spectrum of the standard pure noise:

$$P = S(x(t))^2 - S(q(t))^2 \tag{4}$$

In formula (4) $S(x(t))$ it is the amplitude spectrum of noisy speech; $S(q(t))$ it is a pure speech amplitude spectrum.

Second spectrum subtraction: subtract the square of the amplitude spectrum of the first frame of music noise from the result of the first spectrum subtraction:

$$P' = P - S(f_1(t))^2 \tag{5}$$

In formula (5) $S(f_1(t))$ it is the amplitude spectrum of music noise in the first frame.

Third spectrum subtraction: subtract the square of the amplitude spectrum of the second frame of music noise from the result of the second spectrum subtraction:

$$P'' = P' - S(f_2(t))^2 \tag{6}$$

In formula (6) $S(f_2(t))$ it is the amplitude spectrum of music noise in the second frame.

By analogy, it is called cascading spectral subtraction.

The fluctuation noise is eliminated by the adaptive filter denoising method, and the wavelet threshold denoising algorithm is selected. The algorithm is mainly divided into four steps:

The first step is to select the wavelet basis function, and select a wavelet function to conduct discrete wavelet transform on the speech signal. Generally, db5 wavelet is selected as the wavelet function, which is decomposed into 3–5 layers of wavelet coefficients.

The second step is to determine the threshold value, which determines the final denoising effect of the denoising method [5]. If the threshold value is too small, the denoising effect will be unsatisfactory. If the threshold value is set too large, the useful part of the signal will be removed, resulting in the distortion of the denoised signal. The threshold selection method is as follows: the maximum variance minimizes the threshold to produce an extreme value of the minimum mean square error, which minimizes the maximum mean square error.

In the third step, threshold function is used to process the wavelet coefficients. The threshold function used is hard threshold:

$$l_i = \begin{cases} l_i & |l_i| \geq \gamma \\ 0 & |l_i| < \gamma \end{cases} \tag{7}$$

In Formula (7) l_i represents wavelet coefficients; γ is a hard threshold.

The fourth step is wavelet reconstruction, which uses discrete wavelet inverse transform to reconstruct the wavelet coefficients of threshold processing.

For impulse noise, the denoising method adopted is the optimized VMD decomposition and wavelet threshold denoising algorithm for speech signal denoising. Combining GWO optimized VMD algorithm with adaptive decomposition of signal and wavelet with the best approximation efficiency of one-dimensional signal, a speech signal denoising algorithm based on optimized VMD decomposition and wavelet threshold denoising is proposed. This method can eliminate the impulse noise interference of speech signal to a great extent.

When GWO algorithm is used to optimize VMD parameters, the fitness function uses permutation entropy. The complexity of the signal can be seen from the permutation entropy by calculating the fitness function. If the signal is more complex, the calculated permutation entropy is larger, and vice versa. After speech signal is decomposed by VMD, if there are many noise components included in IMF components, the higher the signal complexity, the greater the permutation entropy; If there are few noise components included in the IMF component, the more regular the signal is, the less complex the signal is, and the lower the permutation entropy is. Once the component K and penalty factor are determined α, VMD is used to decompose it, and the component with the smallest entropy value is the component with the best feature information of voice signal. Therefore, the objective of parameter optimization is to minimize the permutation entropy as the fitness value.

The specific implementation steps of the proposed optimized VMD decomposition and wavelet threshold denoising methods are as follows:

Input: voice signal with noise

Output: clean voice signal after denoising

Start.

Step 1: Use GWO algorithm to find the optimal combination of decomposition mode number and penalty factor for VMD to decompose noisy speech signal [k0, α0];

Step 2: Use the optimal combination found by the GWO algorithm to decompose the noisy voice signal into VMD, and obtain a finite number of modal components:

$$R = \{r_1, r_2, \ldots, r_k\} \tag{8}$$

In formula (8) r_k it refers to the No k Modal components.

Step 3: use correlation coefficient to select noisy modes from finite modal components;

3.1 Calculate the correlation coefficient between each modal component and the original signal;

3.2 Calculate the screening threshold through the screening principle of correlation coefficient f. When the modal component meets the following formula:

$$r_k > f \tag{9}$$

It can be considered that the correlation between the modal component and the original voice signal is good, and it needs to be retained; Otherwise, the corresponding modal components are taken out for wavelet denoising.

Step 4: carry out wavelet threshold processing for noisy modes;

4.1 Select appropriate wavelet bases, wavelet decomposition levels and wavelet thresholds. The wavelet base of sym8 is selected in this paper, and the number of decomposition layers is 6. Wavelet threshold uses the threshold of wavelet threshold de-noising algorithm;

4.2 Use the selected wavelet basis function to carry out wavelet transform on IMF components to obtain the corresponding.

A set of wavelet coefficients;

4.3 Compare the wavelet threshold with the wavelet coefficient obtained. If the threshold is less than the wavelet coefficient, it can be considered that the useful signal constitutes the wavelet coefficient, and then the wavelet coefficient needs to be retained; On the contrary, it is considered that the wavelet coefficient is composed of noise signals, so the wavelet coefficient needs to be discarded [6].

4.4 For the reserved signal, the denoised IMF component can be obtained only after reconstruction;

4.5 The selected IMF components must be operated from 4.2 to 4.4;

Step 5: Reconstruct the mode and effective mode after wavelet threshold processing to obtain the denoised speech signal.

End.

The specific process is shown in the following Fig. 3.

2.3 Feature Extraction

A speech signal feature extraction method based on improved EMD is designed to implement the feature extraction of voice signals in wireless communication networks.

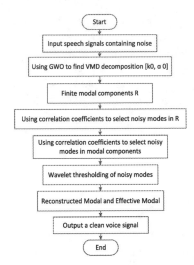

Fig. 3. VMD decomposition and wavelet threshold denoising flowchart

The traditional MFCC extraction algorithm is improved. The speech signal is decomposed by the improved empirical mode decomposition algorithm, and then the eigenmode components obtained by decomposing the speech signal are classified. The eigenmode components with a high proportion of the original signal are reconstructed and their MFCC is extracted. The proposed method saves as many components of the original signal as possible while eliminating noise signals. Since the noise of the signal can be decomposed into IMF components of each layer to enhance the signal-to-noise ratio of the weak signal layer, the improved method can be obtained as follows:

(1) EMD is decomposed into N IMF components and differentiated into reconstructed signals

The processed voice signal is decomposed into several IMF components by EMD method, as follows:

$$V = \{v_1, v_2, \ldots, v_j\} \tag{10}$$

In Eq. (10) v_k it refers to the No j IMF quantity.

Then autocorrelation processing is performed on the IMF components obtained. According to the autocorrelation function waveform, the dividing point between the useful original signal and the noise signal is derived, and the obtained eigenmode components are divided into two categories: the IMF component dominated by the noise signal and the IMF component dominated by the original signal. And reconstruct the main intrinsic mode components of the original signal to obtain the signal $E(t)'$.

(2) Fast Fourier Transform

The transformation of speech signal in the time domain can hardly see the characteristics of the signal. Usually, the speech signal is converted to the frequency domain to observe the characteristics of the speech signal. Therefore, the speech signal spectrum is obtained through fast FFT.

(3) Mel filter bank

The obtained voice signal is then passed through a set of Mel scale triangular filter banks, and the central frequency of the filter bank is $c(b)$ The transfer function satisfying each bandpass filter is:

$$c(b)=\begin{cases} 0 \ u < c(b-1) \\ \frac{u-c(b-1)}{c(b)-c(b-1)} \ c(b-1) \leq u \leq c(b) \\ \frac{c(b+1)-u}{c(b+1)-c(b)} \ c(b) \leq u \leq c(b+1) \\ 0 \ u > c(b-1) \end{cases} \qquad (11)$$

The sampling rate is set to 8000 Hz. Here, the number of filters is set to M = 24.

(4) Take logarithmic energy

It is found that logarithmic energy is suitable for speech signal feature extraction. Compared with traditional entropy, logarithmic energy entropy can describe information more carefully. At present, most of the traditional entropy algorithms are used to describe the global information and column direction information. However, logarithmic energy can not only describe the overall global information of the signal. When describing the signal, the one-dimensional feature vector of the original signal can be changed into a single eigenvalue, so that the dimension of the original signal can be used to analyze and identify future information. Logarithmic energy is insensitive to a certain extent. In practice, logarithmic energy can be obtained by taking logarithm of the processed energy spectrum $o(m)$.

(5) DCT transformation

Discrete cosine transform (DCT) is a transform strongly related to Fourier transform. The discrete cosine transform is similar to the discrete Fourier transform. Because most of the collected information is noisy, if the data is not cleaned, direct prediction will produce data disturbance, greatly reducing the prediction accuracy. The discrete cotransform process uses only real numbers [7]. From the perspective of frequency domain, many natural signals are concentrated in the low-frequency part of the signal after discrete cosine transform. Convert logarithmic energy $o(m)$ MFCC coefficient is obtained through discrete cosine transformation:

$$\delta(z)=\sum_{m=0}^{M-1} o(m)COS\frac{\pi(m-0.5)}{L} \qquad (12)$$

In Eq. (11) M refers to the number of signals; L it refers to the maximum value of the energy spectrum.

(6) Dynamic and static characteristics

The traditional cepstrum parameter MFCC can accurately reflect the static characteristics of speech signals, but the dynamic characteristics of parameters can not be well understood. Therefore, this paper proposes to obtain the dynamic characteristic parameters of MFCC from the difference of static MFCC, and then combine the dynamic characteristic parameters and static characteristic parameters to form a mixed MFCC parameter as the training feature of speech signal.

2.4 Speech Recognition

A speech recognition model is designed based on the regression neural network in deep learning to realize the noise resistant speech recognition in wireless communication networks.

The designed speech recognition model includes three core parts: acoustic model, language model and decoder. In the DNN-HMM based acoustic model modeling, the DNN function is to replace the original GMM model and estimate the verification probability after HMM status.

The Kaldi open source speech recognition system is selected as the carrier to implement the DNN-HMM model. In Kaldi, the script is called quickly to make verification easier [8].

In the DNN-HMM acoustic model, the DNN network structure consists of one input layer, six hidden layers and one output layer.

For the input layer, 39 dimensional MFCC features are obtained by extracting voice information, 11 frames of voice information are used, and there are 429 input nodes. According to the number of triphone states corresponding to the target output of the network, that is, the number of clustered triphone state IDs, the output layer node is set to 1462 [9].

DNN network uses back-propagation algorithm to adjust parameters, so monitoring information is needed for training. For voice signals, it is necessary to know the phoneme state corresponding to each frame. After the forced alignment recognition result, the trisyllon state corresponding to the original voice information is obtained as the training supervision information [10].

The training criteria of the acoustic model should be able to be simply calculated and have a high correlation with the task. The improvement of the criteria should finally be reflected in the completion level of the task. Therefore, the minimum expected loss function should be selected as the training criterion of model parameters [11].

$$\theta_{EL} = U(\theta(\sigma, \varsigma O\rho)) \tag{13}$$

In Eq. (13) U is a statistical expectation operator; $\theta(\sigma, \varsigma O\rho)$ is the loss function, where σ, ς is a model parameter, O to observe the vector, ρ is the corresponding output vector.

Given training criteria, model parameters σ, ς the famous error back propagation algorithm can be used for learning, and the chain rule can be used for deduction. In its simplest form, the model parameters are optimized using the first derivative information. The gradient of the top weight matrix relative to the training criteria depends on the training criteria [12].

The function of the decoder is to extract the decoding map from the trained model, and use this decoding map to search and match the test speech signal, and output the word sequence with the highest probability. The decoding diagram consists of four parts:

1) Grammar, which is the receiver of coding grammar or language model;
2) Phonetic dictionary, which is used to convert phonemes into words;
3) Context relation, which is used to output phonemes according to the most possible combination of phonemes in the context window;
4) HMM is defined to transform PDF ID (PDF index assigned by decision tree clustering) into phonemes representing context.

Firstly, context sensitive HMM acoustic model, ternary grammar language model and voice dictionary are integrated into a weighted finite state converter [13], which is optimized by deterministic algorithm and minimization algorithm to build a search space. Then send the test voice into the search space through the following four steps:

(1) Initialize search path:

$$\xi = \xi_0 \tag{14}$$

(2) Use acoustic model and language model to re judge the path score;
(3) Cut out the path with lower scores;
(4) The optimal path is obtained by backtracking, and the optimal result is obtained after the search.

The language model uses the pre training language model PLMs, which is composed of the Embedding layer, Pre training and Fine tuning. In the pre training process, a large number of unmarked corpus sets are used for unsupervised pre training, and then the weights in Fine tuning are initialized to the weights obtained in the pre training process [14]. All the pre training parameters will participate in the training when fine-tuning, and marked corpus is used in the training process. In different downstream tasks, their respective BERT models will be created as needed.

When using the Pre training Model for feature representation, there are generally two types of strategies: feature based strategy and fine-tuning based strategy. The traditional standard LM is the single way mode, which leads to many limitations when selecting language architecture [15]. BERT attempts to break through the one-way restrictions in the past standard language models by using the masked language model. The specific approach is to use [mask] to mask a random number of characters each time, and predict the words to be [mask] through the objective function. At the same time, BERT is also committed to the NSP task, that is, to determine whether the two sentences given are adjacent sentences in the text segment, so that the model can predict sentence level information [16].

The embedding layer is the embedding layer. BERT has three embedding layers, namely, Token Embedding, Segment Embedding and Position Embedding. The word elements in the input text complete vector conversion, sentence segmentation and word segmentation after three layers of embedding.

3 Experimental Test

3.1 Experimental Data Set

For the designed anti noise speech recognition method based on deep learning in wireless communication networks, its performance is tested through experimental data sets. The voice data, language model and dictionary used in this experiment are all from the "THCHS30 2015" dataset. This Chinese speech dataset contains 25 h of training data (30 speakers), 6 h of test data (10 speakers), and corresponding annotations. The voice data contains a total of 1000 sentences, and the coverage of binary phonemes and ternary phonemes reaches 71.5% and 14.3% respectively. All voice data are voice data under the wireless communication network, the sampling frequency is 16000 Hz, and the number of data bits is 16.

3.2 Experimental Process

First, the experimental data set is preprocessed by anti aliasing filtering, analog-to-digital transformation, pre emphasis, framing and windowing, endpoint detection, etc.

Then, a series of denoising processes are implemented on the experimental data set, and feature extraction is implemented through the speech signal feature extraction method based on improved EMD. In feature extraction, the order of Mel filter is defined as 24, the length of fast Fourier transform is 256, and the sampling frequency is 16000 Hz. Normalize Mel filter bank coefficients, divide speech signals into frames, and calculate MFCC parameters of each frame. Fast Fourier transform is performed to calculate first-order difference coefficient and second-order difference coefficient, and then MFCC parameters and first-order difference parameter MFCC are combined. After MFCC feature extraction, the relationship between the dimension and amplitude of the voice feature, and the relationship between the number of frames and amplitude can be drawn. After feature extraction, the voice feature parameter of one time pronunciation of each English number is 500×24.

Finally, through the speech recognition model based on regression neural network, the anti noise speech recognition of wireless communication network is realized. The script used for acoustic model training is shown in Table 1.

The word error rate and word error rate in the speech recognition of the design method are tested. In the test, the anti noise speech recognition technology based on RBF neural network and the anti noise speech recognition technology based on 3F speech enhancement distortion compensation are used as the comparison test methods, which are represented by technology 1 and technology 2 respectively.

3.3 Experimental Results

Table 2 shows the test results of word error rate in design method and technology 1 and technology 2 speech recognition.

According to the above table, the lowest word error rate of the design method is only 0.156, and the word error rate in speech recognition is far lower than that of technology

Table 1. Script used for acoustic model training

S/N	Script Name	Specific role
1	decode_fmllr.sh	Decoding the speaker adaptive model
2	train.sh	Training depth neural network model
3	pretrain_dbn.sh	Deep neural network pre training foot
4	align_si.sh	Align the specified data as input to the new model
5	decode.sh	Decode and generate word error rate results
6	mkgraph.sh	Establish identification network
7	train_sat.sh	Speaker adaptive training based on maximum likelihood linear regression in feature space
8	train_deltas.sh	Training a context sensitive three phoneme model
9	train_mono.sh	Training monophone hidden markov model

Table 2. Word error rate in speech recognition

Number of tests	Word error rate (%)		
	Design method	Technology 1	Technology 2
5	0.156	0.525	0.317
10	0.157	0.522	0.313
15	0.159	0.524	0.317
20	0.150	0.555	0.315
25	0.141	0.557	0.335
30	0.156	0.553	0.337
35	0.165	0.555	0.335
40	0.162	0.594	0.356
45	0.161	0.596	0.357
50	0.162	0.590	0.359

1 and technology 2, which shows that the design method has better noise resistance and more accurate speech recognition.

The test results of word error rate in the design method and technology 1 and technology 2 speech recognition are shown in Fig. 4.

According to the above figure, the word error rate in the design method speech recognition is also far lower than that in technology 1 and technology 2, which shows that its speech recognition performance is better. At the same time, it can be seen that the word error rate in speech recognition is higher than the word error rate as a whole.

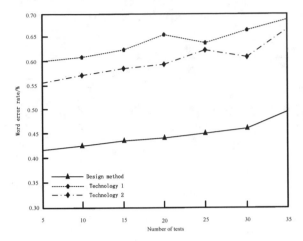

Fig. 4. Word Error Rate Test Results

To further validate the practicality of the design method, comparative tests were conducted using speech recognition time as the experimental indicator, and the test results are as follows (Fig. 5).

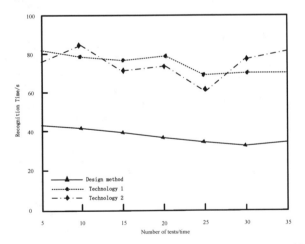

Fig. 5. Identify time test results

According to the above figure, it can be seen that the longest recognition time of the design method is 42 s, while the longest recognition times of the technology 1 and technology 2 methods are 81 s and 84 s, respectively. The recognition efficiency of the design method is significantly higher than that of the comparison method, indicating that the method proposed in this paper is practical.

4 Conclusion

The communication environment of wireless communication network will bring various types of noise to speech, so an anti noise speech recognition method based on deep learning of wireless communication network is designed to achieve anti noise speech recognition in this environment. In the wireless communication network, a series of preprocessing operations are carried out on the speech signal, and the speech signal features are extracted based on the improved EMD. According to the extracted feature signals, the speech recognition model is constructed by using the deep learning regression neural network, and the anti-noise speech recognition task in the wireless communication network is realized. However, the design method also has some limitations and some problems, and will continue to be optimized in the future.

References

1. Michelsanti, D., Tan, Z.H., Zhang, S.X., et al.: An overview of deep-learning-based audio-visual speech enhancement and separation. IEEE/ACM Trans. Audio Speech Lang. Process. **PP**(99), 1 (2021)
2. Goehring, T., Archer-Boyd, A.W., Arenberg, J.G., et al.: The effect of increased channel interaction on speech perception with cochlear implants. Sci. Rep. **11**(1), 1–9 (2021)
3. Datta, H., Hestvik, A., Vidal, N., et al.: Automaticity of speech processing in early bilingual adults and children– CORRIGENDUM. Bilingualism: Language and Cognition **24**(2), 1 (2021)
4. Kapnoula, E.C., Mcmurray, B.: Individual differences in speech perception: evidence for gradiency in the face of category-driven perceptual warping. J. Acoustical Soc. Am. **149**(4), A54–A54 (2021)
5. Kyaw, W.T., Sagisaka, Y.: Studies on association characteristics between vowels and visual colors using multiple speakers' speech. Acoust. Sci. Technol.. Sci. Technol. **42**(4), 161–169 (2021)
6. Abryutina, A., Ponomareva, A.: German-English Interference in the Field of Vocalism (Based on the Speech of Germans who Study English as a Foreign Language). Izvestia of Smolensk State University **1**(53), 128–143 (2021)
7. Qiang, H.: Consumption reduction solution of TV news broadcast system based on wireless communication network. Complexity **2021**(23), 1–13 (2021)
8. Dong, N., Lv, W., Zhu, S., et al.: Anti-noise model-free adaptive control and its application in the circulating fluidized bed boiler. Proc. Inst. Mech. Eng. Part I J. Syst. Control Eng. **235**(8), 1472–1481 (2021)
9. Wang, Q., Jiang, X., Weng, B., et al.: A 3D curvature attribute analysis method with excellent anti-noise property suitable for high steep formation. Geophys. Prospect. Petroleum **56**(4), 559–566 (2022)
10. Guan, Y., Hu, Z., Chen, C., et al.: An anti-noise transmission algorithm for 5G mobile data based on constellation selection and channel joint mapping. Alex. Eng. J. **60**(3), 3153–3160 (2021)
11. Basak, S., Agrawal, H., Jena, S., et al.: Challenges and limitations in speech recognition technology: a critical review of speech signal processing algorithms, tools and systems. Comput. Model. Eng. Sci. **2023**(5), 1053–1089 (2023)
12. Hadwan, M., Alsayadi, H.A., AL-Hagree, S.: An end-to-end transformer-based automatic speech recognition for qur'an reciters. Comput. Mater. Continua **2023**(2), 3471–3487 (2023)

13. El-Bialy, R., Chen, D., Fenghour, S., et al.: Developing phoneme-based lip-reading sentences system for silent speech recognition. CAAI Trans. Intell. Technol. **8**(1), 129–138 (2023)
14. Kamal, M.B., Khan, A.A., Khan, F.A., et al.: An innovative approach utilizing binary-view transformer for speech recognition task. Comput. Mater. Continua **2022**(9), 5547–5562 (2022)
15. Alsulami, N.H., Jamal, A.T., Elrefaei, L.A.: Deep learning-based approach for Arabic visual speech recognition. Comput. Mater. Continua **2022**(4), 85–108 (2022)
16. Nisar, S., Khan, M.A., Algarni, F., Wakeel, A., Irfan Uddin, M., Ullah, I.: Speech recognition-based automated visual acuity testing with adaptive mel filter bank. Comput. Mater. Continua **2022**(2), 2991–3004 (2022)

Acquisition Method of Direct Sequence Spread Spectrum Signal Based on Deep Residual Network

Jia Pan[✉]

Guangxi Science and Technology Normal University, Laibin 546100, China
jiapan_gxkjsfxy@163.com

Abstract. In order to improve the ability of network host to collect spread spectrum signals and accurately deduce the sequence of signal parameters, a direct sequence spread spectrum signal acquisition method based on deep residual network is proposed. The data information is rendered in the depth residual network, and the loss function model is defined through training processing, so as to realize the signal sequence modeling based on the depth residual network. On this basis, the pseudo-random sequence code is determined, and the direct acquisition of the sequence pseudo-code is completed according to the pseudo-code acquisition result of the spread spectrum signal. The experimental results show that the application of deep residual network can greatly enhance the ability of network host to collect spread spectrum signals, and meet the practical application requirements of accurately deriving the expression of signal parameter sequence.

Keywords: Deep residual network · Direct sequence · Spread spectrum signal · Data rendering · Signal training · Loss function · Pseudo-random sequence code

1 Introduction

The difficulty of tracking and acquisition of spread spectrum signal lies in the large amount of algorithm calculation and high resource usage in the process of realizing pseudo code synchronization and carrier synchronization in the signal receiver. However, with the rapid development of FPGA technology in recent years, its practical application scenarios in the field of signal processing technology are more and more extensive. You can choose to use FPGA chips with extremely rich computing resources and storage resources to complete the acquisition and tracking of spread spectrum signals. FPGA has a strong ability to handle parallel computing, especially when dealing with repeated and complex operations such as large point FFT, and its computing speed is significantly higher than that of chips such as DSP [1]. FPGA is suitable for fixed-point number calculation, and it needs to be applied to DSP chip in the process of angular error signal calculation and AGC subsection control requiring floating point number operation. This kind of working mode in which FPGA and DSP chips cooperate with each other can give full play to the advantages of each chip in data processing, and greatly improves the speed

L. Yun et al. (Eds.): ADHIP 2023, LNICST 547, pp. 169–183, 2024.
https://doi.org/10.1007/978-3-031-50543-0_12

of the system's acquisition and tracking of direct sequence spread spectrum signals. The problems of direct sequence spread spectrum signal itself, such as being easily submerged by noise interference and intercepted, have greatly limited the performance of the system, which is difficult to fundamentally solve the problem by upgrading software algorithms and adjusting hardware system parameters.

Direct sequence spread spectrum can be divided into incoherent spread spectrum and coherent spread spectrum. Incoherent spread spectrum is different from coherent spread spectrum in that the information data clock and pseudo-random code clock are asynchronous, and the information data rate is variable within a certain range. The incoherent direct sequence spread spectrum system is an existing TT&C spread spectrum communication system, which has the outstanding advantages of anti-interference, low interception, code division multiple access, etc., and is widely used in telemetry and telecontrol signals in aerospace TT&C communication. Although the incoherent direct sequence spread spectrum has the advantage of flexible application and flexible change of data rate according to the service requirements, due to the uncertainty of the relative relationship between the data bit jump edge and the phase of the pseudo-random code, a coherent integration time may contain data bit polarity jump. The SHF based local signal acquisition method predicts the signal amplitude and phase level by defining the local transmission characteristics of signal parameters, and uses the SHF mixing method to complete the on-demand acquisition of signal objects. The signal acquisition method based on the finite-difference time-domain uses the Lagrange function algorithm to obtain the interference signal, and determines the unit acquisition quantity of the signal object according to the finite-difference time-domain principle. However, the application capability of the two methods mentioned above is limited, which is not enough to improve the acquisition capability of the network host for the spread spectrum signal.

As the deepening of neural network increases the difficulty of training, the deep residual network solves the problem of gradient disappearance and model degradation in the training process of deep neural network by designing a residual learning framework. Each basic residual block is composed of a weight layer, an activation function layer, an identity mapping and a normalization layer [2]. The identity mapping layer neither adds additional parameters nor increases computational complexity. The output of the residual block is obtained by adding the output of the original input through multiple weight layers and the original input elements, and then passing through the activation function. The weight layer can be a convolution operation or a linear function operation. Based on the above research background and the deep residual network, a new method of direct sequence spread spectrum signal acquisition is designed, render the data information in the deep residual network, and construct the Loss function model through training processing to determine the pseudo-random sequence code. According to the pseudo-random code acquisition results of the spread spectrum signal, complete the direct acquisition of the sequence pseudo-random code, and realize the design of the acquisition method of the basic Direct-sequence spread spectrum signal. And the practical application value of this method is highlighted through comparative experiments.

2 Signal Sequence Function Modeling Based on Deep Residual Network

The modeling of spread spectrum signal sequence is the basis for realizing signal acquisition. On the basis of deep residual network, data rendering is completed, and the complete loss function expression is solved by training signal objects. This chapter conducts in-depth research on the above contents.

2.1 Data Rendering in Deep Residual Network

The apparent function used by the depth residual network is the bidirectional reflection distribution function (BRDF). It is an empirical formula used to describe the transmission behavior of signal data in the network host. It is necessary to measure the input rate and output rate of signal parameters and the signal data utilization efficiency in different combinations.

In general, professional equipment is required to measure data samples under various transmission conditions to obtain the BRDF function expression of the object. However, the requirement of data rendering is to restore the performance of the data under various transmission conditions through a signal sample with a specific transmission rate [3]. We hope to obtain the BRDF apparent distribution attribute of a single information under unknown transmission conditions through the depth residual network, so we need to use the network model to render the information parameters of some required samples, and select appropriate rendering programs to express the characteristics of BRDF functions.

The complete depth residual network structure model is shown in Fig. 1.

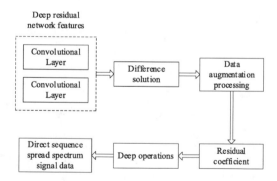

Fig. 1. Deep residual network structure model

In order to suppress the influence of incoherent data modulation on signal acquisition, data preprocessing can be used to eliminate data codes, and then coherent integration can be performed.

(1) First, the signal and the local pseudo-random code are processed by chip spacing differential pre-processing, and then the parallel code phase acquisition based on the deep residual network is performed. The differential pre-processing of the signal

can eliminate the impact of data bit polarity hopping, but the same processing of the pseudo-random code will greatly reduce the spread spectrum gain and reduce the anti noise performance of the acquisition.

(2) Secondly, because the direct sequence signal is a differential input, a differential amplifier chip needs to be selected. The deep residual network uses a single power supply instrument amplifier AD623, which allows the use of a single gain setting resistor for gain programming. Since AD623 itself can provide excellent AC common mode rejection ratio (ACCMRR) that increases with gain, it can maintain the minimum error. Due to the influence of CMRR, the line noise and harmonics can still remain constant up to 200 Hz.

regulations $\dot{\delta}$ represents the interference integration characteristics of direct sequence spread spectrum signals, and its definition formula is as follows:

$$\dot{\delta} = \sqrt{\frac{\chi \times \left(o_{\text{max}}^2 - o_{\text{min}}^2\right)}{\vec{P}}} \tag{1}$$

where, o_{max} represents the maximum assignment result of the data sample bit polarity jump behavior parameter, o_{min} represents the minimum assignment result of the bit polarity jump behavior parameter, χ indicates the spread spectrum gain parameter, $\vec{P}$ represents the transmission eigenvector of the direct sequence spread spectrum signal in the deep residual network.

For signal capture affected by deep residual network, data preprocessing or correction correlator is required before rendering data to eliminate the impact of data bit polarity jump. Data preprocessing algorithms are not suitable for applications in low SNR environments due to the use of nonlinear transformations such as delay multiplication and square transformation, resulting in SNR loss.

Using formula (1), the data rendering expression in the depth residual network can be defined as:

$$I = \sqrt{\frac{\sum_{i=1}^{+\infty} \alpha \left|\frac{\hat{u}}{\vec{p}}\right|^2}{\beta \times \dot{\delta}}} \tag{2}$$

where, i represents the preprocessing parameters of direct sequence spread spectrum signals, α represents the nonlinear transformation parameter, β represents the signal parameter extraction coefficient based on the depth residual network model, $\hat{u}$ indicates the SNR loss characteristics of signal parameters, $\hat{p}$ indicates the delay multiplication parameter of the signal parameter.

Direct sequence spread spectrum signal data rendering can spread spectrum modulation information data code and pseudo-random code with higher transmission rate to achieve the purpose of expanding signal bandwidth. In the deep residual network, the data rendering receiver performs reverse operation on the received signal parameters [4]. In order to demodulate the data signal accurately, the receiver must synchronize with the pseudo code phase and carrier phase of the received signal. This synchronization control process is called direct rendering processing.

2.2 Signal Training

If you want to obtain more signal samples that meet the needs of the deep residual network, you can achieve this by increasing the diversity of training data. If a large number of unlabeled data are used to inject into the network, they often contain some valuable information, such as correlation, integration, phase, etc. In this way, a search space containing apparent signal information is formed for the deep residual network. Through continuous training, the search space will be expanded.

The existing scarce labeled apparent spread spectrum data set is used to train it, so that the corresponding apparent spread spectrum data can be predicted when the direct sequence information is input. With such a model [5], we can then input unmarked general information to obtain its apparent spread spectrum data. By combining parameters with input signal information, we can improve and strengthen this deep residual network.

Acquired from the original depth residual network n unequal common signal information $U_1, U_2, \cdots, U_n$, the value expression is:

$$
\begin{cases}
U_1 = \sqrt{\dfrac{\gamma_1 \times |\Delta Y|}{\dot{y}_1}} \\[2ex]
U_2 = \sqrt{\dfrac{\gamma_2 \times |\Delta Y|}{\dot{y}_2}} \\[2ex]
\vdots \\[1ex]
U_n = \sqrt{\dfrac{\gamma_n \times |\Delta Y|}{\dot{y}_n}}
\end{cases}
\tag{3}
$$

where, ΔY represents the unit cumulative amount of direct sequence spread spectrum signal data in the deep residual network, $\gamma_1, \gamma_2, \cdots, \gamma_n$ respectively n randomly selected sequence information prediction parameters, and $\gamma_1 \neq \gamma_2 \neq \cdots \neq \gamma_n$ inequality value condition of is always true, $\dot{y}_1$ representation and parameters γ_1 matched spread spectrum signal data enhancement vector, $\dot{y}_2$ representation and parameters γ_2 matched spread spectrum signal data enhancement vector, $\dot{y}_n$ representation and parameters γ_n matched spread spectrum signal data enhancement vector.

Signal training first selects a small amount of labeled data and its corresponding sequence parameters to pre train the deep residual network, so that the network has an initial prediction ability and can generate simple apparent information. In this way, unlabeled data samples can be input into the network to obtain their corresponding apparent spread spectrum parameters, and these data can be rendered. A group of channel acquisition samples is obtained, and now they become a new marked data pair with the signal data used to render them previously.

The simultaneous formulas (2) and (3) can express the direct sequence spread spectrum signal training conditions based on the deep residual network as:

$$
R = \frac{U_1 \times U_2 \times \cdots \times U_n}{(\varphi - 1)^2 - (r_1 \cdot r_2 \cdot \cdots \cdot r_n)^2} \times I
\tag{4}
$$

φ represents the initial prediction parameters of the direct sequence spread spectrum signal, r_1, r_2, $\cdots$, r_n respectively representing and U_1, U_2, $\cdots$, U_n relevant signal parameter spread spectrum processing authority.

Signal training also prevents the features extracted from some networks from taking effect only under specific circumstances, otherwise the network will be too sensitive to some special training data. When these features are missing, the prediction effect of the network will decline sharply.

2.3   Loss Function Model

In order to improve the training effect, the loss function model is used to pre train a deep residual network. In order to provide the grid with as much information as possible, the direct sequence spread spectrum signal at each training stage is normalized, because the direct sequence spread spectrum signal is closest to the original appearance of the signal object, more intuitive, and its value representation is closest to the original representation of the signal data.

FIR filters are responsible for generating loss functions related to direct sequence spread spectrum signals. There are four types of FIR filters: direct type, cascade type, linear phase type and frequency sampling type. The linear phase of FIR filters is very important because the system is required to have linear phase for data transmission, and FIR filters can be made into strict linear phase because of their finite impulse response. And the number of times of multiplication is reduced by half by using symmetry relation, which greatly improves the calculation efficiency.

The calculation process of the estimated loss function model is the process of normalizing all direct sequence spread spectrum signals. The specific calculation formula is:

$$E = \left| \frac{\tilde{e}}{\dot{W}} \right|^2 \times \left. \left(R \middle/ |\Delta T| \right) \right|_{\dot{W} \neq 0} \tag{5}$$

where, $\tilde{e}$ represents the normalized processing vector of the direct sequence spread spectrum signal, $\dot{W}$ represents the original arrangement characteristics of signal objects, ΔT represents the unit transmission period of signal data in the deep residual network.

The deep residual network inputs the results of convolution operation of each layer to one layer and transmits them in turn. Its main function is to extract the direct sequence characteristics of input spread spectrum signals. The batch standardization layer (BN) is used to improve the stability, calculation efficiency and performance of the residual network, alleviate the problem of slow function convergence to a certain extent, and reduce the difficulty of training[6]. The specific loss function model operation process is shown in the figure below (Fig. 2).

After normalization, the transmission behavior of direct sequence spread spectrum signals starts to be different from that of the last deep residual network, and the residual network system is composed of a set of bilinear up sampling and convolution layers throughout the transmission process. Its structure is similar to the analysis subnet. Before bilinear sampling, each convolution layer will have a group of batch normalization layers (BN) and activation layers. The activation function is the loss function [7]. In addition,

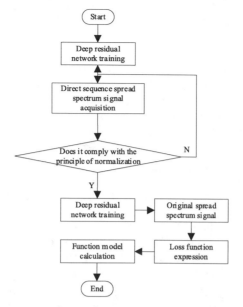

Fig. 2. Calculation flow chart of loss function model

a jump connection is added to each layer to map the corresponding convolutional layers one by one to enhance and supplement the high-frequency details that may be lost in the later analysis layers.

The solution of the depth residual network signal loss function model satisfies the following expression:

$$Q = 1 - \sqrt{\sum_{\substack{\varepsilon=1 \\ \iota=1}} |\phi E| \times \left| \frac{A_\varepsilon}{A_\iota} \right|} \tag{6}$$

where, ε, ι represents the sampling parameters of two direct sequence spread spectrum signals that are not equal to zero, and $\varepsilon \neq \iota$ value condition of is always true, A_ε indicates parameter based ε Signal data activation vector, A_ι indicates parameter based ι signal data activation vector, ϕ indicates the jump connection parameters of the deep residual network.

The loss function model obtains a strong regression model by combining multiple weak classifiers. Its underlying idea is that even if one classifier makes a prediction error, other classifiers can correct it, so as to achieve accurate analysis of the transmission behavior of direct sequence spread spectrum signals.

3   Design of Direct Sequence Spread Spectrum Signal Acquisition Method

In the deep residual network model, the pseudo-random sequence code is determined, and the acquisition and processing of the sequence pseudo-random code are completed according to the captured pseudo-random code parameters of the spread spectrum signal, thus realizing the design of the direct sequence spread spectrum signal acquisition method based on the deep residual network.

3.1   Pseudo Random Sequence Code

Ordinary signals can send information codes to the communication channel directly through carrier modulation, but direct sequence spread spectrum signals need to send information codes to the communication channel after pseudo-random sequence code modulation (spread spectrum modulation), and then through carrier modulation. The period length of the pseudo-random code is usually long, and the receiver can only complete the acquisition and tracking of the spread spectrum signal on the premise of predicting the pseudo-random code used by the sender.

The modulation principle of direct sequence spread spectrum signal is shown in Fig. 3.

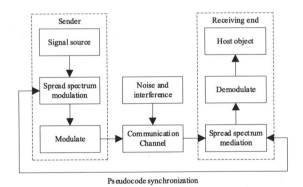

Fig. 3. Modulation principle of direct sequence spread spectrum signal

Pseudo random sequence is a group of binary sequence generated by shift register. The structure of shift register can be divided into simple shift register generator (SSRG) and module tap sequence generator (MSRG).

When the number of register taps is the same, the pseudo-random sequence code rate generated by MSRG is higher. Both SSRG structure and MSRG structure can generate m-sequence, and a set of m-sequence optimal pairs can generate Gold sequence. Both are pseudo-random sequences with excellent performance [8]. M-sequence, also called m-code, is the longest linear shift register sequence, which can be generated by SSRG structure or MSRG structure.

All "0" states of the shift register are removed in each cycle of the m-code, so the number of "1" in the m-code is only one more than the number of "0". The number of

"1" and "0" in m code is basically equal, and the balance is strong. Therefore, the signal modulated by m-code is more hidden and has strong anti interception ability.

The solution of pseudo-random sequence code of direct sequence spread spectrum signal satisfies the following expression:

$$S = \frac{1}{\int\limits_{a=1}^{+\infty} \sqrt{\frac{1}{\lambda}|\hat{d}|Q^{\frac{\mu+\nu}{2}}}} \tag{7}$$

where, a represents the standard length parameter of the m-code of the direct sequence spread spectrum signal, λ represents signal sample storage parameters, $\hat{d}$ represents the signal parameter coding characteristics based on pseudo-random sequence, μ, ν represents two randomly selected concealment parameters, whose values belong to $[1, +\infty)$ Value range of.

Direct sequence spread spectrum (DSSS) works by directly using pseudo-random code to broaden the spectrum of the signal at the transmitter. The received spread spectrum signal is despread at the receiving end using the same pseudo-random code as the transmitting end, and the spread spectrum signal whose spectrum is broadened is restored to the signal before spread spectrum modulation. The direct sequence spread spectrum system has the advantages of low signal transmission power and strong anti-interference ability.

3.2 PN Code Acquisition of Spread Spectrum Signal

The acquisition method based on the deep residual network, referred to as the residual acquisition method, has the structure shown in Fig. 4. The pseudo code generated by the local pseudo code generator is multiplied with the input signal, and then sent to the threshold decision module through integral accumulation processing. If the pseudo code acquisition decision is successful, the output signal will be sent.

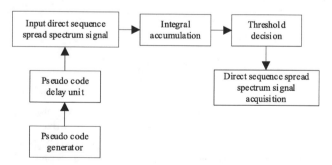

Fig. 4. Block diagram of acquisition method based on depth residual network

If the PN code acquisition decision fails, a control signal is sent to control the PN code delay unit to delay the PN code generated by the PN code generator by one time unit. The length of the time unit is determined according to the PN code acquisition accuracy in the specific design. Multiply the PN code delayed by one time unit with the input signal again and do subsequent processing until the correlation result passes the PN code acquisition decision flag and then outputs the signal to complete the serial PN code acquisition.

The serial PN code acquisition method uses less resources in practical applications and is relatively simple to implement, but its processing speed is slow. The acquisition accuracy of the depth residual network is half a symbol length, and the gold code with a pseudo-random code of 1023 code length is taken as an example. In the worst case, 2046 times of serial processing is required to capture the threshold output signal through the pseudo-random code, and complete the coarse synchronization of the local pseudo-random code phase and the pseudo-random code phase in the input signal [9]. Therefore, the serial PN code acquisition method can only be used when the system has low requirements for processing time.

Using formula (7), the PN code acquisition expression of spread spectrum signal is derived as:

$$F = \sqrt{f \times \left(\frac{S}{\theta - 1}\right) \times \frac{\left(g_{max}^2 - g_{min}^2\right)}{\overline{g}}} \tag{8}$$

where, f represents the coarse synchronization parameters of PN code of spread spectrum signal, θ represents the threshold value, g_{max} indicates the maximum value of the local pseudo code phase, g_{min} represents the minimum value of the local pseudo code phase, $\overline{g}$ indicates the average value of pseudo code phase parameters.

Compared with the serial PN code acquisition method, the parallel PN code acquisition method uses more resources, and the system design complexity is higher, but the processing speed is greatly improved. Under the premise of allowing the transmission amount of direct sequence spread spectrum signal, the acquisition speed and accuracy can be further improved by designing more parallel branches. However, in practical engineering applications, the PN code acquisition module designed according to the parallel acquisition method is often too complex and consumes too much resources. Therefore, this method can only be used when the speed is extremely high and a lot of resources can be consumed.

3.3 Direct Acquisition of Sequence Pseudo Code

To realize direct acquisition of sequence PN code, it is necessary to use numerical control oscillator NCO, low-pass filter LPF, PN code generator, time-sharing selection module, multiplier, loop filter and PN code VCO. The input signal is mixed with the sine and cosine waves generated by the local carrier NCO and processed by a low-pass filter to remove the influence of the center frequency on the signal. The phase deviation of the PN code sent by the local PN code generator in time sharing is two data points, and the sending probability of the lag PN code and the leading PN code in time is half. After multiplying the leading pseudo code and the output signal of the low-pass filter, the actual acquisition signal can be obtained through integral accumulation and square processing.

Regulations $\tilde{k}$ represents the low-pass filtering transmission characteristics of direct sequence spread spectrum signals, $\vec{h}$ represents the local carrier vector of direct sequence spread spectrum signal. The simultaneous formula (8) can express the phase deviation condition of sequence PN code as:

$$J = \hat{l} \times \dot{j}^2 - \left| \frac{\tilde{k} \times \vec{h}}{F} \right|^2 \tag{9}$$

where, $\hat{l}$ represents the data point definition parameter, $\dot{j}$ represents the central frequency characteristics of direct sequence spread spectrum signals.

The digital down conversion can be realized by mixing the input data with the numerical control oscillator and low-pass filtering, in order to remove the influence of the carrier center frequency on the signal. The local pseudo code generator generates three pseudo code signals in turn, namely, leading pseudo code, real-time pseudo code and lagging pseudo code, with a phase length interval of one data point. The leading pseudo-random code branch and lagging pseudo-random code branch signals are processed by integration and accumulation respectively before FFT processing [10, 11]. The reason why FFT is used instead of square processing in correlation processing is that the Doppler frequency offset is large in the process of satellite signal transmission, and square processing is difficult to obtain the peak value accumulated by correlation results. With FFT, the correlation between the leading pseudo code branch and the lagging pseudo code branch and the acquisition signal can be found more quickly.

The performance of A/D acquisition directly affects the overall performance of weak signal acquisition system. Therefore, the selection of A/D devices is particularly important. The acquisition method based on deep residual network uses AD7667 chip to realize A/D conversion. AD7667 is a 16 bit high-precision analog-to-digital converter, which can amplify weak signals and conduct lossless acquisition. It is very suitable for weak signal acquisition.

Using formula (9), the direct acquisition expression of sequence pseudo code is derived as:

$$L = \frac{\xi \cdot J}{\sum\limits_{z=1}^{+\infty} \frac{b|\hat{C}|}{\psi} \times \dot{X}^2} \tag{10}$$

where, ξ represents the lag performance parameter of the sequence pseudo code frequency, z represents the signal transmission coefficient in the deep residual network branch, b represents the cumulative vector of the direct sequence spread spectrum signal, $\hat{C}$ represents the performance characteristics of signal peak, ψ represents the alternative parameter of the sequence pseudo code, $\dot{X}$ represents the real-time acquisition coefficient of the spread spectrum signal.

Compared with the PN code acquisition link whose accuracy is half of the symbol length, the PN code tracking link whose accuracy is one data point can complete more accurate phase adjustment. After the PN code tracking is completed, the input signal can be completely consistent with the real-time PN code generated by the local PN code generator. After removing the influence of PN code modulation, the PN code synchronization link needs to realize Doppler frequency offset compensation in the local digital controlled oscillator at the receiver through the carrier synchronization link to recover the carrier signal with the same frequency as that before PN code modulation. After the input signal data removing the influence of pseudo code is processed by twice integral accumulation, more accurate Doppler frequency offset is obtained through the arc tangent phase discriminator PD and loop filter LP in turn, and corresponding frequency offset adjustment control words are generated at the same time. The frequency offset of the output signal in the numerical control oscillator can be compensated by adding the control word to the numerical control oscillator, and the newly generated signal after the frequency offset compensation can be down converted with the input signal again to form a closed loop system. Because the closed-loop processing speed is fast, it can quickly complete the compensation of the Doppler frequency offset in the signal, so it can be considered that this method can accurately compensate the Doppler frequency offset of the signal in real time, and realize the synchronous acquisition of direct sequence spread spectrum signals.

4 Example Analysis

In order to verify the practicability of the direct sequence spread spectrum signal acquisition method based on the deep residual network, the local signal acquisition method based on SHF, and the signal acquisition method based on the finite-difference time-domain, the following comparison experiments are designed.

4.1 Experiment Preparation

The AGC equipment is used to extract the direct sequence spread spectrum signal in the transmission network, and the obtained signal parameters are used as experimental objects, as shown in Fig. 5.

Fig. 5. Direct Sequence Spread Spectrum Signal Extraction

The specific models of equipment components used in this experiment are shown in Table 1.

Table 1. Selection of Experimental Equipment

Experimental equipment	Component model
SIGNAL PROCESSOR	UltraScale + MPSOC XCZU components
Signal recognition equipment	AT89S51/52 microcontroller
Frequency band sampling device	STC89C52RC-40I-PDIP40
Processor board	AX7Z010 020
Embedded Motherboard	STC89C52RC-40I-PDIP40
Signal trigger	MC68HC908

In order to ensure the accuracy of the experimental results, after the signal sample collection is completed, the control switch must be disconnected until the equipment components return to the initial state, and then the closed loop switch is closed again for the next experiment.

4.2 Principle and Process

The acquisition capability of the network host to the spread spectrum signal can be used to describe the definition accuracy of the signal parameter sequence expression. When the signal output behavior remains stable, the stronger the acquisition capability of the network host to the spread spectrum signal, the higher the accuracy of the derived signal parameter sequence expression.

The specific implementation process of this experiment is as follows:

- Use the equipment elements shown in Table 1 to set up an experimental environment, and start the experiment after ensuring that the elements at all levels remain in a stable connection state.
- Under the effect of the direct sequence spread spectrum signal acquisition method based on the deep residual network, record the specific number of spread spectrum signals collected by the network host, and the results are the experimental group variables.
- Under the SHF based local signal acquisition method, record the specific number of spread spectrum signals collected by the network host, and the result is the control group A variable.
- Under the action of signal acquisition method based on finite-difference time-domain, record the specific number of spread spectrum signals collected by the network host, and the results are the control group B variables.

4.3 Results and Conclusions

The figure below reflects the experimental values of the spread spectrum signal acquisition quantity of the experimental group, control group A and control group B.

It can be seen from the analysis of Fig. 6 that with the extension of the experimental time, the amount of spread spectrum signal acquisition in the experimental group, control group A and control group B keeps increasing.

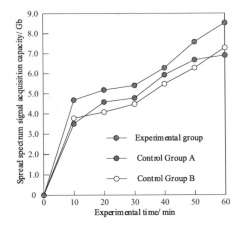

Fig. 6. Acquisition Amount of Spread Spectrum Signal

By the end of the 60 min experiment, the maximum amount of spread spectrum signal acquisition in the experimental group reached 8.5 Gb; The maximum amount of spread spectrum signal acquisition in control group A is 6.8 Gb, which is 1.7 Gb lower than the maximum amount in experimental group; The maximum value of the spread spectrum signal acquisition quantity of control group B is 7.3 Gb, which is 0.5 Gb higher than the maximum value of control group A, and 1.2 Gb lower than the maximum value of experimental group.

5   Conclusion

The conclusion of this experiment is:

(1) The implementation of local signal acquisition method based on SHF has relatively limited application ability in improving the acquisition amount of spread spectrum signal, which is insufficient to enable the network host to accurately define the signal parameter sequence expression.

(2) Compared with the SHF based local signal acquisition method, the application of the signal acquisition method based on the finite-difference time-domain is more capable of improving the acquisition amount of spread spectrum signals, but it is still unable to achieve the accurate definition of the signal parameter sequence expression.

(3) The application of the direct sequence spread spectrum signal acquisition method based on the deep residual network has greatly improved the acquisition capability of the network host for the spread spectrum signal. Compared with the SHF based local signal acquisition method and the time-domain finite-difference based signal acquisition method, it is more in line with the actual application requirements of accurately deriving the signal parameter sequence expression.

The next test work will focus on the field test and other real working environment closer to the signal processing board, and improve the new problems found. We are committed to further optimizing and improving the model structure and improving the acquisition performance and effect. At the same time, more data sets and scenarios are considered to verify the generalization ability and adaptability of the model.

References

1. Deelip, M.S., Govinda, K.: ExpSFROA-based DRN: exponential sunflower rider optimization algorithm-driven deep residual network for the intrusion detection in IOT-based plant disease monitoring. Int. J. Semant. Comput. **17**(01), 5–31 (2023)
2. Akhtar, M.M., Ahamad, D., Shatat, A., et al.: Big data classification in IOT healthcare application using optimal deep learning. Int. J. Semant. Comput. **17**(01), 33–58 (2023)
3. Fang, Z., Zhang, T., Wang, R., et al.: Acquisition of binary offset carrier modulation signals in a highly dynamic environment. J. Signal Process. **38**(10), 2164–2172 (2022)
4. Schmitter, S., Grundmann, S., John, K., et al.: Reynolds stress tensor and velocity measurements in technical flows by means of magnetic resonance velocimetry Data acquisition in the flow field behind the sudden expansion of the FDA benchmark nozzle and identification of systematic errors. tm - Technisches Messen **89**(3), 201–209 (2022)
5. Hosseini-Nasab, H., Ettehadi, V.: Development of opened-network data envelopment analysis models under uncertainty. J. Ind. Manage. Optim. **19**(3), 1963–1982 (2023)
6. Ye, L., Shen, W., Jianliang, X., et al.: A blind estimation method of DSSS signal information sequence and pseudo code sequence based on EVD. J. Signal Process. **38**(7), 1442–1449 (2022)
7. Yang, Y., Ji-Shan, H.E., Di-Quan, L.I.: Energy distribution and effective components analysis of 2 sequence pseudo-random signal. Trans. Nonferrous Metals Soc. China **31**(7), 2102–2115 (2021)
8. Su, N., Chen, X., Guan, J., et al.: One-dimensional sequence signal detection method for marine target based on deep learning. J. Signal Process. **36**(12), 1987–1997 (2021)
9. Avraham, Y., Pinchas, M.: A new approach for the DFT NIST test applicable for non-stationary input sequences. J. Signal Inf. Process. **12**(1), 1–41 (2021)
10. Hong, F., Ao, S., Li, X., et al.: Design of CC2530 wireless data acquisition system based on photovoltaic energy. Comput. Simul. **39**(01), 80–85 (2022)
11. Liu, S., et al.: Human inertial thinking strategy: a novel fuzzy reasoning mechanism for IoT-assisted visual monitoring. IEEE Internet Things J. **10**(5), 3735–3748 (2023)

A Secure Sharing Method for University Personnel Archive Data Based on Federated Learning

Xinwei Li[1(✉)], Yue Zhao[1], and Min Zhou[2]

[1] Human Resources Department, Changchun University of Architecture and Civil Engineering, Changchun 130000, China
lixinwei58666@163.com
[2] Department of Employmeng and Enterpreneurship, Xiamen Ocean Vocational College, Xiamen 361000, China

Abstract. In response to the complex data trust evaluation process in the current process of secure sharing of university personnel archive data, which leads to long data encryption time and poor data sharing and distribution performance, a federated learning based method for secure sharing of university personnel archive data is proposed. Build a data federation learning module to provide a platform for subsequent data processing. Optimize federated learning algorithms and complete incremental federated learning of archive data. Federated incremental learning of archival data. Improve data privacy and security. Apply Kalman filtering technology and data mapping technology to achieve secure sharing of archival data. The experimental results show that this method can effectively reduce data encryption time and provide data sharing and distribution performance.

Keywords: Federated learning · Archive data processing · Data distribution · Data sharing

1 Introduction

With the rapid development of modern computer science and significant improvement of computing power, machine learning is widely used in various fields of production and life, such as intelligent medical treatment, automatic driving and face recognition. These applications of machine learning are changing the way people live, learn and produce. The development of the Internet of Things makes the data of the Internet of things more and more huge, huge data can mine a large number of valuable information. GSMA predicts that by 2025, the number of connected Internet of Things devices in the world will reach 25 billion. The massive data collected by Internet of Things devices can generate meaningful information for human production and life through the mining of artificial intelligence and other technologies [1, 2]. For example, in smart medicine, doctors diagnose and analyze human health information collected by wearable devices to make scientific diagnosis and treatment suggestions for patients. In the scenario of

© ICST Institute for Computer Sciences, Social Informatics and Telecommunications Engineering 2024
Published by Springer Nature Switzerland AG 2024. All Rights Reserved
L. Yun et al. (Eds.): ADHIP 2023, LNICST 547, pp. 184–197, 2024.
https://doi.org/10.1007/978-3-031-50543-0_13

university Internet of Things, personnel file data collected by Internet of Things devices can help universities optimize education effect, improve teaching efficiency and reduce the cost of manual file management through intelligent analysis.

Under the influence of big data technology, the data size and model complexity of university personnel archive data are gradually increasing. In large-scale machine learning applications, the amount of data can be so large that a single computer's computing power cannot afford it. At this point, distributed machine learning has attracted the public's attention. Unlike model training on a single computer, distributed machine learning divides data and computing tasks that were originally concentrated on the server into different nodes, and multiple nodes collaborate to assist the server in training the model, thus supporting large-scale data training [3, 4]. Today's Internet of Things and edge computing, based on distributed training, have accelerated the pace of AI industrialization on the premise of ensuring data security.The effective union of production data between data nodes can realize the joint optimization of archival data, and further develop the technology of personnel data processing and sharing in colleges and universities. Due to the limited computing and storage capacity of Internet of Things devices, it is often necessary to outsource computing and storage requirements to cloud service providers. Therefore, before blockchain technology was widely studied by the academic community, Internet of Things data sharing was mainly realized through cloud services, supplemented by cryptography algorithm to realize data access control and ensure data security. However, the current proposed security sharing method of personnel files data in colleges and universities is complicated to set the link of data credit score, which leads to a long time in data security processing and low accuracy in data distribution. Therefore, a security sharing method of personnel files data in colleges and universities based on federal learning is proposed to optimize the shortcomings of the current security sharing method in the application process.

2 Build a Data Federation Learning Module

Literature research shows that federated learning can ensure the sharing of gradient information among all participants. Currently, in the process of data security sharing, the method of calculating the weighted average of all parties is commonly used. Therefore, the upload of client encryption gradients, aggregation of server encryption gradients, and distribution of global shared models are involved, and the overall operation time is long, and the operation security factor is poor [5, 6]. Therefore, in this study, federated learning algorithms were used to optimize the current method and compensate for its shortcomings. In the specific design, the following considerations were made in this study:

① Improved timeliness of data encryption:

Use Differential privacy technology: Differential privacy is a method to protect individual privacy, which can provide high data encryption timeliness while protecting data privacy. The Differential privacy technology can be used to encrypt and Data anonymization sensitive data.

Optimize encryption algorithms and key management: Choose efficient encryption algorithms and key management strategies to improve the efficiency of encryption and decryption, and reduce the impact of the encryption process on system performance.

② Improve data distribution accuracy:

Introduction of model aggregation and parameter update mechanism: in Federated learning, all participants jointly train models and share updated parameters, which can improve the accuracy of data distribution through model aggregation and parameter update mechanism. Participants share some model parameters rather than complete original data, thus protecting data privacy.

Data standardization and cleaning: Before data distribution, standardize and clean the data to ensure consistency and accuracy. This will help improve the accuracy and availability of data distribution.

③ Building Federated learning module:

Build a secure federated learning platform: Establish a secure and trustworthy federated learning platform to ensure security and privacy protection during data transmission and model training processes.

Design a reasonable participant collaboration mechanism: clarify the roles and responsibilities of each participant, develop a reasonable participant collaboration mechanism, and ensure the smooth progress of data sharing and model training.

When completing personnel file data sharing in colleges and universities, A data set for data sharing processing is usually given, in which each sample is represented as a number pair (E, F) composed of data feature E and data label F, and an initial model parameter u and a given form of loss function $\alpha(E, F, u)$ are given. Then the loss function can be used to calculate the loss of each data sample (E, F) on model u, so as to obtain the gap between the model performance and the training target based on the current data set and the current model parameter training. Finally, minimizing the gap between the reality and the target is taken as an important criterion throughout the process of data sharing to guide the updating direction of the model and realize the training of the model. Take the r sample in the training data set as an example, the feature of this data point is E_r and the label is F_r. If model parameter u is given, the loss function $f(u)$ on the whole training data set is defined as the form of sum of finite terms:

$$f(u) = \frac{\sum_{i=1}^{n} f_r(u)}{n} = \frac{\sum_{i=1}^{n} f_r(E_r, F, u)}{n} \tag{1}$$

where, n represents the number of training data, u represents the parameters of the neural network model, and f_r represents the loss function of a single sample. Use this function to control the loss during data processing. For the other half of the privacy budget δ, in each iteration, δ is divided into two parts, the gradient budget δ_i and the step budget δ_j. δ_i is used to calculate gradients with noise, and δ_j is used to find the optimal gradient descent step size. At the same time, add an initialization weight vector β. At this point, in the second row of federated learning calculation, set t to record the number of iteration rounds initialized to 0, and randomly set δ to δ_i and δ_j. The third line, when the privacy budget is not exhausted, performs a loop, which is model training. On line 4, record the index value of the gradient step with δ_j and initialize it to O. There are:

$$h_t(s) \leftarrow \nabla \alpha(f(u), n) \tag{2}$$

$$h_t(s) \leftarrow \frac{h_t(s)}{\max(1, \frac{\|h_t(s)\|_2}{Y})} \tag{3}$$

$$\widetilde{h_t(s)} \leftarrow \frac{\sum (h_t(s) + N(\frac{0,Y}{\delta_j}))}{N} \tag{4}$$

Y indicates the data processing threshold. The above function can be applied to the data credit score, according to the score results, determine the security level of the data. The above Settings are sorted out and constructed as a federal learning framework, so that archive data users can sort and update data in this framework and realize data cloud interaction.

3   Design of Security Sharing Method of Personnel File Data in Universities

In federated learning, a model is jointly trained among multiple users in a decentralized manner, and the model learning process is transformed to the user side. The user only uploads the model parameters to the cloud platform, which then integrates the parameters and sends them back to each user. Due to the potential involvement of sensitive information in training data, publishing the model during or after training poses a risk for federated learning frameworks to leak user privacy [7, 8]. The personnel files of universities contain a large amount of personal privacy data, and any abnormal encryption will cause irreparable losses. Therefore, based on previous research results and the data

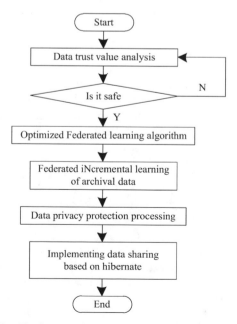

Fig. 1. Security Sharing Process of Personnel Archive Data in Universities

federation learning process mentioned earlier, the secure sharing process of archive data is set in the following form.

Next, follow the process shown in Fig. 1 for data processing and distribution. The specific data sharing operation process is set as follows.

3.1 Federated Incremental Learning of File Data

The object of this study is the highly complex personnel file data of colleges and universities. How to mine the newly generated state data for the exponential growth of new data and combine it into the existing mining mode has become a new focus. The model trained by the traditional data processing algorithm repeatedly extracts data features from the local data for training and learning, but it cannot adapt and incrementally modify the industry association model with the newly added data T in real time, resulting in an increase in time cost and a decrease in model diagnostic accuracy. This study uses incremental weighting to solve the federated incremental learning problem. The number of samples reflects the diversity of samples to some extent, and the model based on high-complexity data training has better scalability [9]. The process of model training can be understood as the "learning" process of the model. Generally, as time goes by, the model gets closer to the optimal solution of the problem, but more new data will increase the distance between the computing cloud and the optimal solution of the learning problem. Therefore, it is obviously unreasonable for the computing cloud with uneven new data to update the model parameters in the file association end in the same amount. Therefore, it is targeted to optimize.

The incremental weight value represents the proportion of the number of newly added samples in the cloud to the total number of original samples. The incremental weight of cloud c can be calculated from the number of newly added samples and the total number of samples:

$$\varepsilon_c = \frac{|U_c|}{|V_c|} \tag{5}$$

where, $|U_c|$ is the number of newly added samples in the computing cloud, and $|V_c|$ is the total number of original samples in the computing cloud. In the process of parameter optimization, there is a certain depth value. Let the parameter depth value be:

$$\varpi_{i+1}^c = \varpi_i^c - (\varepsilon_c * \varpi_i^c) \tag{6}$$

During the time interval between downloading and uploading parameters from the cloud, new training data will be generated, and the parameter depth value will also be updated to a certain extent. The parameter depth value represents the impact of newly added data from the local dataset on model performance during the completion of an iterative learning process in the computing cloud, reflecting the update level of the computing cloud. In order to reduce the parameter weighting of cloud computing with larger parameter depth values and relatively smooth attenuation process, this study chose the arctangent function as the incremental weighted attenuation function:

$$\sigma_{i+1}^c = \frac{2 \arctan \varpi_{i+1}^c}{\pi} \tag{7}$$

Under the above Settings, only the computing cloud in the participating subset is updated in each round. The contribution of the model to the aggregation operation can be determined according to the parameter depth value of the computing cloud model, which can effectively utilize the historical information and distinguish the utilization value of the local model, which is expected to improve the effectiveness of the aggregation operation. Therefore, the parameter weighting of the local model is further paid attention to, and the improved aggregation strategy is proposed as follows:

$$\lambda_{i+1} \leftarrow \sum_{c=1}^{c} \frac{n_c}{n} * \sigma_{i+1}^c * \varpi_{i+1}^c \qquad (8)$$

In the process of federated incremental learning, the model parameters submitted by the computing cloud need to be incrementally weighted to participate in industry joint model optimization. The modified parameters are updated on the data joint processing end based on specific optimization algorithms to update the model parameters. After optimization, the latest industry joint model parameters are obtained from the cloud and covered by local parameters for the next round of iterative learning.

3.2 Data Privacy Protection Processing

The general process of federated learning was set in the previous section. However, the process of encryption and decryption of gradient or local model weights was ignored in the description. If the gradient or current local model weight is not encrypted, the above process can train a model normally. However, it will lead to the disclosure of user privacy data. Although in the general process of federated learning, data does not flow directly out of the local, only out of the gradient or when the local model weight, however, some research results have shown that such information contains user privacy data. Specifically, user data can be backderived through gradient, and the leakage gradient is almost a direct disclosure of user privacy data [10–12]. Therefore, the gradient or local model weights uploaded by users per round need to be protected by relevant techniques. Common protection schemes include security aggregation, homomorphic encryption and differential privacy. In this study, security aggregation is selected to build the protocol of federal learning privacy protection, and the specific main contents are set as follows.

Any user of this agreement needs to communicate with all remaining users, with the purpose of enabling Diffie Hellman key negotiation between any two users to obtain a secret that only both parties know. Any user, in order to ensure the security and robustness of the protocol, will share their secrets twice, that is, split their secrets into "secret fragments" and distribute them to all remaining users[13–15]. Only when they have a certain number of "secret fragments" can the shared secrets be reconstructed. Before sending data to the server, users will apply two masks locally, and the two masks will be removed during the aggregation phase in two queries from the server.

To ensure the security of the communication content on the channel, symmetric encryption can be used to encrypt the communication content. Symmetric encryption means that the same key is used to encrypt and decrypt information. To ensure information security, the key cannot be disclosed and only the communication parties know it. In the public key cryptosystem, the public key is used for encryption and the private key for decryption. A public key is different from a private key. Public keys can be made public, and anyone can use them to encrypt; however, public keys cannot be decrypted. The private key is not public, and only the person with the private key can decrypt it. Let's say we're using a symmetric cryptosystem. Before the communication starts, both parties do not know the key and cannot perform symmetric encryption. Then, there needs to be a way for both parties to know this key securely. Therefore, in this study, a temporary secure private channel was established to transfer the key used by both parties for the convenience of subsequent processing.

3.3   Sharing File Data Securely

Based on the above settings, a new method for secure sharing of archival data is proposed. Due to the fact that archive data is mostly heterogeneous, Hibernate will be used in this study to achieve data sharing. Through the research on Hibernate based MCDMS heterogeneous data sharing technology, starting from the heterogeneous data source of archives, this paper explores the data mapping of each archive and the inheritance relationship of each archive's data table after saving the heterogeneous data source data to the Oracle9i database. In addition, this study will also describe the underlying data filtering technology of a single discipline and the Lucen data indexing technology currently used in the system. By using open source components and Java Web programming knowledge, the data indexing function will be implemented, providing a data level shared data query solution for the sharing of personnel archives data in universities.

Set the data in the file to $O = \{\mu1, \mu2, \mu3, ..., \mu n\}$; Inherit the dependency relationship as the parent class PriorDiscipline, all derived subclasses as Derived Discipline N, and $N \in$ as any natural number. Then, in the relational database, you can use additional fields to represent the recorded ID of a specific subclass in the PrioritDiscipline table, and save the data segments that all subclasses need to share. When doing SQL or HQL (Hibernate Query Language) CRUD, you can use this ID to read the specific subclass, obtain the attribute values of the subclass saved in the parent class PrioritDiscipline, and complete the inheritance mapping. According to this setting, an additional field filter can be used for the PriorDiscipline parent class in the archive database (for example, the filter can be equal to the numerical value 1 to identify the parent class, and if it is NULL, it is not the parent class). The PriorDiscipline table does not store the data segments that all subclasses need to share. Instead, by passing the ID field of the PriorDiscipline parent class to the Derived DisciplineN, the ID is both the primary key of the Derived DisciplineN and the foreign key of the Derived DisciplineN. When using SQL or HQL for CRUD, the first step is to determine whether the filter is NULL. If it is not, the ID of PriorDiscipline is taken to compare all subclass IDs. If a subclass with a matching ID value is found, the attribute values of the subclass are obtained to complete the inheritance mapping.

Some noise is generated in the mapping processing process. In order to avoid the impact on the archive data and increase the process, Kalman filtering technology is used in this study to complete this part of operation. Kalman filter can deal with normally distributed noise in linear system. The noise added to the model parameters after differential privacy processing is Gaussian noise, which meets the condition of normal distribution, and the state update equation is linear. Therefore, Kalman filter can be used to de-noise the model parameters to reduce the impact of differential privacy on model accuracy. The data update parametric equation can be expressed as:

$$
\begin{cases}
g_{t+1} = g_t - \rho \sum h_i \\
g_{t+1,1} = g_{t+1} - \rho g_{t+1} + \wp(0, \zeta^2 D) \\
\ldots \\
g_{t+1,N} = g_{t+N} - \rho g_{t+N} + \wp(0, \zeta^2 D)
\end{cases}
\tag{9}
$$

Among them, g_t represents the global model parameters for the t round of global training, and g_{t+N} represents the model parameters for the target data sent to different data nodes in the t round. The noise added during the differential privacy process uses the Gaussian mechanism, which satisfies the Gaussian distribution $\wp(0, \zeta^2 D)$. Transform the parameter update equation of model training into matrix form. In the t round of federated learning, there are N sending nodes with model parameters $g_{t+1,1}$ and global model parameters $g_{t+1,N}$. The parameter vector composed of all node model parameters is $G = \begin{bmatrix} g_{t,1} & g_{t,2} & \ldots & g_{t,n} \end{bmatrix}$, and its gradient value can be expressed as $g_{t,n}$. At this point, the gradient vector composed of the gradient values of all data distribution nodes is $g_{t,n}$, and the data filtering linear coefficient matrix can be written as:

$$
L = \begin{bmatrix}
1 & 0 & 0 & \ldots & 0 \\
0 & \frac{G'}{N} & 0 & \ldots & 0 \\
0 & 0 & \frac{G'}{N} & \ldots & 0 \\
\ldots & \ldots & \ldots & \ldots & \ldots \\
0 & 0 & 0 & \ldots & \frac{G'}{N}
\end{bmatrix}
\tag{10}
$$

Through this formula, the noise in the data is filtered and the content set above is followed to realize the safe sharing of archive data. So far, based on the federal learning university personnel file data security sharing method design is completed.

4   Experimental Analysis

This study proposes a secure sharing method for university personnel archive data based on federated learning, and improves the data user credit scoring section of the method. In this stage, the method will be tested and analyzed, including data encryption testing, simulation application performance testing, and credit scoring performance testing.

4.1   Experimental Data

This experiment uses two commonly used open source datasets, MNIST and FMNIST, as experimental data. Below are brief introductions.

The MNIST dataset records handwritten archive images from 100 different individuals, with numbers ranging from 0 to 9 and 10 categories. The statistical purpose is to train recognition models for handwritten digits. The MNIST dataset contains a total of 10000 grayscale images with channel 1, of which 5000 were divided into training datasets and the remaining 5000 were used for this experiment. The data size of each handwritten image is 56 pixels * 56 pixels.

The EMNIST dataset is an extension of the MNIST dataset. Based on handwritten numerical pictures of different people, the relevant attribute data is recorded. The handwritten digits are 0 to 9, the attribute data characters are set as lowercase letters a to z and uppercase letters A to Z, and the number of categories is 62. The FMNIST dataset contains 10000 pieces of data. There are a large number of categories in the FMNIST data set, and the upper and lower case forms of some handwritten letters are similar, so the FMNIST data set can be divided into five categories, which are detailed as shown in Table 1.

Table 1. Classification results of FMNIST dataset

Category name	Training set data quantity/item	Test set data quantity/item	Total number of data sets per/item
Z1	458	1042	1500
Z2	2000	1000	3000
Z3	4500	1500	6000
Z4	1000	1000	2000
Z5	1350	1150	2500

Organize the above content and use it as the data source for this experiment to analyze the application effect of the methods in the text.

4.2 Experimental Environment

All experiments in this chapter were performed on four high-performance CPU servers with the same configuration. Table 2 shows the specific configurations of each server.

Table 2. Configuration information of the experimental server

Information sequence number	Message name	Specific content
1	CPU	8 nuclei, Intel Xeon Processor (Skylake, IBRS)@2194.848 MHz
2	Memory	32 GB
3	Disk	1TB
Tb	Network card	Red Hat, Inc. Virtio network device
5	Operating system	Centos 7.2
6	Kernel version	3.10.0?1160.15.2.e17.x86–64

In terms of software, all implemented applications in this experiment were deployed in the form of containers using Docker software; The development language includes Java and Golang; The IBPRE algorithm is implemented based on the official JPBC library of Java (Java Pairing BasedCryptography); The blockchain platform uses Tendermint and Hyperledger Fabric. Fabric is used for comparison with Tendermint; The functional testing software used Postman performance testing software and Apache's open-source JMeter project.

4.3 Encryption Timeliness Test

On the premise of fixed experimental data, compare and analyze the data encryption effects of the basic method, deep learning method, and the method in the text. The encryption effect is reflected by the encryption time, as shown in Fig. 2.

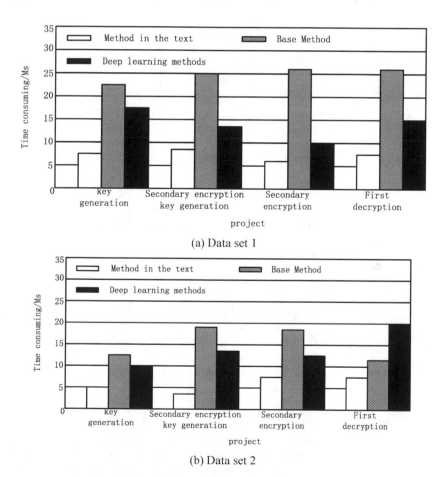

(a) Data set 1

(b) Data set 2

Fig. 2. Encryption experiment results

Analyzing Fig. 2, it can be seen that there are significant differences in encryption time among the three methods for different databases. For database 1, the encryption process of this method takes significantly less time, and it can complete the encryption of university personnel file data within 10ms. Compared with the methods in this article, the encryption efficiency of the two comparison methods is relatively low. Compared to Database 1, the three methods generally take less time to encrypt Database 2, but the encryption time of this method is still less than that of the two comparative methods. Therefore, the encryption timeliness of this method is higher.

4.4   Simulation Application Performance Test

This experiment sets the performance testing index of the secure sharing method as the accuracy of data distribution, and the specific calculation formula is set as follows:

$$A = \frac{a+d}{a+b+c+d} \tag{11}$$

where, a represents the correct number of samples distributed; b represents the correct number of negative samples distributed; c represents the number of samples wrongly distributed; d represents the negative number of samples distributed incorrectly.

According to formula (11), complete 50 data sharing and distribution tests. The accuracy of data distribution using different methods is shown in Fig. 3.

Analyzing Fig. 3, it can be seen that the data distribution capability of the method proposed in this paper is far superior to the other two methods. During 50 tests, the method proposed in this article can maximize the accuracy of data distribution, with a distribution accuracy rate consistently above 90%, significantly higher than the two comparative methods. This indicates that the method proposed in this article can effectively avoid the problem of abnormal data sharing and ensure the reliability of archival data applications.

4.5   Credit Score Performance Test

In this study, the credit score part of the security sharing method is optimized. This experiment will compare the data transaction processing performance after the sharing method is applied in the case of malicious nodes and credit scoring algorithm, so as to verify the effectiveness of the method proposed in this study (Fig. 4).

Comparing the content in the above figure, it can be seen that when the data volume is constant, the data processing capacity of each node after the application of the method in the article is relatively large, indicating that this method can effectively control the credit value of the data and avoid data distribution and sharing problems caused by abnormal credit scores. The credit scoring process of the other two methods has poor application results when there are malicious nodes, and the credit scoring process affects the efficiency of data processing and transactions. Based on the above experimental results, it can be determined that the credit scoring stage setting of the method in the article is relatively reasonable.

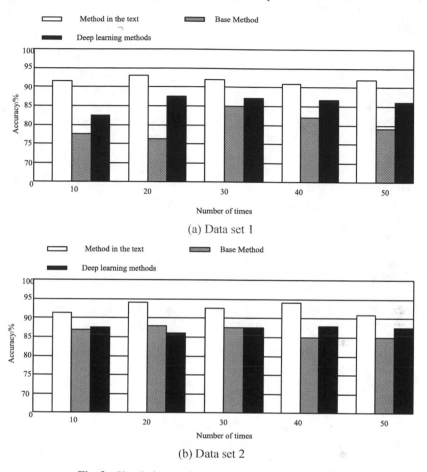

(a) Data set 1

(b) Data set 2

Fig. 3. Simulation application performance test results

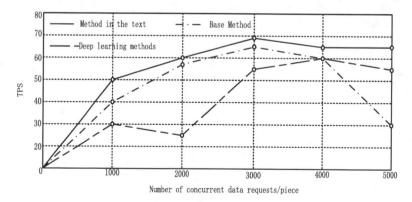

Fig. 4. Credit score performance test results

5   Conclusion

In this study, from the part of data encryption and data security management, the current sharing method of personnel files in colleges and universities was optimized, and the federal learning module was added to conduct credit assessment on the original data, so as to improve the security of data and the ability to resist malignant interference.

The method proposed in this article has achieved good application results at the current stage. This method can not only effectively shorten the encryption time, but also improve the distribution accuracy and credit rating of the data sharing process. However, during the testing process, it was found that there is still room for further improvement in the scalability of this method. Therefore, in future research, it is necessary to optimize the application process of this method, including increasing the number of supported users, data scale, and flexibility of the architecture, in order to further ensure the application effect of this method.

References

1. Guo, L., Zhu, Q., Zheng, J., et al.: The research on data sharing and security protection for students' comprehensive literacy evaluation based on federated learning. China Educ. Technol. **10**, 56–63 (2022)
2. Lu, C., Deng, S., Ma, W., et al.: Clustered federated learning methods based on DBSCAN clustering. Comput. Sci. **49**(z1), 232–237 (2022)
3. Liu, W., Xu, X., Zhang, X., et al.: Federated learning based method for intelligent computing with privacy preserving in edge computing. Comput. Integr. Manuf. Syst.. Integr. Manuf. Syst. **27**(9), 2604–2610 (2021)
4. Wen, Y., Chen, M.: Medical data sharing scheme combined with federal learning and blockchain. Comput. Eng. **48**(5), 145–153,161 (2022)
5. Mo, H., Zheng, H., Gao, M., et al.: Multi-source heterogeneous data fusion based on federated learning. J. Comput. Res. Dev. **59**(2), 478–487 (2022)
6. Zhao, Y., Wang, L., Chen, J., et al.: Network anomaly detection based on federated learning. J. Beijing Univ. Chem. Technol. (Natural Science Edition) **48**(2), 92–99 (2021)
7. Zheng, J., Li, W., Liu, X., et al.: Research on the federal sharing technology of remote sensing image artificial intelligence datasets. Spacecraft Recovery Remote Sensing **43**(4), 12–24 (2022)
8. Su, Y., Zhang, H., Liu, J.: Federated recommendation algorithm based on equilibrium learning. J. Hefei Univ. Technol. (Natural Science) **45**(5), 625–632 (2022)
9. Liu, X., Yin, Y., Chen, W., et al.: Secure data sharing scheme in Internet of Vehicles based on blockchain. J. Zhejiang Univ. (Eng. Sci.) **55**(5), 957–965 (2021)
10. Yang, Y., Lin, D., Huang, F., et al.: Blockchain based secure data sharing system for internet of things. J. Fuzhou Univ. (Natural Sci. Edition) **49**(6), 739–746 (2021)
11. Jia, Z., Zhang, J., Yi, K.: Mobile education resource sharing method for wireless broadband connection. Secur. Commun. Networks **16**(5), 127–135 (2021)
12. Noreen, H.S., Thar, B., Ghulam, A., Haq, A.Z.: MACRS: an enhanced directory-based resource sharing framework for mobile ad hoc networks. Electronics **11**(5), 725–732 (2022)
13. da Costa, L.A.L.F., Kunst, R., de Freitas, E.P.: Intelligent resource sharing to enable quality of service for network clients: the trade-off between accuracy and complexity. Computing **14**(7), 1–13 (2022)

14. Erqi, Z.: Application of conditional random field model based on machine learning in online and offline integrated educational resource recommendation. Math. Probl. Eng.Probl. Eng. **20**(1), 551–5526 (2022)
15. Qian, X., Li, F.: Evolutionary game analysis of information sharing in closed loop supply chain based on cloud platform. J. Univ. Shanghai Sci. Technol. **43**(06), 606–616 (2021)

Research on Software Test Data Optimization Using Adaptive Differential Evolution Algorithm

Zheheng Liang[1,2](✉), Wuqiang Shen[1,2], and Chaosheng Yao[1,2]

[1] Joint Laboratory On Cyberspace Security of China Southern Power Grid, Guangzhou 510000, China
liangzhaoheng23213@163.com, shenwuqiang@gdxx.csg.cn
[2] Guangdong Power Grid, Guangzhou 510000, China

Abstract. In order to improve the coverage of the target path corresponding to the generated software test data after optimization, and make the data better adapt to the software test process, the adaptive differential evolution algorithm is introduced to design the optimization method for software test data. Using the PSO algorithm to simulate the biological evolution mechanism in nature, and with the help of computer programming, the generated software test data are preliminarily trained; Draw on the path correlation based regression test data evolution of adaptive differential evolution algorithm to generate the relevant path representation method, mark the program to be tested, and construct the test data fitness function based on this; A hybrid model MPSO is proposed to select the best individual data in software test data; The selection criteria of scaling individuals are introduced into the adaptive scaling factor to cluster the optimal data individuals, so as to realize the design of optimization methods. The comparison experiment results show that the designed method has a good effect in practical application. This method can improve the target path coverage corresponding to the generated software test data on the basis of controlling the time length and evolution times required for software test data optimization.

Keywords: Adaptive differential evolution algorithm · Data selection · Individual clustering · Fitness function · Software test data · Optimization

1 Introduction

In order to ensure the quality of software, not only need advanced technical means, perfect R & D process, but also need continuous testing, no matter how advanced the current technology, how perfect the R & D process, can not 100% guarantee the software developed zero defects. Software testing can help scientific research and technology developers find out the errors or potential defects in the process of software development, and reduce the damage caused by software errors and potential defects as much as possible. In order to ensure the high reliability and integrity of software products, it is necessary to design enough test cases. However, the scale of software products is increasing day by day, the structure is becoming more and more complex, and the update

L. Yun et al. (Eds.): ADHIP 2023, LNICST 547, pp. 198–209, 2024.
https://doi.org/10.1007/978-3-031-50543-0_14

cycle is becoming shorter and shorter. It is a huge challenge for software testing to carry out comprehensive testing to find the problems and potential defects in the increasingly tight software development cycle. Therefore, it is necessary to use effective test cases within a limited test time to find out the errors and potential defects of the product as far as possible to ensure the accuracy and reliability of the software product, and the basis of this test is a high degree of target path coverage.

In recent years, the use of artificial intelligence algorithms such as ant colony algorithm, particle swarm optimization algorithm and genetic algorithm to automatically generate test cases has attracted more and more researchers' attention, and has achieved good research results. For example, Li Lu et al. [1] studied the defect optimization method of high-performance traffic analysis software based on particle swarm optimization algorithm, which introduced particle swarm optimization algorithm to analyze the current defect status of high-performance traffic analysis software, reset defect parameters, and build the defect optimization model of high-performance traffic analysis software. However, this method has the problem of long running time. Yang Bo et al. [2] studied a software fault location assisted test case generation method using improved genetic algorithm. Based on genetic algorithm, this method assisted the generation of test cases in the process of software fault location through the ranking of suspected faults in software fault location, but the marking path coverage of this method was low. The research shows that the adaptive differential evolution algorithm has the advantages of fast yield with fewer parameters and fast yield in high dimensions [3, 4]. Therefore, this paper takes the program code of unit test in software testing as the research object, introduces adaptive differential evolution algorithm to calculate all the tested paths, and uses the fitness function of the tested paths to improve the process of output optimal solution by optimizing the mutation operator, and improves the adaptive differential evolution algorithm by improving the mutation operator. Test cases covering each path are generated automatically to optimize software test data. This paper makes use of the convenient and efficient characteristics of artificial intelligence algorithm, which not only reduces the test cost, but also obtains a high efficiency of automatic case generation, moreover, the target path coverage of the generated software test data is more than 90%, which is more than 10% higher than that of the comparison method, which is of great significance to the improvement of product quality.

2 Software Test Data Generation

A lot of work has proved that PSO algorithm can solve the problem of software test data generation better than GA algorithm. Moreover, the PSO algorithm is simple, with fewer parameters to configure and easy to implement [5]. Therefore, this chapter introduces PSO algorithm to design the generation of software test data.

In this process, with the help of computer programming, the problem to be solved is expressed into strings (or chromosomes), that is, binary codes or digital strings, by using the PSO algorithm to simulate the biological evolution mechanism in nature, so as to form a group of strings, and they are placed in the problem solving environment. According to the principle of survival of the fittest, the strings that adapt to the environment are selected for replication, and the two gene operations are crossover and mutation, Create a new generation of clusters that are more adaptable to the environment. After such continuous changes from generation to generation, finally converge to a string that is most suitable for the environment, and obtain the optimal solution of the problem [6]. That is, the training of test software requirement data is preliminarily realized.

On this basis, it should be clear that the essence of PSO algorithm is an optimization algorithm based on swarm intelligence, and its idea comes from artificial life and evolutionary computing theory. The algorithm simulates the behavior of birds flying to find food, and makes the group achieve the optimal goal [7] through the collective cooperation between birds. When using the PSO algorithm to generate software test data, it is necessary to input the software test data into the test environment first, and express the test environment as D, on D. In the target retrieval space of dimension, each software test data is regarded as a particle, and the particle is represented as m, then m Particles form a population, where the i Particles in the d. The position of the dimension is x_{id}, x_{id} The corresponding flight speed in space is v_{id}, the optimal location searched by this particle is p_{id}, the current maximum value of the entire particle swarm p_{gd} In this way, the particle position in space is updated, and this process is taken as the data update process, as shown in the following calculation formula.

$$v_{id}^{t+1} = wv_{id}^t + c_1r_1(p_{id} - x_{id}^t) + c_2r_2(p_{gd} - x_{id}^t) \tag{1}$$

In formula (1): v_{id}^{t+1} Means on $t + 1$ The corresponding flight speed of particles in space at time; w Is the inertial factor, w The larger the value of, the more suitable for large-scale exploration of the solution space, w The smaller the value of is, the more suitable it is to explore the solution space in a small range; v_{id}^t Means on t The corresponding flight speed of particles in space at time; x_{id}^t Means on t At the moment i Particles in the d. Dimension position; c_1, c_2 Represents normal number, called acceleration factor; r_1, r_2. Represents a random number with a value between 0 and 1 [8]. On this basis x_{id}^{t+1} The calculation formula is as follows:

$$x_{id}^{t+1} = x_{id}^t + v_{id}^{t+1} \tag{2}$$

On the basis of the above contents, it should be clear that, d The value range of is $1 \le d \le D$, for the d Position of dimension x Constrain the range of change and make it clear x The value range of is $[x_{d\,min}, x_{d\,max}]$, for the d Dimensional velocity v Constrain the range of change and make it clear v The value range of is $[v_{d\,min}, v_{d\,max}]$.

In the iteration, if the position and speed exceed the boundary range, the boundary value shall be taken. The particle position in the space can be updated according to Fig. 1 below. In this way, the spatial position of the software test data can be updated to generate more complete and global adaptive test data [9].

The termination condition of the PSO algorithm takes the maximum number of iterations or the predetermined minimum fitness threshold that the optimal position searched by the particle swarm optimization meets according to the specific problem.

Because p_{gd} It is the optimal position of the whole particle swarm. Therefore, the software test data can be generated according to the above steps, or it can be used as the retrieval process of global particles. The global PSO algorithm has a fast convergence speed, but sometimes it falls into a local optimum; The local PSO algorithm

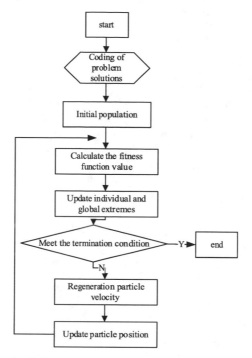

Fig. 1. Software test data generation and update process

converges slowly, but it is relatively difficult to fall into the local optimal value. There-fore, corresponding measures can be taken for data processing according to the specific requirements of software test data generation in practical applications.

3 Construct Test Data Fitness Function Based on Adaptive Differential Evolution Algorithm

On the basis of the above contents, in order to achieve the extraction of the optimal data in the generated software test data and master the fitness of different data in the generated data and the software test process, the adaptive differential evolution algorithm is introduced to construct and design the fitness function of the test data [10].

In this process, it should be clear that the establishment of the mathematical model of the multi-path coverage test data generation problem is closely related to the path rep-resentation method, and the tested program generally has multiple paths. In structured programming, the program usually consists of three structures: sequence, selection, and cycle. Among them, selection and cycle structure determine the trend of differ-ent branches of the program; For cyclic structure, by introducing Z Path coverage can decompose it into multiple selection structures. The existing path representation methods include: using the statement number sequence of the program to represent the path, using the branch node sequence to represent the path, and using Huffman code to represent the target path. Using the path correlation based regression test data evolution generation correlation path representation method in adaptive differential evolution algorithm for reference, the program to be tested is marked.

Assume that the program to be tested is Q, insert piles into each branch node in the program, and set the value of "true" to be 1 when crossing the corresponding branch node. If it is false, it is - 1, and if it does not pass through the branch node, it is 0. Therefore, the code sequence of 1, - 1, and 0 can be used to represent the path of program execution. Run the test program after inserting piles with a set of input data, and the branch node sequence is $D_1, D_2, \cdots, D_n$ The corresponding execution path is expressed as $q(T_i)$, where n Is the path $q(T_i)$ Length of. In addition, the program to be tested Q The target path of is $q_1, q_2 \cdots, q_N$, N Is the number of target paths.

Hypothetical procedure Q The input field of is α, input data is $T_1, \cdots, T_N$, then $T_N \in \alpha$, the path will be executed $q(T_i)$ Compare the coding sequence of and the target path from left to right, find the first different coding position, record the number of the same coding before, and record it as $|q(T_i) \cap q_j|$, take the ratio between it and the target path code, that is, the total length of the target path, to obtain the path similarity, and express it as $f_i(T)$, then $f_i(T)$ It can be calculated by the following formula.

$$f_i(T) = \frac{|q(T_i) \cap q_j|}{\max\{q(T_i), q_j\}} \tag{3}$$

In formula (3): $f_i(T)$ Indicates the path similarity. After the above calculation is completed, the problem of solving test data generation can be transformed into solving $f_1(T), f_2(T), f_N(T)$. The fitness function of the maximum value problem is as follows.

$$H = \max(f_1(T), f_2(T), \cdots, f_N(T)) \tag{4}$$

In formula (4): H Represents the fitness function of test data. According to the above method, the research on the construction of test data fitness function based on adaptive differential evolution algorithm is completed.

4 Selection of Optimal Individual Data in Software Test Data

After completing the above design, in order to select the best individual data in the software test data and seek the best test data, it should be clear that in the GPSO, particles evolve in the direction of the global optimal particles, and more obviously converge to the current optimal solution. Each time the particle position is updated, the characteristics of all particles in the population are integrated, and the information transmission speed is fast, but the population diversity is easy to lose, and the probability of falling into the local extreme value is large. In the local model LPSO, each particle only shares information with its neighbor nodes, which slows down the speed of information transmission, but the diversity of the population is guaranteed and it is not easy to fall into local extremum.

Therefore, based on the analysis of the impact of different fixed forms of neighbor patterns on the algorithm performance, it is found that the higher the average connectivity of social interactions between particle individuals, the faster the information transmission speed in the population, and the premature convergence phenomenon is more prone to occur. In view of this, a hybrid model MPSO is proposed, that is, by observing the diversity index of particle swarm, each generation of particles selects GPSO or LPSO for evolutionary optimization, so as to realize the selection of the best individual data in software test data. In this process, describe the diversity of particles, that is, analyze the diversity of software test data in space. This process is shown in the following calculation formula.

$$I = \frac{1}{|k| \times |L|} \sum_{i>1}^{N} \sqrt{\sum_{d>1}^{D} (x_{id}^{t} - x_{d}^{t})^2} \tag{5}$$

In formula (5): I Represent the diversity of software test data in the space; k Represents the size of particles in the search space; L Indicates the maximum diagonal length in the search space. After calculation, output I. The calculation results of I It represents the dispersion level of each particle in the community, I The larger the value, the more dispersed the population; I The smaller the value, the more concentrated the population, and the lower the diversity of the population.

Therefore, by observing the group I The feedback information automatically adjusts the topology of particle swarm to maintain the diversity of the community, maintain a certain population density, and avoid the algorithm falling into the local optimal value. On the basis of the above I. The population with higher values will be globally and locally optimized, and the process is shown in the following calculation formula.

$$W(I) = \frac{I_i - l_{\min}}{e^{l_{\max} - l_{\min}}} \tag{6}$$

In formula (6): $W(I)$ Represents the best individual data in software test data; $l_{\min}$ Represents the minimum branch nesting depth; $l_{\max}$ Represents the maximum nesting depth; e Indicates the depth of the current branch. Complete the selection of the best individual data in the software test data according to the above method.

5 Optimal Individual Clustering and Optimization of Software Test Data Set

On the basis of the above contents, extract the optimal individual data from the software test data selected in the above way, introduce the selection criteria of scaling individuals into the adaptive scaling factor, and cluster the optimal data individuals. The process is shown in the following calculation formula:

$$r = \sum_{i>1}^{n} Mr_i / \sum_{i>1}^{n} r_n \cdot v \tag{7}$$

In formula (7): r Represents the optimal individual clustering; M Represents the cluster center. After completing the above calculation, the adaptive differential evolution algorithm is applied to automatically generate the model of test data, as shown in Fig. 2 below.

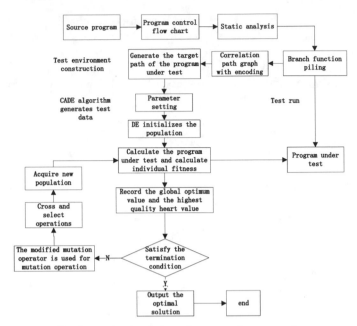

Fig. 2. Optimal solution of output software test data

The construction of the test environment is the basis for the generation of test data. In this stage, the source program is statically analyzed, and the coded test program flowchart generated by branch function instrumentation is used to generate the target path. This method is more concise than the traditional data flowchart.

However, as far as the current research is concerned, the existing methods assume that the mass of each particle in the population is the same, and do not take into account the differences between the particles. Therefore, in the optimization process, the process of outputting the optimal solution is improved. On the basis of the existing algorithms, a mutation operator based on the center of mass is proposed. The improved mutation operator can be calculated by the following formula.

$$A(t) = x_i(t) + F(r + c_1 r_1 + c_2 r_2) \tag{8}$$

In formula (8): $A(t)$ Represents the improved mutation operator; F Represents the mass of individual particles. After the selection of the improved mutation operator is completed, in order to break the stagnation of evolution, jump out of the local best, let the operator fluctuate in size, and by limiting its value range, make the difference dynamic scaling, and search for optimization in a large range, theoretically it can improve the ability of global optimization. In the early stage of evolution, the operator tries to maintain the diversity of the population, and takes the small value first; in the later stage, the drill takes the large value to speed up the convergence speed. In this paper, the corresponding generation operator of normal distribution random number is used as follows:

$$\beta = normrnd(U \times F) \cdot A(t) \tag{9}$$

In formula (9): β Represents the generating operator process; *normrnd* Represents the deviation compensation value; U Represents a normally distributed random number. By compensating the deviation of the operator generation process, the software test data can be better optimized within the optimal range, which is helpful to jump out of the region where the local extreme value is located quickly.

The cross factor plays a fine tuning role, and the appropriate value can maintain the diversity of the population. If the crossover probability decreases, the candidate will contain more target individuals. If the crossover probability increases, the weight of variant individuals in the newly generated candidate will increase.

At the early stage of the iteration, the diversity of the population is relatively high, so cross operation with a small probability can maintain the diversity of the population. As the number of iterations increases, the diversity of the population is gradually losing and approaching the extreme point. Therefore, local search should be increased to accelerate the convergence of the algorithm, so as to approach the extreme point at a faster speed.

In this process, it should be clear that generating test data is the core part of the whole model based on the adaptive differential evolution algorithm. In this stage, the population is initialized according to the target path and test data range constructed by the test environment, combined with the setting of parameters required by the algorithm, and the population is updated by running the improved mutation operator to guide the population to evolve towards the optimal value; And use the test data to run the program to be tested after pile insertion to calculate the fitness value. When the fitness value meets the needs of software testing, complete the data optimization design.

According to the above method, complete the optimal individual clustering and software test data set optimization, and realize the design and research of software test data optimization method based on adaptive differential evolution algorithm.

6   Comparison Experiment

The research on optimizing software test data using adaptive differential evolution algorithm has been completed from four aspects above. In order to test the application effect of software test data optimization methods, the following will take the tested software program provided by a teaching and research institute as an example to design a comparison experiment and carry out the research as shown below.

In order to ensure that the designed method can play its expected role in the test, the environment in which the software program under test runs is described before the experiment, and the relevant contents are shown in Table 1 below.

According to the above contents, after completing the design of the technical parameters of the comparative experimental environment, the method designed in this paper is used to optimize the data of the tested software.In the optimization process, it assists modern intelligent algorithms such as GA to generate software test data.On this basis, the adaptive differential evolution algorithm is introduced to construct the fitness function of the test data. In order to ensure that the constructed fitness function can be used as the basis for evaluating the software test data, the parameters of the algorithm in the computer can be set according to Table 2 below during the iteration of the adaptive differential evolution algorithm.

Table 1. Technical Parameters of Comparative Experimental Environment

S/N	project	parameter
1	Experimental programming language	Java Language
2	Experimental program running environment	Eclipse environment
3	Microcomputer environment	Windows operating system Intel (R) Core (TM) i3 CPU 2.0GHz
4	Computer running memory	2GB RAM

Table 2. Training parameter settings of adaptive differential evolution algorithm

S/N	project	parameter
1	Population iterations (times)	1000
2	Operator crossover probability (%)	0.6
3	Operator compilation probability (%)	0.01
4	Number of independent runs for each case of different programs (times)	50
5	Data cycle branch (s)	6
6	Select nesting quantity (pcs)	3
7	search space	3
8	Learning factor	1.5
9	Inertia weight	0.9
10	Diversity threshold in algorithm	0.1

After setting the experimental parameters, select the best individual data in the software test data, and optimize the software test data set by clustering the best individual data.

On the basis of the above content, software test data optimization method based on particle swarm optimization (reference [1] method) and software test data optimization method based on genetic algorithm (reference [2] method) are introduced, and the introduced method is regarded as traditional method 1 and traditional method 2. The method in this paper and the two traditional methods are used to optimize software test data.

In the optimization process, three software test data populations of different scales are selected, and three methods are used to evolve the software test data respectively. The average length of time required by the three methods to optimize the software test data is compared, which is the key basis for testing the application effect of the method in this paper. The statistical experimental results are shown in Table 3 below.

From the experimental results shown in Table 3 above, it can be seen that the average time required for optimization of software test data using this method is less than 1s,

Table 3. Average time required for three different methods to optimize software test data

S/N	Population size	Average duration (s)		
		Methods in this paper	Traditional method 1	Traditional method 2
1	50	0.125	0.569	0.639
2	100	0.156	0.693	0.678
3	150	0.269	0.854	1.069
4	200	0.489	0.910	1.598
5	300	0.526	1.025	1.756
6	400	0.963	1.156	1.963
7	500	0.956	1.569	3.051

while the average time required for optimization of software test data using traditional methods is significantly higher than the average time required for optimization of data using this method.

After completing the above experiment, set the input range of software test data to [0,64], [0512], [1024], use three methods to optimize the software test data, and compare the average evolution times of the three methods on the input data during the optimization process. The results are shown in the following figure:

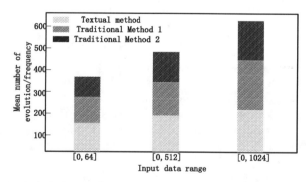

Fig. 3. Average evolution times of three methods for input data in different ranges during optimization

It can be seen from Fig. 3 above that the three methods optimize the training of input data in different ranges. With the increase of input data range, the average evolution times of the three methods show an increasing trend. Use this method to optimize [1024] software test data, and the average number of evolutions in the optimization process is less than 300;The traditional method 1 is used to optimize software test data, and the average evolution times are less than 500;The traditional method 2 is used to optimize the software test data, and the average evolution times are less than 700.Based on the above results, it can be seen that among the three methods, only using this method to optimize software test data can ensure that the average evolution times are at a relatively low level, while using traditional method 1 and traditional method 2 to optimize software test data, the corresponding method has more average evolution times, that is, the data optimization process is more complex.

After the above design is completed, several software test paths are set according to the software test requirements. It is known that different software test paths require different software test data sizes and categories. After comparing the three methods to optimize the software test data, the generated test data can cover the extent of the target test path, that is, compare the availability of the optimized corresponding data in the software testing process. According to the above method, carry out a comparison experiment and make statistics of the experimental results, as shown in Fig. 4 below.

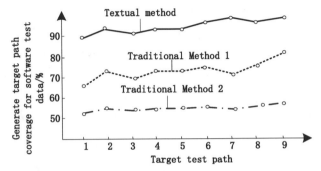

Fig. 4. Target path coverage corresponding to software test data generated by the three optimized methods (%)

From the experimental results shown in Fig. 4 above, it can be seen that the method in this paper is used to optimize the software test data, and the target path coverage corresponding to the generated software test data after optimization is > 90%, while the target path coverage corresponding to the generated software test data after optimization in the traditional method 1 is between 60% and 80%;The target path coverage corresponding to the software test data generated after optimization of traditional method 2 is between 50% and 60%.

Based on the above experimental results, the following experimental conclusions are obtained: compared with the traditional methods, the software test data optimization method designed in this paper using adaptive differential evolution algorithm has a good effect in practical application. This method can improve the target path coverage corresponding to the generated software test data on the basis of controlling the time required for software test data optimization and the number of evolutions. In this way, it provides further support and guidance for the standardized implementation of software program testing and other related scientific research work.

7 Conclusion

In order to find potential software defects as much as possible and improve the coverage of target paths corresponding to software test data generation, this paper conducts testing research on the optimization method of software test data through software test data generation, construction of test data fitness function based on adaptive differential evolution algorithm, selection of optimal individual data in software test data, optimal individual clustering and optimization of software test data set.After completing the design of this method, taking the software program under test provided by a teaching and research unit as an example, a comparative experiment was designed to prove that the method designed in this paper has a good effect in practical application. This method can improve the coverage of the target path corresponding to the generated software test data on the basis of controlling the length of time required for software test data optimization and the number of improvements.

References

1. Lu, L.: Defect optimization method of high-performance traffic analysis software based on particle swarm optimization. J. Anhui Tech. College Water Resources Hydroelectric Power **21**(4), 56–59 (2021)
2. Bo, Y., He, Y., Xu, F., et al.: IGA: software fault location assisted test case generation method using improved Genetic Algorithm. J. Beijing Univ. Aeronautics and Astronautics **48**(3), 1–13 (2022)
3. Zhang, X., Liu, Q., Qu, Y.: An adaptive differential evolution algorithm with population size reduction strategy for unconstrained optimization problem. Appl. Soft Comput. J. **138** (2023)
4. Niu, D., Liu, X., Tong, Y.: Operation optimization of circulating cooling water system based on adaptive differential evolution algorithm. Int. J. Comput. Intell. Syst. **16**(1) (2023)
5. Li, W., Sun, Y., Huang, Y., et al.: An adaptive differential evolution algorithm using fitness distance correlation and neighbourhood-based mutation strategy. Connection Sci. **34**(1) (2022)
6. Kumar, G.S., Kumar, M.A.: A self-adaptive differential evolution using a new adaption based operator for software cost estimation. J. Inst. Eng. (India) Series B **104**(1) (2022)
7. Chen, J., Chen, X., Zan, T., et al.: An automatic generation of software test data based on improved Markov model. Web Intelligence **20**(4) (2022)
8. Ferreira, V.R., et al.: Bio-inspired optimization to support the test data generation of concurrent software. Concurrency and Computation: Practice and Experience **35**(2) (2022)
9. Kattuva, M.A.K., Narayanan, N., Mangottiri, V., et al.: Parameter evaluation of a nonlinear Muskingum model using a constrained self-adaptive differential evolution algorithm. Water Practice Technol. **17**(11) (2022)
10. Mehdi, E., Hossein, D.A.: Automation of software test data generation using genetic algorithm and reinforcement learning. Expert Syst. Appl. **183** (2021)

Personalized Scheduling of Distributed Online Educational Resources Based on Simulated Annealing Genetic Algorithm

Xiaotang Geng[1(✉)] and Yan Huang[2]

[1] Heilongjiang Polytechnic, Harbin 150086, China
gxt7809@163.com
[2] School of Tourism and Humanities, Heilongjiang Polytechnic, Harbin 150086, China

Abstract. The arrival of the information age has accelerated the development of information education, and distributed online education resources have increased exponentially. However, due to the low level of scheduling service, the application effect of education resources is poor, which hinders the follow-up development of information education. A personalized scheduling method of distributed online education resources based on simulated annealing genetic algorithm is proposed. The membership relationship between knowledge points and educational resources is calculated using fuzzy logic method, and the corresponding educational resource model is constructed. Based on this, the purpose and key problems of personalized scheduling of educational resources are analyzed, and the objective function of personalized scheduling of distributed online educational resources is constructed. The objective function is solved based on simulated annealing genetic algorithm, Obtain the final personalized scheduling scheme of distributed online education resources, and realize the personalized scheduling of distributed online education resources. Experimental data shows that after the proposed method is applied, the minimum response time of distributed online education resource scheduling is 6s, and the maximum precision of distributed online education resource scheduling is 96%, which fully confirms that the proposed method has better application performance.

Keywords: Distributed · Individualization · Dispatch · Online education resources · Simulated annealing genetic algorithm

1 Introduction

With the rapid development of information science and technology, mankind has entered the information age. The State Council, based on China's basic national conditions, put forward the overall idea of informatization development. Information resources and talents, information technology application, information network, information technology industry, and information policies, laws and regulations are important elements of the national information system. However, from the perspective of the development of an

L. Yun et al. (Eds.): ADHIP 2023, LNICST 547, pp. 210–225, 2024.
https://doi.org/10.1007/978-3-031-50543-0_15

information industry, information resources are the core, while information networks are the foundation, the use of information resources and the application of information technology are the purpose, and the information technology industry, information talents, and information policies, regulations and standards are its guarantee. One of the most basic components of the education system is information. The reform and development of education requires the informatization of education to be promoted by the comprehensive and in-depth use of modern information in the entire educational field. The inevitable result of educational informatization is to form a new form of education, that is, information-based education [1]. Informational education is characterized by the globalization of educational resources, the multimedia of course materials, the autonomy of learning, the personalization of teaching, the cooperation of activities, the virtualization of the environment and the automation of management, among which educational resources are the core of all applications. At the same time, educational resources are also the top priority of educational informatization construction, which plays an extremely important role in improving the quality of education and teaching and promoting the development of education.

At present, the three characteristics of educational resources, namely, heterogeneity, dynamics and autonomy, make educational resources have many problems in sharing and collaboration, which greatly affects the efficiency of users in obtaining information related to teaching resources. However, the ultimate goal of the grid is to realize the sharing and distributed collaboration of a large number of teaching resources distributed in various geographical regions. Grid refers to the connection of all major clusters, supercomputers, database servers, large-scale storage systems and other geographically dispersed special instruments and equipment, as well as all resources including personal computers (in this case, computing, software, information, storage, communication, knowledge and other related resources) to be used as a completely unified resource. A parallel distributed computing platform with large capacity, high performance and high speed transmission is the inevitable result of the development of grid technology to a certain extent. Grid technology mainly emphasizes the realization of comprehensive sharing of resources and application services. In order to solve the problems in teaching resource sharing and teaching resource distribution services, a scheduling strategy for teaching resources is proposed. It is mainly aimed at the scheduling of distributed online teaching resources, which can improve the application efficiency and effect of educational resources and provide assistance for the follow-up development of online education [2].

The construction of educational resources is a systematic project. The design, development, management, application and evaluation of educational resources are interlinked and interdependent steps in the resource construction project. After the resources are designed and developed, they need to be delivered to the resource users in a timely and effective manner to serve students, teachers and other educators, so as to reflect the ultimate purpose and value of educational resources. The purposefulness, dynamics, timeliness and individuation of educational resource construction fully reflect the significance of resource distribution in resource construction. The long-term and stable continuous updating and maintenance is the guarantee of the lasting vitality of educational resources. The effective solution to the problems existing in the construction plays a very important role in promoting the construction of educational resources, promoting the development of educational informatization, and realizing the balanced and leapfrog development of education. The new resource distribution service system will bring great application value. As far as the existing research results are concerned, the more frequently used scheduling methods are the educational resource scheduling method based on genetic algorithm and the educational resource scheduling method based on P2P technology. The above two methods have the problems of low bandwidth utilization, long response time and poor scalability, which cannot meet the development needs of today's education field, Therefore, a personalized scheduling method of distributed online education resources based on simulated annealing genetic algorithm is proposed. By establishing a user base model and interest model, users can be segmented and classified, providing personalized recommendation services, recommending content, products or services that meet their interests and preferences, and improving user satisfaction and experience. Using fuzzy logic method, calculate the membership relationship between knowledge points and distributed online education resources, and construct a distributed online education resource model. This paper describes the personalized scheduling problem of distributed online education resources, constructs the personalized scheduling objective function of distributed online education resources, solves the objective function based on Simulated annealing genetic algorithm, and obtains the personalized scheduling scheme of distributed online education resources, which effectively shortens the scheduling response time and improves the scheduling accuracy.

2   Research on Personalized Scheduling of Online Education Resources

2.1   Construction of User Base Model and Interest Model

In order to improve the accuracy of educational resource scheduling, the first step is to build a user base model and interest model, which will lay a solid foundation for subsequent research.

The establishment of user model is not only conducive to personalized resource scheduling, but also conducive to the construction of new resources. The role of the user model on the system is mainly reflected in the following three aspects:

According to the user model and resource model, matching can realize personalized resource scheduling based on content;

Clustering user interest groups according to user model to achieve collaborative resource scheduling for users;

When the interests of some users with higher concerns in the user interest model do not match the corresponding resources, the headquarters resource development center can develop new resources according to these users' interests, which is conducive to better understanding the needs of resource users and improving the pertinence and practicality of the developed resources.

The user base model is mainly described in terms of school name, school type, characteristic theme, natural situation, informatization level, etc. The natural conditions include the economic development level of the school location, the school area, the number of classes, the number of students, the number of teachers, the composition of teachers' qualifications and so on; The informatization level includes teachers' information literacy level, students' information literacy level, campus network construction, computer room configuration, teacher machine configuration, etc. The detailed user basic information questionnaire is shown in Table 1.

Table 1. Basic User Information Questionnaire

Project	Range	Value Description
School name	-	Full name
Type of school	A	Primary school
	B	Junior middle school
	C	High school
Featured activities	A	Music
	B	Calligraphy and painting
	C	Performance
	D	Science and technology
	E	Sports
	F	Other
School area	A	>40000 m^2
	B	10000–4000 m^2
	C	<10000 m^2
Number of students	A	>2500 people
	B	1000–2500 persons
	C	<1000 persons
Number of teachers	A	>150 persons
	B	60–150 persons
	C	<60 persons
Configuration of machine room	A	Excellent
	B	Secondary
	C	Low or none
Campus network construction	A	With LAN and Internet access
	B	There is a LAN, but no access to the Internet
	C	No network
Information literacy level of students	A	Most students can operate computers skillfully
	B	Some students can operate computer skillfully
	C	A few students can operate computers skillfully
Teachers' information literacy level	A	Most teachers have received information technology training
	B	Some teachers have received information technology training
	C	A few teachers have received information technology training

The user interest model should have strong adaptability and robustness. When describing the user interest model, the following two factors should be considered:

First, users with different background knowledge of the content represented by a keyword will have different understandings, such as "firewall". Users with computer knowledge background will understand it as a kind of software; Users with architectural knowledge background will understand it as a wall for fire prevention. Therefore, when modeling users, the user's background information needs to be considered [3].

Second, users' interests will change over time. Some topics that users were originally interested in will be gradually forgotten, and new interest topics will gradually emerge. This gradual process of user interest is also called "interest drift". The existence of interest drift makes the user model should also change, otherwise the user model will not reflect the user's interest well. Therefore, the forgetting and updating mechanism of user interest should be considered in the design of user modeling.

In order to reflect user interest more realistically, user interest model is represented by vector space model based on background and temporal. That is to say, disciplines and learning stages are introduced into the model as background constraints, and interest weight functions based on temporal changes are introduced into vector space $\omega_n(\overline{T}_n)$ to calculate the attenuation and update of user interest weight.

At a certain moment t, the user interest model expression is

$$U = \{S, G, K\} \tag{1}$$

In formula (1), U represents the user interest model; S refers to the collection of disciplines (such as mathematics, Chinese, biology, etc.); G refers to the collection of learning stages (primary school one year, primary school two years, primary school three years, etc.); K represents the user interest keyword vector space, recorded as $K = \{(k_1, \omega_1(\overline{T}_1)), \cdots, (k_n, \omega_n(\overline{T}_n))\}$, where, k_n refers to the n keywords describing interest, $\overline{T}_n$ represents a keyword k_n time set of each submission, $\omega_n(\overline{T}_n)$ represents a keyword k_n the time related weight function of.

The attenuation and update of user interest are calculated based on the time window mechanism, that is, at a certain time window Δt the keyword is submitted, the weight will be increased; Otherwise, the weight is attenuated.

Assumptions are as follows:

1. Each time window Δt inner keywords k_n for each submission, the interest weight will be increased by units a;
2. Each time window Δt inner keywords k_n if not submitted, the interest weight attenuation unit b;

So, at some point t, keyword k_n the interest weight function of is expressed as:

$$\omega_n(\overline{T}_n) = \begin{cases} \sum (f(t_{n1} + (i-1) \cdot \Delta t) \cdot a - c \cdot b) & \omega_n(\overline{T}_n) > 0 \\ 0 & \omega_n(\overline{T}_n) \leq 0 \end{cases} \tag{2}$$

In formula (2), $f(t_{n1}, t)$ means $[t_{n1}, t]$ time window within time period Δt number of; $f(t_{n1} + (i-1) \cdot \Delta t)$ means that $[t_{n1} + (i-1) \cdot \Delta t, t_{n1} + i \cdot \Delta t]$ number of keywords submitted in the time window; c represents a constant, and the value is 0 or 1. The value rule is as follows:

$$c = \begin{cases} 0 \ f(t_{n1} + (i-1) \cdot \Delta t) > 0 \\ 1 \ f(t_{n1} + (i-1) \cdot \Delta t) \leq 0 \end{cases} \tag{3}$$

The user interest model is described in tree structure as shown in Fig. 1.

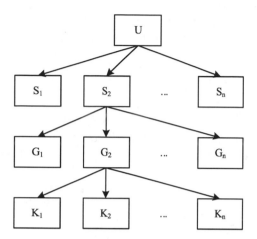

Fig. 1. User interest tree

The above process completed the construction of user base model and interest model, providing support for the subsequent construction of online education resource model.

2.2 Construction of Online Education Resource Model

How to accurately model educational resources is the key to realize personalized scheduling of educational resources. In order to speed up the development and utilization of high-quality educational resources and realize the sharing and benefit of educational resources in many ways, it is necessary to accurately describe and model educational resources [4]. In order to realize the recommendation service of educational resources based on learners' personality and meet learners' personalized learning needs, the primary task is to reasonably classify, label and manage educational resources, so as to realize the integration and clustering reconstruction of educational resources. In this study, knowledge points are used as a link to annotate educational resources at the semantic level, and fuzzy logic method is used to calculate the membership relationship between knowledge points and educational resources, and the corresponding educational resource model is constructed.

In the practical application of distributed online education resources, a resource often belongs to multiple categories at the same time in different degrees. For example, a movie may belong to comedy, love, adventure and other types at the same time, and a learning resource may belong to multiple knowledge point categories at the same time. Therefore, this paper uses fuzzy set theory to model educational resources, that is, a resource will belong to multiple knowledge points in varying degrees. In the process of educational resource scheduling, educational resources correspond to projects to be scheduled I_j the knowledge points associated with educational resources correspond to categories x_k, thus, you can use $\mu_{x_k}(I_j)$ represents a resource I_j belongs to knowledge points x_k degree, and the association between resources and knowledge points uses two-dimensional vector $(x_k, \mu_{x_k}(I_j))$ express. Common forms of membership function mainly include triangle, trapezoid, Gaussian function, exponential function, etc. The membership function defined in this paper $\mu_{x_k}(I_j)$ the calculation formula of is based on the form of exponential function, and the expression is

$$\mu_{x_k}(I_j) = \begin{cases} \left(\frac{1}{2}\right)^{\lambda|L_j|(r_k-1)} & 1 \leq r_k \leq |L_j| \\ 0 & r_k > |L_j| \end{cases} \tag{4}$$

In formula (4), λ indicates the adjustment parameters; $|L_j|$ represents the relationship between resources and I_j number of associated knowledge points; r_k represents knowledge points x_k in all resources I_j the arrangement position of the associated knowledge points.

It should be noted that, $\mu_{x_k}(I_j)$ meet the following conditions: the higher the degree of association between resources and categories, the greater the corresponding degree of membership; Otherwise, the degree of membership is smaller; If the resource is not associated with a category, the corresponding membership is 0.

The education resource model is defined as:

$$RM = (RI, KS, RS, DD, PT, OI) \tag{5}$$

In formula (5), RM represents the educational resource model; RI represents the basic information of educational resources, mainly including the name, capacity, author, upload time, storage path, etc. of the resources; RS represents the style type of educational resources, which is defined by $\{\langle r_s, r_t \rangle\}$ composition. Among them, r_s refers to the media type of educational resources, r_t represents the content form of educational resources; KS refers to the set of knowledge points associated with the educational resource, including the knowledge points themselves and their association degree, which is represented by membership; DD, PT and OI respectively represents the difficulty coefficient, present time and target content of educational resources. Among them, the difficulty coefficient is the degree of difficulty in learning and mastering the educational resources; The present time is the approximate time needed to learn the educational resources: the target content indicates the learning objectives contained in the educational resources. The learning objectives in the target content, which have been described in the previous definition, mainly include knowledge points and the learning level requirements for knowledge points, and are divided into three levels, namely memory learning, understanding learning and extended learning.

The online education resource model constructed breaks through the relationship between resources and knowledge points in the previous education resource model. The fuzzy logic method is introduced to express the relationship between resources and knowledge points with membership, which more truly depicts the relationship between resources and knowledge points.

2.3   Personalized Scheduling Problem Description of Online Education Resources

Based on the user model and online education resource model constructed above, the purpose and key issues of personalized scheduling of education resources are analyzed, and the objective function of personalized scheduling of distributed online education resources is constructed to provide a basis for the determination of the final personalized scheduling scheme of education resources[5].

Scheduling matching problems are common in all walks of life, such as the famous traveling salesman problem, secondary allocation problem, sequence scheduling problem, vehicle routing problem, job shop scheduling problem, grid resource scheduling, weapon target allocation problem, and so on[6]. The personalized scheduling of distributed online education resources is mainly realized by the scheduling server, and its specific structure is shown in Fig. 2.

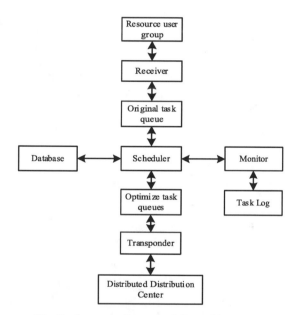

Fig. 2. Structure diagram of dispatching server

As shown in Fig. 2, the scheduling server consists of receiver, scheduler, repeater, monitor, original task queue, optimization task queue, task log, and database. The specific components are as follows:

Receiver: It is responsible for receiving requests from resource users, collecting information about resource identification requested by users, connection speed between users and distribution centers, and forming the original task queue.

Scheduler: According to the original task queue and the corresponding information in the database, such as resource size, distribution center load, etc., tasks are re queued to form an optimized task queue according to the allocation scheme optimized by some algorithm.

Forwarder: assign the corresponding distribution center to provide services for user tasks according to the distribution center allocation scheme in the optimization task queue.

Monitor: According to the task log information, fire the task scheduler to determine whether to adjust the scheduling policy and re optimize the task queue.

From a macro perspective, the scheduling server balances the tasks of each resource distribution center, tries to avoid excessive requests for a single center, and makes the system load tend to balance. At the micro level, the reasonable allocation scheme between the distribution center and resource users is obtained as far as possible through the optimization algorithm to optimize the system performance. An early warning mechanism is set up for the prevention of abnormal problems, which can monitor the health of the system in real time during the system operation, and automatically trigger the emergency processing program to protect the normal operation of the system when abnormal conditions occur.

At the same time, scheduling server load balancing is also crucial. The flow chart of load balancing is shown in Fig. 3.

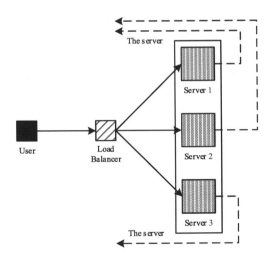

Fig. 3. Load Balancing Flow Chart

As shown in Fig. 3, the general process of load balancing is: when a user's request arrives, the load balancer selects the most appropriate server in the cluster according to the status information of each server in the cluster, using the given load balancing policy, and redirects the request to that server, which then provides services for the user [7]. Load balancing strategy is the core part of load balancing.

Assume there is m distribution center servers $C_1, C_2, \cdots, C_m$, for n users $U_1, U_2, \cdots, U_n$ provide distributed online education resource distribution services. Section j centers C_j the maximum number of connected users is MAX_{C_j}, section i users U_i the download task is M_{U_i}, Task M_{U_i} the physical size of is SM_{U_i}, completed on TM_{U_i}, user U_i the maximum number of connected distribution centers is MAX_{U_i}, user U_i and center C_j download speed between $V_{U_i-C_j}$.

Constraints: The maximum number of connected users in the distribution center is MAX_{C_j} upper limit of connecting distribution center with users MAX_{U_i}. The way of optimizing scheduling in the system mainly considers downloading tasks M_{U_i} select the distribution center of. The purpose of optimization is to minimize the completion time of all download tasks.

Then the objective function of personalized scheduling of distributed online education resources is expressed as:

$$
\begin{cases}
\min E = \sum_{i=1}^{n} TM_{U_i} \\
TM_{U_i} = \dfrac{SM_{U_i}}{\sum_{j=1}^{m} V_{U_i - C_j} \cdot x_{U_i - C_j}} \\
\sum_{i=1}^{n} x_{U_i - C_j} \leq MAX_{C_j} \\
\sum_{j=1}^{m} x_{U_i - C_j} \leq MAX_{U_i} \\
x_{U_i - C_j} = \begin{cases} 1 & C_j \to U_i \\ 0 & other \end{cases}
\end{cases}
\tag{6}
$$

In formula (6), E represents the personalized scheduling objective function of distributed online education resources; $x_{U_i - C_j}$ represents the real-time quantity of educational resource scheduling; $C_j \to U_i$ means C_j by U_i provide services.

From the above mathematical model, it can be seen that this is a typical multi-objective optimization problem. The total objective function is determined by each sub objective function, and each sub objective may conflict. The optimization of one sub objective function is often accompanied by the degradation of other sub objective functions.

2.4 Determination of Personalized Scheduling Scheme of Educational Resources

Based on the objective function of personalized scheduling of distributed online education resources determined above, the objective function is solved based on simulated annealing genetic algorithm [8–10] to obtain the final personalized scheduling scheme of distributed online education resources.

Simulated annealing algorithm is a random optimization algorithm. Its starting point is based on the similarity between the cooling process of high-temperature solid materials in physics (also called annealing) and the solving process of general combinatorial optimization problems. Therefore, simulated annealing algorithm is often used to solve combinatorial optimization problems. The basic steps of the simulated annealing algorithm are:

Step 1: Determine the initial temperature T_0;

Step 2: Given the cooling function, the expression is

$$
T_{k+1} = \upsilon T_k
\tag{7}
$$

In formula (7), T_{k+1} and T_k respectively represent the time $k + 1$ and k temperature value of; υ represents the cooling coefficient.

Step 3: When the temperature is close to 0, the algorithm is terminated.

Genetic algorithm is a stochastic optimization algorithm based on the mechanism of "survival of the fittest" to solve combinatorial optimization problems. Genetic algorithm uses chromosome to express the solved problem[11]. Through the continuous evolution of population replication, crossover and mutation operations, the best individual is the optimal solution of the problem. After nearly half a century of development, genetic algorithm has developed into one of the most widely used algorithms in the field of combinatorial optimization. The basic steps of genetic algorithm are:

Step 1: Randomly generate the initial population $P(0)$;

Step 2: Individual fitness evaluation;

Step 3: If the fitness meets the design requirements, output the results; if not, continue the following steps;

Step 4: Select, cross and mutate;

Step 5: Jump back to Step 2 until the maximum number of iterations is reached.

Both simulated annealing algorithm and genetic algorithm have certain defects, and can not obtain the best personalized scheduling scheme of distributed online education resources, so they are effectively integrated - simulated annealing genetic algorithm [12, 13]. Simulated annealing genetic algorithm combines the characteristics of simulated annealing algorithm and genetic algorithm in optimization operation, principle and other aspects, making the search behavior in the optimization process more perfect, enhancing the ability of global search and local search, and effectively controlling the emergence of "premature" phenomenon, which can theoretically better solve the problem of educational resource scheduling.

The process of obtaining personalized scheduling scheme of distributed online education resources based on simulated annealing genetic algorithm is as follows:

(1) Encoding and decoding:

The algorithm uses operation based coding, and uses constraints to generate initial chromosomes. The decoding process is to first convert the chromosome into an ordered operation table, and then process each operation one by one according to the constraints to generate a scheduling scheme.

(2) Objective function:

The objective function is to minimize the scheduling completion time, which means that the scheduling scheme is better in this evaluation index. Its mathematical expression is as follows:

$$f = \min E = \sum_{i=1}^{n} TM_{U_i} \tag{8}$$

(3) Genetic operator operation:

a. Select operator. The elitist retention strategy and proportional selection method can make individuals with high adaptability inherit with greater probability, thus improving the efficiency of the algorithm.

b. Crossing operator. By improving the conventional crossover operation and designing a simple sequential selection method based on complementary sets, crossover operation has the ability of rapid evolution.

c. Mutation operator. Using the exchange operation, the genes at two different places in the chromosome are exchanged randomly.

(4) Algorithm parameter setting:

Population size, generally 10–200; Crossover probability P_c and probability of variation P_m; Termination algebra, generally 50–200; The initial temperature is determined by the following formula, the expression is

$$t_0 = -\frac{c_w - c_b}{\ln P_r} \tag{9}$$

In formula (9), t_0 the initial temperature; c_w and c_b respectively represent the most adaptive and least adaptive target values of the initial population; P_r represents the relative acceptance probability.

Attemperation function, as shown in Formula (7); The state receiving function refers to the passing probability $\min\left(1, \exp\left(-\frac{\Delta}{T_k}\right)\right)$ accept the new value.

The above process determines the personalized scheduling scheme of distributed online education resources. The implementation of the determined scheme can realize the personalized scheduling of distributed online education resources and provide assistance for the subsequent application of education resources.

3   Experiment and Result Analysis

3.1   Experiment Preparation Stage

In this experiment, three distribution centers provide 10 users with personalized scheduling services for distributed online educational resources as an example to test the application performance of the proposed method. The servers are as follows: the three servers are respectively represented as: C_1, C_2 and C_3; The server C_1 the maximum number of connections for users is 5, and the server C_2 the maximum number of connections for users is 8, and the server C_3 the maximum number of connections for users is 7; The 10 users are represented as: U_1, U_2, U_3, U_4, U_5, U_6, U_7, U_8, U_9, U_{10}; Assume that the maximum number of server connections for 10 users is 1, and the same resource data package is downloaded. The file size is 5000K, and the download speed between each user and the server is limited to 100k/s. The specific data is shown in Table 2.

The personalized scheduling operation interface of distributed online education resources is shown in Fig. 4.

The above process has completed the experimental preparation and provided some convenience for the subsequent experiments.

3.2   Analysis of Experimental Results

Based on the contents of the above experiment preparation stage, the comparison experiment of personalized scheduling of distributed online education resources is carried

Table 2. Download speed between server and user/k/s

	C_1	C_2	C_3
U_1	300	500	400
U_2	300	900	500
U_3	500	300	800
U_4	700	600	200
U_5	100	800	500
U_6	100	100	900
U_7	800	800	300
U_8	400	500	900
U_9	200	700	700
U_{10}	800	100	300

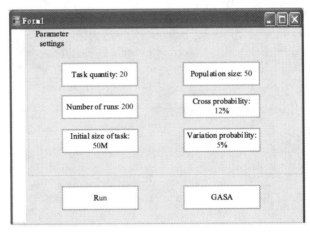

Fig. 4. Example of distributed online education resources personalized scheduling operation interface

out with the educational resource scheduling method based on genetic algorithm and the educational resource scheduling method based on P2P technology as comparison method 1 and comparison method 2,The application effect of the proposed method is visually displayed through the response time of distributed online education resource scheduling and the scheduling accuracy of distributed online education resources.

The response time of distributed online education resource scheduling obtained through experiments is shown in Fig. 5.

As shown in the data in Fig. 5, under different user backgrounds, the response time of distributed online education resource scheduling obtained after the application of the proposed method is less than that of comparison methods 1 and 2, and the minimum value reaches 6s.

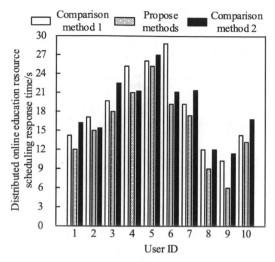

Fig. 5. Schematic diagram of response time of distributed online education resource scheduling

The scheduling accuracy of distributed online education resources obtained through experiments is shown in Fig. 6.

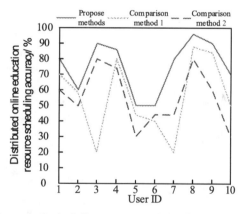

Fig. 6. Schematic diagram of scheduling accuracy of distributed online education resources

As shown in the data in Fig. 6, under different user backgrounds, the scheduling accuracy of distributed online education resources obtained after the application of the proposed method is greater than that of comparison methods 1 and 2, and the maximum value reaches 96%.

Further validate the performance of distributed online education resource scheduling, using throughput as an evaluation indicator. The higher the throughput, the more efficient the resource scheduling can handle a large number of requests or tasks.

The throughput of distributed online education resource scheduling obtained through experiments is shown in Table 3.

Table 3. Distributed Online Education Resource Scheduling Throughput/MB/s

Number of educational resources/MB	Propose methods	Comparison method 1	Comparison method 2
200	189	179	152
400	391	357	342
600	573	541	420
800	779	749	725
1000	985	974	955

According to Table 3, as the number of educational resources increases, the throughput of distributed online educational resource scheduling for different methods also increases. When the number of educational resources is 1000MB, the distributed online educational resource scheduling throughput of Comparison method 1 and Comparison method 2 is 974MB/s and 955MB/s, respectively. The distributed online education resource scheduling throughput of Proposal methods is as high as 985MB/s. From this, it can be seen that the distributed online education resource scheduling throughput of Proposal methods is high and can efficiently handle a large number of requests or tasks.

4   Conclusion

With the rapid development of information technology, teaching resources will become more and more abundant, especially the research on audio and video teaching resources, teaching resources sharing and teaching resources distribution services will be the most urgent requirements at present or in the future. Because of the heterogeneity of the Internet, network resources cannot be fully shared, and the traditional campus network structure also makes it difficult to share educational resources, and almost all educational resources use the C/S service mode, which makes the server face huge load pressure when facing large-scale user access, and it is difficult to ensure the quality of service. Therefore, this paper proposes a personalized scheduling method for distributed online education resources based on simulated annealing genetic algorithm. Build a user base model and interest model to improve user satisfaction and experience. Using fuzzy logic method, calculate the membership relationship between knowledge points and educational resources, and construct an educational resource model. This paper describes the key issues of personalized scheduling of educational resources, constructs the objective function of personalized scheduling of distributed online educational resources and solves it based on Simulated annealing genetic algorithm, which realizes personalized scheduling of distributed online educational resources, thus shortening the response time of distributed online educational resources. The proposed method greatly shortens the response time of distributed online educational resource scheduling, improves the scheduling accuracy of distributed online educational resources, and provides more effective method support for educational resource scheduling. But online education resource scheduling involves a large amount of student data and personal privacy information. Future research should focus on how to protect the privacy and security of student data, and establish reliable data management and sharing mechanisms.

References

1. Liu, S., Xu, X., Zhang, Y., et al.: A reliable sample selection strategy for weakly supervised visual tracking. IEEE Trans. Reliab.Reliab. **72**(1), 15–26 (2022)
2. Bauer, M.N., Probert, M., Panosetti, C.: Systematic comparison of genetic algorithm and basin hopping approaches to the global optimization of Si(111) surface reconstructions. J. Phys. Chem. A **126**(19), 3043–3056 (2022)
3. Shao, R., Zhang, G., Gong, X.: Generalized robust training scheme using genetic algorithm for optical neural networks with imprecise components. Photonics Res. **10**(8), 1868 (2022)
4. Nasrabadi, A.M., Moghimi, M.: Energy analysis and optimization of a biosensor-based microfluidic microbial fuel cell using both genetic algorithm and neural network PSO. Int. J. Hydrogen Energy **47**(7), 4854–4867 (2022)
5. Mishra, M., Dash, M.K., Sudarsan, D., et al.: Assessment of trend and current pattern of open educational resources: a bibliometric analysis. J. Acad. Librariansh.Librariansh. **48**(3), 102520 (2022)
6. Chen, Z., Liu, Y., Hou, H.: Do they really know what we need?" exploring learners' versus universities' views on open educational resources in Chinese universities. Int. J. Educ. Res. **109**(3), 101817 (2021)
7. Yang, S., Lee, J.W., Kim, H.J., et al.: Can an online educational game contribute to developing information literate citizens? Comput. Educ.. Educ. **161**(4), 104057 (2021)
8. Liang, Z., Liu, M., Zhong, P., et al.: Hybrid algorithm based on genetic simulated annealing algorithm for complex multiproduct scheduling problem with zero-wait constraint. Math. Probl. Eng.Probl. Eng. **2021**, 1–21 (2021)
9. Han, B.: Water saving control of turfgrass irrigation robot using genetic simulated annealing algorithm. Mob. Inf. Syst. **2021**, 1–7 (2021)
10. Hou, X., Ji, Y., Liu, W., et al.: Research on logistics distribution routing problem of unmanned vehicles based on genetic simulated annealing algorithm. In: 2021 2nd International Conference on Artificial Intelligence and Information Systems, pp. 1–7 (2021)
11. Archambault, L., Shelton, C., Harris, M.A.: Teachers beware and vet with care: online educational marketplaces. Phi Delta Kappan **102**(8), 40–44 (2021)
12. Zaikov, K.S., Saburov, A.A., Tamitskiy, A.M., et al.: Online education in the Russian arctic: employers' confidence and educational institutions' readiness. Sustainability **13**(12), 6798 (2021)
13. Wang, Z., Yao, N., Liu, Z.: Research on key technology of edge-node resource scheduling based on linear programming. J. Adv. Manuf. Syst. **22**(01), 85–96 (2022)

Accurate Recommendation of Personalized Mobile Teaching Resources for Piano Playing and Singing Based on Collaborative Filtering Algorithm

Xiaojing Wu[✉]

Tianshui Normal University, Tianshui 741000, China
wuxiaojing2313@163.com

Abstract. With the rapid development of education informatization, online education resources are growing explosively. In order to avoid the waste of resources and enable piano playing and singing learners to accurately and quickly find the mobile teaching resources courses they are interested in the massive resources, this paper proposes a precise recommendation method of personalized piano playing and singing mobile teaching resources based on collaborative filtering algorithm. Through the context awareness method, we can obtain the demand information of learners in real time, store it in the database, and calculate the degree of interest of piano playing and singing learners on this basis. Based on the collaborative filtering algorithm, we can constantly optimize the accuracy of the algorithm through the analysis of the degree of interest of learners and other information, accurately recommend learning resources for piano playing and singing, and improve the learning effect and interests of learners. The experimental results show that the proposed method has achieved good application results in practice, and has certain reference and guidance value for enhancing the learning interest of piano playing and singing learners and cultivating autonomous learning ability.

Keywords: Collaborative Filtering Algorithm · Individualization · Piano Playing And Singing · Mobile Teaching Resources · Situational Awareness · Gray Level Correlation Algorithm · Similarity

1 Introduction

The rich network resources facilitate users to search for target resources from massive resources through retrieval, and it has become the main way for people to obtain target resources. However, the explosive growth of network resources has brought convenience to users while also causing a series of problems. For example, when faced with a large amount of learning materials, learners face certain difficulties in finding learning resources that meet their own needs. Even if target resources are found, they may not necessarily be suitable for their own learning. How to recommend massive learning

L. Yun et al. (Eds.): ADHIP 2023, LNICST 547, pp. 226–238, 2024.
https://doi.org/10.1007/978-3-031-50543-0_16

resources to learners is currently a challenge faced by traditional education. The research on precise recommendation methods for personalized piano playing and singing mobile teaching resources [1, 2] can recommend piano playing and singing mobile teaching resources based on users' personalized needs and interests, thereby improving learning efficiency and results, and meeting users' needs and expectations [3, 4]. In addition, it is also of great significance for the development of the education field, providing reference and support for the intelligence and informatization of piano teaching. This paper mainly introduces the research status, recommendation algorithms, advantages and disadvantages of collaborative filtering, which is the most common recommendation algorithm. Let us understand the main idea of collaborative filtering through simple examples. Personalized learning involves analyzing, processing, and mining a large amount of student learning log data, and recommending the mining results. Based on learners' basic knowledge mastery, interests, learning abilities, and other characteristics, personalized chemistry learning models are designed to support teaching, providing personalized learning resources and paths for piano and singing learners.

Reference [5] aims to improve the efficiency of online learners in selecting appropriate high-quality courses from a vast amount of similar learning resources. On the basis of fully mining the value of online comment data, corresponding learning resource profiling and recommendation methods were designed and proposed: based on Apriori algorithm and text sentiment analysis, information such as frequent term itemsets, course features, user emotional tendencies, and online learning platform service quality hidden in online comment information were mined, establish a multi-dimensional feature system for learning resources, and then use the Topsis method to conduct a comprehensive analysis of multiple indicators for alternative courses, exploring the proximity of each object to the optimal ideal solution and the distance from the worst ideal solution, ultimately completing the ranking and recommendation of courses.

Reference [6] proposes a learning resource recommendation method based on multidimensional association ontology. Construct a multi-dimensional association ontology model (MCOM) for learning resource recommendation, and achieve the association of learning resource ontology, learner ontology, and situational ontology through semantic relationships. Then, a dynamic self-balancing binary particle swarm optimization algorithm (DSEBPSO) is designed, and the MCOM ontology model and DSEBPSO algorithm are integrated and applied to implement a learning resource recommendation method based on multi-dimensional associated ontology (MCOM-LROM), which provides learners with optimal learning resources or learning paths.

Reference [7] proposes a multi granularity cloud manufacturing resource combination recommendation method based on self-organizing mapping. By clustering and analyzing the scheduling logs of manufacturing resources from the requester, manufacturing resources are classified into different types based on QoS indicators; Then, sliding window analysis is used to statistically analyze various types of resource scheduling methods, calculate the proportion of different resource scheduling methods in the entire resource scheduling process, and obtain the commonly used scheduling combinations for the requester in the manufacturing process. This is used as a recommended resource combination for the requester to achieve cloud manufacturing resource combination recommendation.

In the current era of information overload, people are facing a massive amount of learning resources, and how to quickly and accurately find suitable teaching resources has become an important issue for learners. However, ordinary learning resource recommendation algorithms lack specificity and cannot effectively meet the personalized needs of learners. Therefore, it is of great significance to study the precise recommendation method of personalized piano playing and singing mobile teaching resources based on collaborative filtering algorithm. By using context awareness method and combining collaborative filtering algorithm to constantly analyze and optimize learners' interests and preferences, learners can find their own learning resources more quickly and accurately, thus improving learning effects and interests. At the same time, the application value of this study is not limited to the learning field of piano playing and singing, but can also provide certain reference and reference for personalized mobile teaching in other disciplines.

2 Design of Personalized Piano Playing and Singing Mobile Teaching Resource Recommendation Method

2.1 A Method for Collecting Information of Piano Singing Learners Based on Contextual Perception

Context awareness refers to the process of perceiving and understanding information such as objects, people, and events in the surrounding environment, enabling machines to understand and adapt to different contexts [8]. Based on this, they decide what data to collect and what methods to use to collect data, and establish a close connection between the collected data and the context.

The data collection process of researchers is an advanced context awareness process [9, 10], This process essentially belongs to the category of human cognition and can be described using models in cognitive psychology. Many models have been proposed in cognitive psychology to describe human cognitive processes. Although these models have significant differences in details, they all describe the cognitive process as three basic stages: first, the brain receives stimuli; Secondly, handle the stimulus; Finally, make a response. Based on these three basic stages, the data collection process of researchers can also be described as three stages:

(1) Perceiving the context related to user activities;
(2) Filter and organize these situations in the brain to identify the situations they are interested in;
(3) Choose appropriate data collection methods. The current situational awareness process in the computer field can also be roughly divided into three stages:
(4) Context acquisition: perceiving and collecting contexts;
(5) Scenario processing: Formalize contextual representations and use contextual reasoning to construct a complete user context;
(6) Service invocation: Using context to trigger the invocation of specific services. This paper will model the situational awareness process of researchers using three processes of situational awareness technology, as shown in Fig. 1.

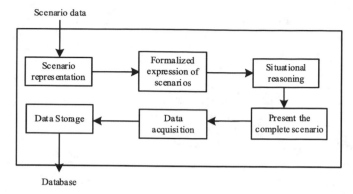

Fig. 1. Context aware data collection model

2.2 Calculation of Interest for Piano Singing Learners

Based on the data collection model, assign an initial value of interest to the target user, their own favorite resources, and the labels used, that is, assign an initial interest to the corresponding vertices in the tripartite graph [11]. These vertices with initial interest will be used as the diffusion origin for the first diffusion.

In the study, it is assumed that the initial interest of the target user towards themselves is 1, and the interest towards other users is 0. Therefore, if y is the target user, the initial interest vector for all users is Y_0. Assuming that users have no difference in their interest in resources, initially, y is only interested in his own favorite resources and has a zero interest in other resources, then y's interest in his favorite resources z can be expressed as:

$$X = \frac{Y_0 z}{n} [z \in I(y)] \tag{1}$$

In the formula, $I(y)$ is the collection of resources that target user y has collected, and n is the number of resources that target user y has collected.

In public classification systems, users often use the same label multiple times to label different resources. To this end, drawing inspiration from the TF-IDF method, a metric was defined to measure the user's interest in different tags, that is, the target user's interest in the tags they use.

Initially, assuming that the target user is only interested in the tags they use and has a zero interest in other tags, the user's interest in tags can be calculated by the following equation:

$$B = y' \times \lg Y_0 z \tag{2}$$

In the formula, y' represents the number of users using labels.

2.3 Collaborative Filtering Based Recommendation Method for Piano Playing and Singing Teaching Resources

In traditional piano playing and singing teaching, the problem of randomization and arbitrariness in the application of online resources is relatively obvious, leading to a decrease in the efficiency of resource application, and the separation of online and offline teaching. Therefore, a precise recommendation method of personalized mobile teaching resources for piano playing and singing based on a collaborative filtering algorithm is proposed.

The main idea of collaborative filtering recommendation method [12, 13] is to use the past behavior of existing user groups to analyze and predict which users will be interested in such items in the future. Generally speaking, collaborative filtering recommendations are divided into three types: user based collaborative filtering, item based collaborative filtering, and model based collaborative filtering.

The implementation of user based collaborative filtering technology mainly includes three steps, namely, finding a set of users with similar interests to the target users, calculating the similarity between items using Pearson correlation coefficient, cosine similarity, Jaccard and other methods, and filtering out the nearest neighbor of the target users by TOP-N method, which refers to taking the first N data after sorting according to certain indicators (such as recommendation, quantity, etc.). This method is usually used in scenarios such as data analysis and leaderboard production, or to filter the nearest neighbor of the target user by setting a threshold. The processing flow of this algorithm is shown in Fig. 2.

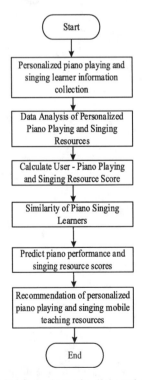

Fig. 2. Recommended flow chart of collaborative filtering algorithm

From Fig. 2, it can be seen that this algorithm collects learner information and conducts data analysis. Based on this, it calculates the score of playing and singing resources, obtains the similarity of learners, predicts scores, and achieves high-precision recommendation of learning resources.

In order to achieve the goal of personalized learning recommendation, students can be divided into two categories. For students without foundation, they can complete knowledge learning in an orderly manner according to the requirements of the curriculum outline. By mastering basic knowledge, they can have a more comprehensive understanding of the content they have learned.

The higher the quality of recommended learning resources, the better the recommendation technology of personalized learning resources [14, 15]. This paper proposes personalized learning resources recommendation based on user preference collaborative filtering algorithm. The algorithm implementation process is shown in Fig. 3.

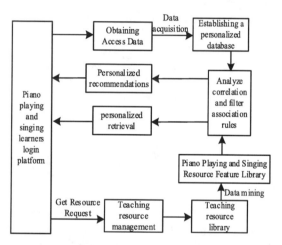

Fig. 3. Flow Chart of Recommended Teaching Resources for Piano Playing and Singing

The core content of teaching resource recommendation is the feature extraction of user data and teaching resources, as well as the analysis or rule association of their features. Feature extraction quantifies and digitizes the user data and teaching resource attribute features of online teaching platforms through data mining, and conducts correlation analysis or association analysis on the user feature library and resource feature library composed of user features and teaching resource features to achieve accurate matching and recommendation of teaching resources. Select user and resource characteristics based on the characteristics of online teaching platforms. The main algorithms are as follows:

Firstly, assuming that the internal and external keywords (requirement features) are a_1 and a_2, with text lengths of δ_1 and δ_2, respectively, where the length of similar strings is δ. Therefore, the similarity $Z(a_1, a_2)$ between the two can be defined as:

$$Z(a_1, a_2) = \begin{cases} \frac{\delta}{\delta_1 + \delta_2 - \delta} & \delta > 2 \\ 0 & \delta \leq 2 \end{cases} \tag{3}$$

From the above equation, it can be seen that $Z(a_1, a_2) \in [0, 1]$. This formula simulates the similarity between the "requirements vs. resources" part inside and outside the school, thereby improving the matching accuracy of similarity between requirements and resource keywords. All resources in the system have several personalized feature words that describe them. Combining the frequency of feature words (keywords), the following processing can be performed:

Firstly, obtain the frequency of feature words for a certain resource:

$$R(\delta_{n+1}) = [R(\delta_1), R(\delta_2), \ldots, R(\delta_n)] \tag{4}$$

From this, Eq. (5) can be obtained:

$$Z(\delta_{n+1}) = [Z(\delta_1), Z(\delta_2), \ldots, Z(\delta_n)] \tag{5}$$

Subsequently, take the maximum value in formula (5) and define it as $Z(\delta_n)_{max}$ to obtain the final value calculated by similarity weighting:

$$H(\delta_n) = Z(\delta_n)_{max} \times (Z(\delta_{n+1}) + 1) \tag{6}$$

Secondly, assuming there are several teaching resources with a total of m feature words, the expression form of their space vector matrix is as follows:

$$A = \begin{bmatrix} a_{11} & a_{12} & \cdots & a_{1n} \\ a_{21} & a_{22} & \cdots & a_{2n} \\ \vdots & \vdots & \vdots & \vdots \\ a_{n1} & a_{n2} & \cdots & a_{nn} \end{bmatrix} \tag{7}$$

The spatial feature component of a resource is the similarity between its feature words and the core feature words (requirements) multiplied by the frequency of the feature words appearing in the abstract, then weighted by 1, and the maximum value is taken. Next, sort them to obtain the recommended resources that are currently ranked high. Subsequently, this system adopts a direct scoring measurement method to describe its evaluation values:

$$\Upsilon(\delta) = a_{nn}H(\delta_n)d + M_t^g v_t, t = 1, 2, \ldots, n \tag{8}$$

In the formula, M_t^g is the evaluation value of the resource by the g-th evaluator (faculty) at time t, where d is the true value, and correspondingly, v_t is the evaluation noise generated by subjective factors at time t. There is unknown prior knowledge in the formula. In the subsequent processing, the gray correlation algorithm [16, 17] is used to calculate the gray correlation degree of the two. This algorithm is a multivariable non parametric statistics statistical analysis method, which can be used to find the relationship between multiple input variables and an output variable. It is mainly used to establish mathematical models and carry out prediction analysis. It is suitable for data with different scales, distributions and intervals standardize the data of distribution and interval, then calculate the correlation degree and analyze it, and provide its similarity

[18, 19]. The grey correlation degree L_{ij} between the recommendation sequence i and the demand sequence j at time t in the system can be expressed as:

$$
L_{ij} = \begin{cases} 1 & , i = j \\ \displaystyle\sum_{i=1, j=1} \dfrac{\min_i \min_j |\Upsilon(\delta) + M_t^g|}{\max_i \max_j |\Upsilon(\delta) + M_t^g|} & , i \neq j \end{cases} \tag{9}
$$

From this, it can be concluded that the similarity between the recommendation sequence and the requirement sequence (grouping of teachers, managers, students, etc.) at time t constitutes a matrix W_l, as shown in formula (8):

$$
W_l(t) = \begin{bmatrix} 1 & w_{12}(t) & \cdots & w_{1n}(t) \\ w_{21}(t) & 1 & \cdots & w_{2n}(t) \\ \vdots & \vdots & \vdots & \vdots \\ w_{n1}(t) & w_{n2}(t) & \cdots & 1 \end{bmatrix} \tag{10}
$$

If there is a large number of elements in matrix W_l, it indicates that it is close to most requirements at time t in the system. From this, it can be further defined that the correlation $T_i(t)$ between the i-th recommendation sequence at time t and the sequence is:

$$
T_i(t) = \frac{\displaystyle\sum_{i=1} w_{ij}(t)}{n} \tag{11}
$$

In the formula, $w_{ij}(t)$ represents the similarity between the recommendation sequence i and the requirement sequence j at time t. At this point, sorting their correlation values and selecting a recommendation sequence with higher correlation can obtain more accurate recommendation targets.

3   Experiment and Analysis

In order to verify the application effect of the personalized piano playing and singing mobile teaching resources precise recommendation method based on the collaborative filtering algorithm, experimental tests were carried out. The experimental environment is shown in Table 1.

The above configuration is a common configuration for online course resource recommendation systems, ensuring the universality of the experiment. The experiment used the proposed method, the method of reference [4], and the method of reference [5] to test recommendation coverage, recommendation time, and recommendation accuracy in sequence. The test content is as follows:

3.1   Recommended Coverage Test

In order to test the recommendation level of different recommendation methods, 1000 students were selected as the recommendation target to test the recommendation coverage

Table 1. Experimental environment

Configuration	Parameter
CPU	Intel(R) Core(TM) i5–9400
Frequency	2.90GHz
The server	Associate SR550
Operating System	Windows 10
Version	18362.1082 Professional Edition
Digit	64bit
Hard disk	8TB
Database	MySQL

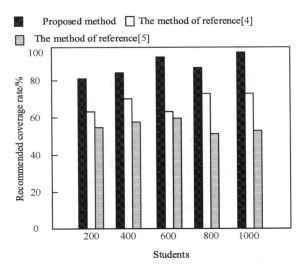

Fig. 4. Recommended Coverage Test Results

of the three methods. The higher the coverage, the better the recommendation ability of the method. The test results are shown in Fig. 4.

From Fig. 4, it can be seen that the recommendation coverage rate using the comparison method is relatively low, while the recommendation coverage rate using the proposed method is always higher than 80%, which can effectively achieve accurate recommendation of personalized piano playing and singing mobile teaching resources.

3.2 Recommended Time Test

In order to test the resource recommendation efficiency of different recommendation methods, 15 sets of tests were conducted on three different methods using recommendation time as the testing indicator. The test results are shown in Table 2.

Table 2. Resource recommendation time for different methods

Number of experimental groups/group	Recommended time/ms		
	Proposed method	The method of reference [4]	The method of reference [5]
1	20.5	46.7	39.9
2	21.2	45.3	39.6
3	20.8	46.6	39.9
4	20.1	46.7	40.2
5	20.8	46.2	39.4
6	20.4	45.9	39.6
7	21.7	46.8	39.8
8	20.9	46.1	40.1
9	21.1	45.8	39.4
10	20.5	45.4	38.8
11	21.8	46.3	40.1
12	20.6	44.5	40.5
13	19.9	45	39.9
14	20.3	44.6	41
15	21	46	40.2

From Table 2, it can be seen that during the experimental process of 15 groups, the recommended time of the method of reference [4] and method of reference [5] fluctuated around 45.7 ms and 39.5 ms, respectively. However, the resource recommendation time of the proposed method was always less than 22 ms, which was lower than the comparison method, indicating a higher resource recommendation efficiency.

3.3 Recommended Accuracy Testing

In order to further test the resource recommendation effectiveness of the three methods, using recommendation accuracy as an evaluation indicator, the resource recommendation accuracy of the three methods for 1000 students was tested. The calculation formula for recommendation accuracy is as follows:

$$\varpi = \frac{\varpi_1}{\varpi_2} \times 100\% \tag{12}$$

In the formula, ϖ_2 represents the total recommended resources, and ϖ_1 represents the actual received recommended resources.

The higher the recommendation accuracy, the better the recommendation accuracy and resource recommendation effect. The test results are shown in Fig. 5.

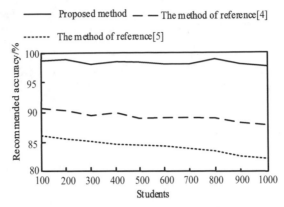

Fig. 5. Recommended Accuracy Test

Figure 5 shows the recommendation accuracy of different methods. From the test results, it can be seen that as the number of students increases, the recommendation accuracy of each method continues to decrease. When the number of students reaches 500, the recommendation accuracy of the method of reference [4] and the method of reference [5] decreases to 87% and 83%, respectively. Although the resource recommendation accuracy of the proposed method shows a decreasing trend, its resource recommendation accuracy always exceeds 97%, it can effectively improve the precise recommendation effect of personalized piano playing and singing mobile teaching resources.

4   Conclusion

This paper proposes a precise recommendation method for personalized mobile teaching resources of piano playing and singing based on collaborative filtering algorithm. Collect the information of piano playing and singing learners based on context awareness method, and calculate the learners' interest in recommended resources. Aiming at the problems of low resource utilization and poor recommendation effect, accurate recommendation of piano playing and singing resources is achieved through Collaborative filtering algorithm. The experimental results indicate that:

1) In the recommendation coverage tests of the three methods, the proposed method consistently achieved a recommendation coverage rate of over 80%, which can effectively complete the precise recommendation task of personalized piano playing and singing mobile teaching resources and has good practicality.
2) In the recommendation time test, the resource recommendation time of the proposed method is always less than 22ms, indicating high recommendation efficiency.
3) When the number of students reaches 500, the recommendation accuracy of the comparison method is less than 90%. Although the resource recommendation accuracy of the proposed method shows a decreasing trend, the overall resource recommendation accuracy is always higher than 97%, indicating good recommendation results.

From the above results, it can be seen that this method has high recommendation coverage and accuracy, and low recommendation time, can quickly and accurately recommend learning resources that learners are interested in, effectively improving learning effectiveness and hobbies. In addition, this method also has certain universality, and the corresponding algorithm can be applied to other mobile teaching fields, which has certain theoretical and practical significance. In summary, this method provides a new approach and approach for personalized recommendation of mobile teaching resources, which can better meet the personalized needs of learners and provide a certain reference for the development of mobile education.

Acknowledgement. Natural Science Fund Project in Shandong Province (2013ZRE27312)

References

1. Xia, Y.: Resource scheduling for piano teaching system of internet of things based on mobile edge computing. Comput. Commun. **158**, 73–84 (2020)
2. Shi, Y., Yang, X.: A personalized matching system for management teaching resources based on collaborative filtering algorithm. Int. J. Emerg. Technol. Learn. **15**(13), 207–220 (2020)
3. Liu, M., Huang, J.: Piano playing teaching system based on artificial intelligence–design and research. J. Intell. Fuzzy Syst. **40**(2), 3525–3533 (2021)
4. Luo, Y., Yang, H.: Teaching applied piano singing while playing based on Xindi applied piano pedagogy: Taking Fujian vocational college of art as an example. J. Contemp. Educ. Res. **6**(8), 123–135 (2022)
5. Zhao, J., Zhou, S., Zhang, J.: Recommendation methods of online learning resource based on online comments. Math. Practice Theory **52**(09), 260–270 (2022)
6. Li, H., Wu, J., Dai, H.: A method of learning resource recommendation based on multidimensional correlation ontology. J. Zhejiang Univ. Technol. **49**(04), 374–383 (2021)
7. Zou, Y., Zhao, X.: Multi-granularity cloud manufacturing resource combination recommendation based on SOM. J. Wuhan Univ. **67**(06), 555–560 (2021)
8. Liu, S., Li, Y., Fu, W.: Human-centered attention-aware networks for action recognition. Int. J. Intell. Syst. **37**(12), 10968–10987 (2022)
9. Kirci, P., Arslan, D., Dincer, S.F.: A communication, management and tracking mobile application for enhancing earthquake preparedness and situational awareness in the event of an earthquake. Sustainability **15**(2), 970 (2023). https://doi.org/10.3390/su15020970
10. Borcoci, E., Vochin, M.C.: A quality-of-service scenario awareness for use-cases of open-source management and control system hub in edge computing. In: 2021 IEEE International Black Sea Conference on Communications and Networking (BlackSeaCom), pp. 1–5. IEEE (2021)
11. Taheri, M., Farnaghi, M., Alimohammadi, A., et al.: Point-of-interest recommendation using extended random walk with restart on geographical-temporal hybrid tripartite graph. J. Spat. Sci. **68**(1), 71–89 (2023)
12. Wang, X., Dai, Z., Li, H., et al.: A new collaborative filtering recommendation method based on transductive SVM and active learning. Discret. Dyn. Nat. Soc. **2020**, 1–15 (2020)
13. Fu, L., Ma, X.M.: An improved recommendation method based on content filtering and collaborative filtering. Complexity **2021**, 1–11 (2021)
14. Bulathwela, S., Kreitmayer, S., Pérez-Ortiz, M.: What's in it for me? Augmenting recommended learning resources with navigable annotations. In: Proceedings of the 25th International Conference on Intelligent User Interfaces Companion, pp. 114–115 (2020)

15. Zheng, H.U.: Multi level recommendation system of college online learning resources based on multi intelligence algorithm. J. Phys. Conf. Ser. **1873**(1), 12078–12085 (2021)
16. Hao, W.: Research on gray correlation algorithm of factors for college students' mental health early warning. Electron. Des. Eng. **30**(11), 12–16 (2022)
17. Sultana, U., et al.: Determination of green spots (trees) for google satellite images using MATLAB. Procedia Comput. Sci. **171**, 1634–1641 (2020). https://doi.org/10.1016/j.procs.2020.04.175
18. Li, L., Zhang, Z., Zhang, S.: Knowledge graph entity similarity calculation under active learning. Complexity **2021**, 1–11 (2021). https://doi.org/10.1155/2021/3522609
19. Wang, H., Wang, W.: Social network behavior inference method based on collaborative filtering recommendation. Comput. Simul. **38**(02), 427–431 (2021)

Research on Fault Signal Reconstruction of Treadmill Equipment Based on Deep Neural Network

Lingling Cui[1](✉) and Juan Li[2]

[1] Department of Physical and Health Education, Wuxi Vocational Institute of Commerce, Wuxi 214153, China
cl_ilikemouse@163.com
[2] College of Intelligent Equipment and Automotive Engineering, Wuxi Vocational Institute of Commerce, Wuxi 214153, China

Abstract. There are a large number of noise components in the fault signals of treadmill equipment, which leads to increased difficulty in signal reconstruction. Therefore, a new method for reconstructing fault signals of treadmill equipment is proposed by introducing deep neural networks. Based on the community structure, fault source localization is achieved through two stages: partitioning fault areas and predicting fault propagation paths. A fault signal acquisition platform is designed based on the fault source localization results, and the collection of fault signals from the treadmill equipment is implemented. A denoising model based on a dual-layer recurrent neural network is constructed using deep neural networks to perform denoising processing on the collected fault signals. The signal reconstruction of treadmill equipment faults is completed using a matching tracking algorithm. The test results show that the reconstruction time of this method is less than 6000 ms, and the minimum signal-to-noise ratio of the reconstructed signal reaches 49.30 dB, demonstrating good practical application effects.

Keywords: Deep Neural Network · Community Structure · Treadmill Equipment · Fault Signal Reconstruction · Orthogonal Matching Pursuit

1 Introduction

At present, busy work makes it difficult to regularly go to the gym for exercise, so exercising at home has become an increasing choice for people. As a kind of household fitness equipment, the Treadmill has obvious advantages that consumers can use the Treadmill to exercise enough even in their own narrow space. At the same time, it is convenient to install, simple to control, and can freely control the intensity and time of movement. Therefore, it has become a very popular sports equipment for consumers [2]. However, under frequent startup, shutdown, and speed regulation modes, the stator and rotor winding circuits of asynchronous motors are prone to generate huge heat, which can lead to thermal aging if they cannot dissipate heat in a timely manner. As a result, the

L. Yun et al. (Eds.): ADHIP 2023, LNICST 547, pp. 239–254, 2024.
https://doi.org/10.1007/978-3-031-50543-0_17

performance of asynchronous motors decreases sharply and faults are prone to occur. Therefore, fault signal reconstruction can improve the accuracy and efficiency of fault diagnosis.

In this context, scholars have conducted certain research work on fault signal reconstruction. So far, fault signal reconstruction methods mainly include methods based on multivariate statistical analysis, observer based methods, and machine learning methods. Among them, some scholars proposed a motor fault signal reconstruction method based on dyadic wavelet transform and triangular Spline interpolation. This method uses the same function - triangular spline wavelet function in dyadic wavelet transform and interpolation reconstruction, which can reflect the essence of signal processing. The proposed triangular spline wavelet has the dual function of serving as both a wavelet function and an interpolation function, greatly improving the effectiveness of the algorithm. It was applied to the reconstruction process of motor fault signals and compared with the Mallat algorithm in terms of signal-to-noise ratio and relative error, with significant results. Some scholars have proposed a multi-sensor rolling bearing fault signal reconstruction method based on the correlation principle. In order to compensate for the defect of incomplete signal collection by a single sensor, a multi sensor array is used to non-contact collect the periodic signal of rolling bearing faults and eliminate excess information. The correlation algorithm is used to reconstruct the incomplete signal fragments into a complete periodic acoustic signal of rolling bearing faults. And a calculation example of multi-sensor waveform data reconstruction is provided to verify the effectiveness of this method. The problems of long reconstruction time and low signal-to-noise ratio of reconstruction signal existing in the above methods are taken as research objectives, and a fault signal reconstruction method of Treadmill equipment based on deep neural network is designed.

2   Design of Fault Signal Reconstruction Method for Treadmill Equipment

2.1   Fault Source Location

Based on the community structure, fault source localization is achieved through two stages: dividing fault areas, predicting fault propagation paths, and locating fault sources. Design a fault source localization method based on the community structure. When a community node is detected to be in a fault state, it cannot be directly determined that the node has malfunctioned. The fault may also be caused by adjacent, highly capable fault nodes through fault propagation. Therefore, a retrospective approach is adopted to achieve fault source localization. At the same time, in order to quickly traverse the fault path, depth first traversal is used for path search [4].

The fault location process is divided into two stages: Dividing the fault area and predicting the fault propagation path to locate the fault source.

(1) Division of fault area

This stage is responsible for obtaining fault set α from the system that detected the fault, thereby dividing the network β where the fault alarm occurred into normal area χ

and fault area α, namely:

$$\chi \cup \alpha = \beta \tag{1}$$

The depth first traversal search and backtracking technology is mainly used for fault area division. The specific process is as follows:

Step 1: Initialize the fault set. Add the detected faulty node to α, randomly select a node from the fault set and start a reverse radial search to determine whether the node has been searched. If not, set it as starting node W and mark it; Otherwise, reselect a node that has not been searched for, set it as the starting node, and mark it as [5].

Step 2: Searches for the fault propagation path. Use depth first traversal and backtracking techniques for reverse search to determine the possible propagation path and corresponding node set of the node fault. Search for nodes with fault propagation impact on the current node, assuming E nodes, and determine whether E node has been searched. If node E is not searched, add that node to the fault propagation path set $H(W)$, record the propagation path, and mark that node as searched.

Step 3: Determines whether the backtracking condition is met. Repeat Step 2 until there is no fault propagation trend at node E. If it has been searched, it is determined that all nodes affected by the fault have been searched; On the contrary, search for other propagation paths and corresponding nodes. If both the path and the node have search marks, it will be traced back to the previous node.

Step 4: Determine whether to end the current search. Repeat Step 3 until the start node is traced back, and judge whether the nodes connected to it have search marks. If not, start searching again from the node that has not been searched; If yes, the affected fault set search for the failed node has been completed.

Step 5: Determine whether to end the fault area division. Determine whether all nodes in α have been searched. If not, set one of the nodes as the starting node W and return to step 2; If all faulty nodes are searched, add the intersection of the fault propagation path sets of each node to α, and the fault area division ends [6].

(2) Predict propagation path and locate fault source

After determining the fault area α, analyze the structural characteristics of the area, and the specific fault source localization process can be divided into the following steps:

Step 1: Determine the size of the fault area. For fault area α, first determine the following equation:

$$|\alpha| = 1 \tag{2}$$

When the above formula is true, it indicates that there is only one fault source at present. Output the node to end the positioning; Otherwise, it indicates that there are multiple fault nodes. Enter Step 2.

Step 2: Determines whether there are independent subnets in α. If it exists, it indicates that multiple fault alarms have occurred in different regions. It is necessary to locate the fault source α of each independent sub network as the fault area and return to Step 1; Otherwise, store the nodes within α into the fault prediction set U and enter Step 3.

Step 3: Sort the nodes of the fault prediction set. Calculate the fault occurrence rate g_v for each node c_v in fault prediction set U, sort from high to low, and select the node with the highest fault occurrence rate as c_{max}.

Step 4: Predicts the fault propagation path and locates the fault source. Calculate the fault occurrence rate of each path node in the fault propagation path set $H(c_{max})$ of c_{max}, comprehensively consider the impact of fault propagation on the path node and the historical changes in fault occurrence rate, determine whether the fault propagates between each node, and obtain the set K of fault propagation paths. Select the node that is most likely to fail in set K as the expected fault source node and store it in fault source set V.

Step 5: Outputs the results. If there are independent subnetworks in the fault area α of Step 2, the predicted fault source nodes found in all subnetworks will form the predicted fault source set V, and the corresponding propagation path will be recorded to complete fault localization and output the localization results.

2.2   Fault Signal Acquisition

The fault signal acquisition platform is designed based on the fault source location results, and the fault signal acquisition of Treadmill equipment is implemented. The platform selects Altera Cyclone IV FPGA as the control core. Firstly, after the acceleration sensor is assembled as required, the platform starts to collect the fault signal of the treadmill equipment. The collected signals are filtered and amplified by the signal conditioning module, and then converted into voltage signals and sent to the AD conversion module. FPGA is responsible for controlling AD chip to convert data, and then performing FIR filtering pretreatment on the converted digital signal. Then, control the two pieces of RAM to complete FIFO data cache in ping-pong structure and send it to SDRAM for real-time storage. It is connected to the upper computer through RS232 serial port or Ethernet, and the serial port and TCP protocol are used for data transmission. Finally, the control network interface chip transmits the data to the upper computer [7]. After the hardware system is connected to the server created by the upper computer, wait for the upper computer to receive data. The upper computer will carry out system analysis and data processing for these data respectively.

The selected sensor is an acceleration sensor. In view of the comprehensive consideration of various parameter ranges of the sensor and its installation environment and other factors, we have adopted the YK-YD20 acceleration sensor of IEPE type. The sensor has a built-in charge amplifier, which has the advantages of large dynamic range, wide frequency response, strong anti-interference ability, etc., and greatly suppresses the increase of noise. The parameters of YK-YD20 acceleration sensor are shown in Table 1.

In order to ensure the normal operation of IEPE type acceleration sensor YK-YD20, it is generally necessary to stimulate (2–20 mA) with a quantitative constant current source to output a constant current to effectively avoid damage to the sensor caused by excessive current. In order to achieve the effective output of constant current excitation, we have chosen LM334 chip for design after comprehensive consideration. Constant current source is greatly affected by temperature, and its output current will change with the difference of temperature factors. Therefore, it can be regarded as a constant current source only when it maintains constant temperature. In order to prevent the impact of temperature, it is necessary to add a resistor and diode on the original basis, which can effectively reduce the current drift [8] caused by temperature changes.

Table 1. YK-YD20 acceleration sensor parameters

S/N	Project	Parameter
1	Measuring range	100 g
2	Frequency range	0.2–5 kHz
3	Supply voltage	12 V–28 V (DC)
4	Maximum allowable acceleration	5 * 100 m·s-2
5	Sensitivity	50 mV/g
6	Working temperature	−50 °C− + 120 °C
7	Transverse sensitivity ratio	≤8%
8	Working current	1–10 mA

The operational amplifier designed with LM358 chip is a co directional proportional amplifier circuit. The two internal amplifiers have high gain and are independent of each other. Through sensor sensitivity PK, measured vibration acceleration a, and charge amplifier sensitivity AK, it can be calculated that the sensor output voltage is about 1 V. Since the input voltage of the AD chip used in the platform acquisition module is ±5 V, ensure that the sensor is normally connected to the AD conversion circuit and can only amplify the input signal.

This circuit is a typical voltage series negative feedback circuit. The acceleration sensor YK-YD20 outputs a voltage of v_q and is connected to the in-phase input terminal of the amplification circuit through a resistor. The amplified output voltage of v_p is connected to the inverting input terminal of the operational amplifier circuit using two resistors. Due to the negative feedback circuit of LM358 and its gain approaching 100 dB, it is considered a linear state. The two input terminals of the operational amplifier can be regarded as equal voltages, namely:

$$v_1 = v_2 \tag{3}$$

The input current is zero, that is, the definition of virtual short and virtual break. According to the resistance voltage division formula:

$$v_2 = \frac{R_a}{R_b + R_a} v_q \tag{4}$$

In Eq. (4), R_a and R_b are external resistors.

$$v_1 = \frac{R_c}{R_d + R_c} v_p \tag{5}$$

In Eq. (5), R_c and R_d are internal resistances. Due to the need to design an amplification circuit with a voltage amplification factor of 5, 5k resistors of R_c, R_d, and 1k resistors of R_a and R_b will be selected, and finally substituted into the above formula to calculate the ratio of v_q to v_p as 5.

FPGA is used as the control chip, mainly because it has rich I/O pin resources. It includes 8 user I/O blocks, and the number of I/Os available to users is 179, which makes it easier to achieve logical control. In this platform, FPGA is mainly responsible for the acquisition and processing of the signal module and the control of the signal transmission to the upper computer. When selecting FPGA, it is necessary to consider the low cost and low power FPGA architecture designed, and select according to the design requirements. Therefore, this paper selects the fourth generation Cyclone IV series FPGA devices produced by ALTERA Company, and selects EP4CE10F17C8 chip as the platform FPGA device chip [9].

According to the chip manual, EP4CE10F17C8 is powered by three different voltages. In order to meet the demand of the power circuit, the voltage conversion chip is used to convert 5 V voltage into different voltage values of 1.2 V/2.5 V/3.3 V, so as to ensure the normal operation of the FPGA control module. It is understood that 1.0 V/1.2 V voltage is required to be supplied to the internal logic voltage (VCCINT) and PLL digital voltage (VCCD_PLL), 2.5 V is required to supply PLL analog voltage (VCCA), and 3.3 V voltage is required to be supplied by IO for the rest.

Cyclone IV series FPGAs are configured in active serial (AS), passive serial (PS), JTAG, etc. FPGA device adopts SRAM architecture, which belongs to volatile memory architecture. Therefore, nonvolatile memory chips must be added to the platform. The 16Mbit nonvolatile memory chip EPCS16 designed by Altera Company was selected. Configure the pins in the Table 2 below for the chip.

Table 2. Function table of configured circuit pins

S/N	Pin name	Function
1	ASDI	Data writing
2	nCS	Chip Select
3	DCLK	Clock signal input
4	DATA	data output

AD acquisition circuit controls the reception of signals, and converts the received analog signals into data signals for the analysis and processing of upper computer data. Because the A/D conversion module has complex performance, when designing the hardware of the A/D module, all the pins on the chip are connected to the FPGA. Through the design of the logical A/D conversion module, the corresponding functions are controlled and realized. This paper uses the AD7606 A/D conversion chip designed and produced by an American company. The chip can support 16 bit synchronous sampling, with a sampling rate of up to 200K, and also has the characteristics of 40 dB anti aliasing suppression signal.

The data conversion of AD7606 chip is controlled by two pins CONVST A and CONVST B, so connecting CONVST A and CONVST B can effectively complete eight channel synchronous sampling. Among them, CONVST A controls the four channels V1–V4, while the other pin controls the remaining four channels.

The memory selected is W9812G6KH-75 as the cache chip. The maximum clock of this chip can reach 133 MHz, and data reading and writing can be completed in a short time. Moreover, the memory capacity of the chip is 4 Banks * 1 Mbits * 16, or 64 Mbit, which largely meets the data storage requirements of the platform.

The selected UART device is RS232 interface, which is mainly used for the communication between the data acquisition module and the PC, and RS232 can also complete two-way data transmission and full duplex communication, so it is widely used in short distance data transmission. Since the input and output of FPGA pin is TTL level, the platform uses serial port conversion core MAX3232 that supports RS232 protocol for data conversion.

W5300 chip is selected to realize the network data upload function. W5300 chip supports TCP/IP protocol stack. The processor only needs to set network parameters such as IP address and mask to perform TCP/IP connection. Its advantage is not only fast, stable and reliable, but also effectively saves the ROM resources of the processor. Connect W5300 to FPGA through the connection mode of control pin. See Table 3 for details.

Table 3. Control pin and function

S/N	Pin	Function description
1	OP_Mode[2:0]	Operation mode selection of internal PHY chip
2	TestMode[3:0	PHY chip mode selection
3	RD	Read enable signal
4	WR	Write enable signal
5	CS	Chip select
6	DATA[15:0]	Data bus signal
7	ADDR9	Address bus signal
8	BIT16EN	Data bit selection
9	RESET	Hardware reset signal input

RJ45 Ethernet interface is adopted, and HX1188 is selected as Ethernet isolation transformer.

2.3 Signal Noise Reduction Processing

Based on the deep neural network, a noise reduction model of a two-layer recurrent neural network is constructed to denoise the collected fault signals [10]. In the input layer, the data set is divided, and the ratio of test set and training set is 7:3. Define the input data as:

$$Y = \{y_1, y_2, \ldots, y_m\} \tag{6}$$

In formula (6), m represents the input data length.

Predict the next fault data 45 based on the fault data $\{y_1, y_2, \ldots, y_{100}\}$ of the first y_{101} moments of the asynchronous motor. Then use $\{y_1, y_2, \ldots, y_{101}\}$ data to predict y_{102}, and so on [11]. The entire model framework consists of four parts, and the network model has four layers. The input layer is responsible for formatting the data, and the output layer is used to output noise reduction results. In order to prevent overfitting, the hidden layer is built with LSTM-GRU and double-layer LSTM, and connected through the Dropout layer [12].

When the neural network unit core is at time t, it receives xtof the current state and output ht − 1 of the hidden layer state of the previous time (t − 1) as the input of the entire neural network unit, which is combined with the weight and bias of the input gate, output gate and forgetting gate respectively to obtain the signal values of the input gate, output gate and forgetting gate. Then update the memory cell c and the current time ht according to the current signal, and use them as the input of the next layer of neural network core unit [13].

In the process of noise reduction for error data, the selection of model parameters determines the quality of network performance and the accuracy of noise reduction results. In this model, there are two main types of parameters. One is the parameters automatically updated and learned through the network model in the training process, and its settings are shown in Table 4.

Table 4. Model parameter settings

S/N	Model parameter	Parameter setting
1	Neuron inactivation ratio dropout	0.1
2	Optimization algorithm	Dropout layer import
3	Initial learning rate	0.005
4	Number of batches	30
5	Number of hidden layers	2
6	Output layer node	1
7	Number of hidden layer nodes	8
8	time step	100

The other is the setting of super parameters, such as learning rate, iterations, and loss function selection.

(1) Super parameter setting

The mean square error in the network model is taken as the loss function, and the mean square error formula is as follows:

$$MSE = \frac{1}{M} \sum_{j=1}^{M} \left(\bar{s}_j - h_j \right)^2 \tag{7}$$

In the above equation, M represents the number of samples used in the iteration; $\bar{s}_j$ represents the j-th data denoising value; h_j represents the original drift error of the j-th data.

(2) Weight parameter update

After the super parameters are set, the double-layer network model learns the error data characteristics through the training of the sample set. The Adam algorithm is used to continuously update the weight and adjust the weight parameters through back propagation. When the loss function is not decreasing or the whole training process is over, the weight and offset are saved. The noise reduction model reduces the noise of data at the next time through the weight and deviation of training.

The Adam algorithm execution flow chart is shown in Fig. 1.

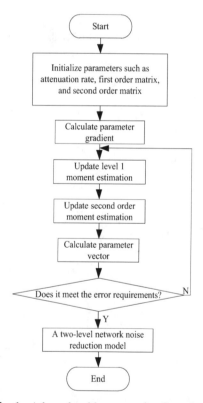

Fig. 1. Adam algorithm execution flow chart

The noise reduction of fault signal is completed through this model.

2.4 Signal Reconstruction

Based on the signal after noise reduction, the signal reconstruction of Treadmill equipment fault is completed by using Matching pursuit algorithm. The regularized orthogonal matching pursuit algorithm of the compressed sensing reconstruction algorithm is an improvement on the orthogonal matching pursuit algorithm [14]. In each iteration, the orthogonal matching pursuit algorithm will only select the column most related to the residual, but it is problematic to select only one column. To solve this problem, a regularized orthogonal matching pursuit is proposed.

Basic symbol description: ω is the original signal, which is ξ dimensions, and ψ is the observation vector, which is ζ dimensions. The general formula is established as follows:

$$\zeta < \xi \tag{8}$$

In reality, ω is generally not sparse, but in a certain transformation domain υ is sparse, that is:

$$\omega = \vartheta \upsilon \tag{9}$$

In formula (9), ϑ represents the sparse coefficient of O, which means that the absolute values of only O non zero terms or O element coefficients of ϑ are much greater than zero.

The observation vector in compressed observation satisfies the following equation:

$$\psi = \varpi \omega \tag{10}$$

In formula (10), ϖ represents an empty set.

In a certain transformation domain υ, the following equation holds:

$$\psi = \varpi \vartheta \upsilon \tag{11}$$

The following formula is established:

$$\varpi \upsilon = B \tag{12}$$

The following formula is established:

$$\psi = B\vartheta \tag{13}$$

In the ROMP reconstruction algorithm, there are also the following symbols, as shown in Table 5.

Table 5. Other symbols

S/N	Symbol	Significance
1	t_r	Residual
2	r	Iterations
3	μ_r	Index found in r-th iteration
4	Δ_r	Index set for the r-th iteration
5	B_i	Column i of matrix B
6	ρ_r	Column vector of $r * 1$
7	B_r	Column set of matrix B selected by index μ_r

The ROMP reconstruction algorithm process is described in detail as follows:

(1) Input:
 1. Sensing matrix $B = \varpi \upsilon$ of $\zeta \times \xi$;
 2. $\xi*1$-dimensional observation vector ψ;
 3. Signal sparsity H.
(1) Firstly, initialize each quantity:

$$t_0 = \psi \tag{14}$$

In formula (14), t_0 represents the initial value of the residual.

$$\Delta_0 = \lambda \tag{15}$$

In formula (15), λ represents an initial value set; Δ_0 represents the initial value of Δ_r.

$$B_0 = \lambda \tag{16}$$

In formula (16), B_0 represents the initial value of B.

$$r = 1 \tag{17}$$

(2) Calculate $\langle t_{r-1}, B_i \rangle$, select the H maximum values or all non zero values in the result, and form a set I with column numbers i corresponding to B.
(3) Regularization: Find subset I_0 in set I, which satisfies the following formula:

$$\begin{cases} |R(I_1)| \leq |R(I_2)| \\ I_1, I_2 \in I_0 \end{cases} \tag{18}$$

Select the I_0 with the highest energy among all subsets I_0 that meet the requirements.
(4) The following formula is established:

$$\begin{cases} \Delta_r = \Delta_{r-1} \cup I_0 \\ B_r = B_{r-1} \cup B_i \\ i \in I_0 \end{cases} \tag{19}$$

(5) Find the least squares solution of $\psi = B_r \vartheta_r$.
(6) Update the residuals using the results from the previous step.
(7) Increase the number of iterations by one. If the number of iterations is less than or equal to H, return to step (2), otherwise proceed to the next step.
(8) The sparse representation coefficients of the reconstructed signal are estimated to have non-zero terms at Δ_r locations, with their values being the least squares solutions obtained from the last iteration [15].
(9) Output:
 1. Estimation of signal sparse representation coefficient;
 2. $\xi * 1$ dimensional residual

3 Signal Reconstruction Test

3.1 Experimental Process

For the designed method of fault signal reconstruction of treadmill equipment based on deep neural network, use it to reconstruct the fault signal of a treadmill equipment, and test the performance of the method. First, the fault source location method based on community structure is used to implement the fault location of the treadmill equipment, and then the fault signal acquisition platform is used to implement the fault signal acquisition. The collected fault signal is shown in Fig. 2.

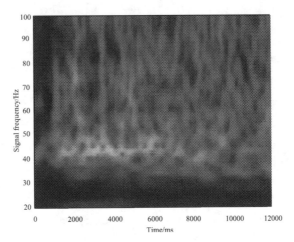

Fig. 2. Fault signal collected

The denoising model based on the two-layer recurrent neural network is used to implement the signal denoising. The signal after noise reduction is shown in Fig. 3.

Finally, the signal reconstruction of treadmill equipment fault is realized by matching pursuit algorithm. The reconstructed signal is shown in Fig. 4.

The time spent in testing the reconstructed signal and the signal-to-noise ratio of the reconstructed signal. In the test Motor fault signal reconstruction method based on dyadic wavelet transform and triangular spline interpolation. Compared with the reconstruction method of multi-sensor rolling bearing fault signal based on correlation principle, the method is tested together and represented by method 1 and method 2.

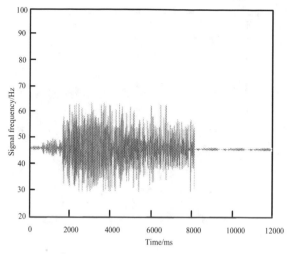

Fig. 3. Signal after noise reduction

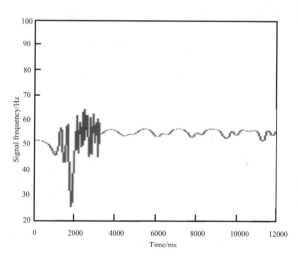

Fig. 4. Reconstructed signal

3.2 Reconstruction Time Test Results

The design method and the reconstruction time test results of method 1 and method 2 are shown in Fig. 5.

The reason is that this method constructs a noise reduction model based on the double-layer Recurrent neural network according to the deep neural network, realizes the noise reduction processing of the collected fault signal, and completes the signal reconstruction of treadmill equipment fault by combining the signal noise reduction processing results with the matching pursuit algorithm. Therefore, this method has lower reconstruction time and higher reconstruction efficiency.

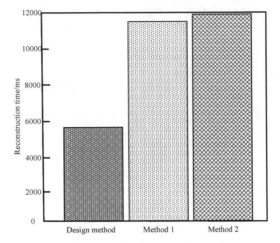

Fig. 5. Reconstruction time test results

3.3 Signal to Noise Ratio Test Results of Reconstructed Signal

The reconstructed signal SNR test results of the three methods are shown in Table 6.

Table 6. Signal to noise ratio of reconstructed signal of three methods

Iterations	Signal to noise ratio (dB)		
	Design method	Method 1	Method 2
5000	50.52	20.36	24.36
6000	50.30	20.01	24.18
7000	50.05	19.96	24.05
8000	49.92	19.64	23.94
9000	49.56	19.38	23.84
10000	49.30	18.97	23.64

According to the test results in the table above, the reconstructed signal signal-to-noise ratio of the design method reaches a minimum of 49.30 dB, while the signal-to-noise ratios of Method 1 and Method 2 are only 18.97 dB and 23.64 dB, respectively. The signal-to-noise ratio of the design method is lower. The signal to noise ratio of the reconstructed signal of the design method is up to 50.52 dB, while the signal to noise ratio of the method 1 and method 2 is only 20.36 dB and 23.64 dB. The reason is that the design method builds a noise reduction model based on the double-layer recurrent neural network according to the depth neural network to achieve the noise reduction of the collected fault signal, so the signal to noise ratio of the reconstructed signal of this method is always kept at a high level.

4 Conclusion

The fault signals of treadmill equipment contain a large amount of noise components, which increases the difficulty of signal reconstruction. Therefore, deep neural networks are introduced to design a new method for reconstructing fault signals of treadmill equipment. This method mainly involves constructing a denoising model based on a dual-layer recurrent neural network using deep neural networks, performing denoising processing on the collected fault signals, and completing the signal reconstruction of treadmill equipment faults using a matching tracking algorithm. The superiority and effectiveness of this method are highlighted through relevant comparative experiments. In summary, the reconstruction of fault signals in treadmill equipment not only improves fault diagnosis accuracy and maintenance efficiency but also promotes the development of intelligent maintenance technology. However, there are inherent limitations such as limited knowledge and insufficient research depth. Therefore, there are still many shortcomings in related research work that need to be further studied and optimized.

References

1. Wang, H., Cheng, Y., Tian, Z.: A new signal fault detection algorithm for vector tracking loop in strong noise environments. J. Northwest. Polytech. Univ. **40**(2), 323–329 (2022)
2. Jiang, J., Cong, X., Li, S., et al.: A hybrid signal-based fault diagnosis method for lithium-ion batteries in electric vehicles. IEEE Access **9**, 19175–19186 (2021)
3. Hu, H.Y., Wang, C., Liu, R.D., et al.: Speed sensorless control of induction motor based on stator current series model. Comput. Simul. **38**(04), 177–181 (2021)
4. Gao, F., Li, F., Wang, Z., et al.: Research on multilevel classification of high-speed railway signal equipment fault based on text mining. J. Electric. Comput. Eng. **2021**(2), 1–11 (2021)
5. Li, F., He, Z., Zhang, L., et al.: Analytical model and spectral characteristics of acoustic emission signal produced by localized fault of rolling element bearing. J. Northwest. Polytech. Univ. **39**(4), 831–838 (2021)
6. Teguia, J.B., Kammogne, A.S.T., Ganmene, S.G.T., et al.: Fuzzy-enhanced robust fault-tolerant control of IFOC motor with matched and mismatched disturbances. Math. Found. Comput. **5**(4), 295–314 (2022)
7. Shi, L., Zhu, Y., Zhang, Y., et al.: Fault diagnosis of signal equipment on the Lanzhou-Xinjiang high-speed railway using machine learning for natural language processing. Complexity **2021**(8), 1–13 (2021)
8. Ren, Z., Wu, G., Wu, Z., et al.: Hybrid dynamic optimal tracking control of hydraulic cylinder speed in injection molding industry process. J. Ind. Manag. Optim. **19**(7), 5209–5229 (2023)
9. Ives, Z.G.: Solving the signal reconstruction problem at scale: technical perspective. Commun. ACM. ACM **64**(2), 105 (2021)
10. Li, Z., Wu, H., Cheng, L., et al.: Infrared and visible fusion imaging via double-layer fusion denoising neural network. Digit. Sig. Process. **123**(1), 103433–103445 (2022)
11. Wen, Z., Wang, H., Gong, Y., et al.: Denoising convolutional neural network inspired via multi-layr convolutional sparse coding. J. Electron. Imaging **30**(2), 023007–023027 (2021)
12. Han, Z.: Nonlinear model predictive control of single-link flexible-joint robot using recurrent neural network and differential evolution optimization. Electronics **10**(19), 2426–2437 (2021)

13. Chen, C., Tian, Y., Gao, Q., et al.: Frequency-division abnormal amplitude attenuation after data reconstruction based on random function and its application in the very thick loess tableland area, ordos basin. Geophys. Prospect. Petrol. **58**(5), 741–749 (2022)
14. Han, C.: The analysis about compressed sensing reconstruction algorithm based on machine learning applied in interference multispectral images. Adv. Multimedia **21**(Pt.1), 1–6 (2021)
15. Polat, N., Kayhan, S.K.: FPGA implementation of LSD-OMP for real-time ECG signal reconstruction. Turk. J. Electr. Eng. Comput. Sci. **29**(4), 1887–1907 (2021)

Intelligent Library Educational Information Digital Resources Retrieval Based on Ant Colony Algorithm

Xu Wang[1]($\boxtimes$) and Mingjie Zheng[2]

[1] Tianjin Maritime College, Tianjin 300150, China
seohee1428@163.com
[2] College of Humanities and Information, Changchun University of Technology, Changchun 130122, China

Abstract. In order to improve the service quality of the smart library, the ant colony algorithm is used to optimize the design of the retrieval method of educational information digital resources of the smart library. Collect educational information digital resources of smart libraries, complete preprocessing and integration operations for different types of digital resources, and extract the characteristics of digital resources. Establish a search engine for educational information digital resources. With the support of the search engine, use ant colony algorithm to determine the optimal path of resource retrieval, and get the final retrieval results of educational information digital resources. Through the performance test experiment, it is concluded that the precision and recall of the optimized design method are increased by 7.35% and 10.55% respectively, and the retrieval response time is significantly shortened compared with the traditional method.

Keywords: Ant colony algorithm · Smart library · Educational information · Digital resource retrieval

1 Introduction

Smart library is a virtual electronic library based on computer and Internet technology. It provides readers with books, documents and other materials by means of online communication on the Internet. Smart libraries are characterized by digitalization of collection, computerization of operation, networking of transmission, liberalization of information storage, resource sharing and structural connectivity. The elements of digital library are: digital resources are the basis of materials, network access is the basis of existence, and distributed management is the advanced stage of development. The smart library is based on the idea of people-oriented public welfare and benefiting the people, so that every reader can get the same space reading and learning solutions, and enjoy the convenience and convenience brought by the smart library [1]. The establishment and development of smart libraries have brought about earth shaking changes in the study and work of readers and librarians. Therefore, the smart library based on library + Internet of Things

L. Yun et al. (Eds.): ADHIP 2023, LNICST 547, pp. 255–271, 2024.
https://doi.org/10.1007/978-3-031-50543-0_18

+ cloud computing + intelligent equipment can, on the one hand, achieve more efficient management for librarians on the basis of intelligence and autonomy, and can also provide quick and convenient information query, reading and other comprehensive services for readers. The retrieval of educational information digital resources is one of the important functions of smart libraries. Digital resources are Literature information. One form of expression of computer technology, communications technology and multimedia technology. The sum of information resources released, accessed and utilized in digital form formed by mutual integration. Resource information retrieval is a process in which people search for the required information from the World Wide Web, local area network or specific resource information database using specific search instructions, search terms and search strategies on computers or computer search network terminals, and then display or print the information by terminal devices.

At present, the retrieval methods of educational information digital resources used in smart libraries mainly include: intelligent library resource retrieval methods based on deep learning, resource retrieval methods based on Internet of Things technology, and library resource retrieval based on CGSP. However, with the support of the above resource retrieval methods, the current smart library has low hit rate of resource retrieval Ant colony algorithm is introduced to solve the problems of incomplete resource retrieval and slow retrieval speed.

Ant colony algorithm, also known as ant algorithm, is a probabilistic algorithm used to find the optimal path in the graph. As a general stochastic optimization method, the ant colony algorithm initially only randomly selects the search path and does not need any prior knowledge. With the understanding of the solution space, the search becomes regular, and gradually approaches to the global optimal solution. Ant colony algorithm is a kind of simulated evolutionary algorithm. Preliminary research shows that the algorithm has many excellent properties. The ant colony algorithm is used to optimize the retrieval method of educational information digital resources of the smart library, so as to improve the retrieval performance of digital resources and improve the service quality of the smart library. Using a smart library as a platform, collect local resources, cooperative resources, and non cooperative resources, and segment, encode, and compress the collected resources to complete resource integration and improve retrieval efficiency. Information gain technology is used to evaluate the amount of information and extract resource features. In order to further improve the effectiveness of resource retrieval, a search engine is composed of information collector, analysis indexer, searcher and query interface to analyze the extracted information features, and Ant colony optimization algorithms is used to determine the digital resource retrieval path to achieve efficient resource retrieval.

2   Design of the Retrieval Method of Library Educational Information Digital Resources

The operation process of optimizing and designing the retrieval method of educational information digital resources of the smart library is shown in Fig. 1.

First, collect all the retrieval objects and build a centralized local text document library. Then we extract the text from the local text document library [2], extract the text

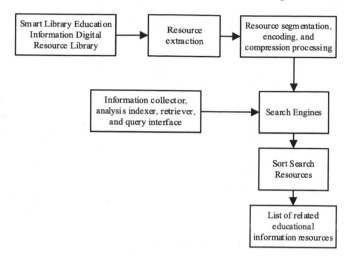

Fig. 1. Library Educational Information Digital Resources Retrieval Flow Chart

string, and preprocess the text, then we can build the index of the document. After the text document is indexed, it can be retrieved [3]. The user first submits the query to the retrieval terminal, which then directly accesses the index. After the retrieval terminal quickly obtains the collection of documents related to the query through the index, the sorting system evaluates the sorting and returns the results to the user.

2.1 Collecting Digital Educational Information Resources of Smart Library

Take the smart library as the resource collection platform, collect different types of educational information digital resources, and take this as the goal of resource retrieval. The educational information digital resources of smart libraries can be divided into three types: local resources, resources with cooperative relationships, and resources without cooperative relationships. The collection of local resources is mainly manual, and resources with cooperative relationships are mainly targeted at database suppliers. The database suppliers should build metadata that meets the specifications and standards on the premise of meeting the interoperability agreement. It should also be able to open metadata to buyers based on cooperation agreements, so as to provide retrieval functions and complete local indexing and storage operations [4]. Resources without partnership can be generated by combining manual filling and automatic generation. Finally, the collection results of educational information digital resources of the smart library are as follows:

$$E = \frac{E_{cpu} + E_{ram}}{\omega_{cpu} + \omega_{ram}} \tag{1}$$

Among E_{cpu} and E_{ram} Digital resources in CPU and memory environments, ω_{cpu} and ω_{ram} It corresponds to the weight value of CPU and memory. According to the above method, we can get the collection results of educational information digital resources of the smart library.

2.2 Integration and Processing of Educational Information Digital Resources of Smart Library

According to the resource storage format, the initially collected digital educational information resources of the smart library can be divided into text resources, image resources, audio resources, video resources and other types. Before resource integration, the digital educational information resources need to be preprocessed first. The pre-processing of text resources mainly includes Chinese word segmentation, text coding, text compression, etc. It is assumed that the collected text digital resources are B In the process of Chinese word segmentation, the string variable storing the segmentation result is recorded as C, temporary string is A, first compare the length of the Chinese string to be segmented and the size relationship of the set maximum comparison length, get the smaller value of the two, and record it as C' [5]. If the length of the string to be segmented is 0, the final segmentation result of the string will be returned directly; otherwise, the length from the head of the Chinese string to be segmented is C' Substring of A, search in the dictionary. If the string cannot be found in the dictionary C' Value minus 1, remove A For the rightmost word, use the formula 2 pairs after completing the above steps B Reassign, and the assignment result is:

$$B = B - A \tag{2}$$

The string to store the result C The assigned value is:

$$B = B + C' + " /" \tag{3}$$

Reassign the temporary storage string as an empty string, return the final segmentation result of the string, and the word segmentation of the digital text resources of the education information of the smart library is completed. In addition, the process of text encoding and compression can be expressed as follows:

$$\begin{cases} B_{code} = B \cdot \kappa_{code} \\ B_{compress} = B \cdot \kappa_{compress} \end{cases} \tag{4}$$

Variables in Formula 4 κ_{code} and $\kappa_{compress}$ They are the coding and compression coefficients of text resources. The processing of image information resources includes rotation, scaling, mirroring, smoothing, enhancement, restoration, compression, reconstruction, feature extraction and recognition. Taking rotation operation as an example, the processing results are as follows:

$$\begin{cases} x' = a + (x - a) \cos\theta - (y - b) \sin\theta \\ y' = b + (y - b) \cos\theta - (x - a) \sin\theta \end{cases} \tag{5}$$

among (a, b) and (x, y) Are the position coordinates of the rotation center and the initial image resource, θ Is the rotation angle. Image scaling is to reduce or enlarge the image or part of the image area. The inspection standard for image scaling is to minimize the spatial distortion of the transformed image. When scaling, it is generally to find one or several pixels [6] in the original image corresponding to each pixel in the target image by reverse mapping. Then, the adjacent pixel replacement method can be used to complete the addition or reduction, or the interpolation calculation method can be used to process. Similarly, the pre-processing of audio resources and video resources can be realized. On this basis, the integration result of digital resources of educational information in smart libraries is as follows:

$$E_{\text{integration}} = \sum E'_i \cdot \omega_i \qquad (6)$$

Among E_i and ω_i Respectively represent the i Preprocessing results and weight values of educational information digital resources of smart libraries. Thus, the integration of digital resources of educational information in the smart library is completed.

2.3 Extracting the Characteristics of Educational Information Digital Resources

In order to facilitate the retrieval of educational information digital resources, the characteristics of integrated information digital resources are used as retrieval tags, so it is necessary to extract the characteristics of all educational information digital resources. Among them, the feature extraction objects of digital text resources are document frequency, information gain, etc. The document frequency of an entry refers to the number of documents in which the entry appears in the training corpus [7]. The adoption of feature extraction is based on the following basic assumptions: entries whose values are lower than a certain appendix value are low-frequency words, which do not contain or contain less category information. Removing such terms from the original feature space can not only reduce the dimension of the feature space, but also improve the accuracy of classification. The extraction result of document frequency characteristics is:

$$\tau_{\text{document frequency}} = \frac{m_\tau}{m_{total}} \qquad (7)$$

In the above formula m_τ and m_{total} Corresponds to the number of documents with feature words and the total number of documents in the training set. Information gain is an important concept in information theory, which is widely used in text classification and information retrieval of machine learning. Information gain can be used to evaluate the amount of information reflected by word attributes in the process of sample classification. Suppose that the current educational information digital text resources can be divided into r Categories, then entries t The formula for calculating the information gain characteristics of is as follows:

$$\tau_{\text{Gain}} = -p(\bar{t}) \sum_{i=1}^{n_r} p(r_i|\bar{t}) \lg p(r_i|\bar{t}) \qquad (8)$$

Variables in the above formula $p(\bar{t})$ The probability that the word t does not appear, $p(r_i|\bar{t})$ Indicates that the text belongs to the r_i Probability of class, n_r Is the number of extracted words. All terms in the training sample set should calculate their information gain value in advance, and select the attribute with the highest information gain as the feature to determine the importance of the attribute set for classification. The larger the value, the more important it is to classification and the greater its contribution. If the information gain of lexical attributes is greater, the classification of classes will also play a greater role in the process of text information classification [8]. Therefore, in the process of feature selection, words with large information gain value are usually selected as the basis for feature selection. Similarly, we can get the feature extraction results of digital image, audio and video resources of educational information of smart library, and finally mark the comprehensive features of each educational information resource as $\tau_{con}(i)$.

2.4 Establish a Search Engine for Educational Information Digital Resources

The search engine is mainly composed of four parts, namely information collector, analysis indexer, searcher and query interface. The structure of the search engine built in the optimized design of the intelligent library's educational information digital resource retrieval method is shown in Fig. 2.

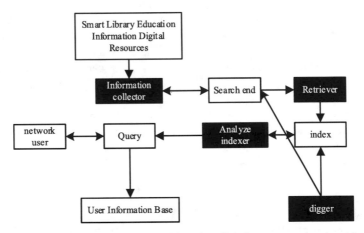

Fig. 2. Structure of educational information digital resources search engine

The operation of the information collector can be abstracted as a directed graph traversal process. When the information collector is running, as long as a small number of user configured initial smart library web pages are provided, the information collector can obtain new smart library web pages and hyperlinks according to certain algorithms, and roam in the smart library along these hyperlinks to collect information resources. The information collector visits the smart library periodically, usually once or several times a month, and the number of visits depends on the update frequency of the smart library. The performance of the information collector greatly affects the scale of the search engine site. The main function of the analysis indexer is to analyze the collected information and establish an index library for query. Analysis indexer can be divided into two parts, namely analyzer and indexer [9]. First, according to the characteristics of online data, the analyzer analyzes the collected web pages and hyperlink information according to specific algorithms, extracts the web page description information related to user retrieval, and then the indexer extracts the index items from the abstract data of the analyzed web pages to build an index. The specific index creation process is shown in Fig. 3.

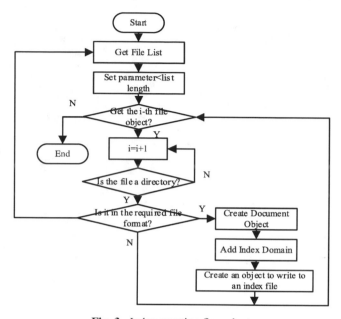

Fig. 3. Index creation flow chart

Index items are divided into objective index items and content index items. Objective index items are independent of the semantic content of the document, such as author name, URL, update time, code length, link frequency and other content index items reflect the content of the document, such as keywords and their grade values, phrases, words, etc. Content index entries include single word index entries and phrase index entries. The function of the searcher is to receive and interpret the user's search request. According to the user's query, quickly check out the document in the index library, calculate the relevance between the web page and the search request, and sort the results to be output to achieve the user relevance feedback mechanism [10]. Query interface is an interface for users to use search engines.Its main function is to input user queries, display query results, and provide user relevance feedback mechanism, so that users can use the search engine to obtain useful information efficiently and variously. Query interface technology mainly includes search request technology, search result representation technology and user behavior analysis technology. In a modular way, the composition modules of the above search engine are integrated, and the results of the establishment of the educational information digital resources search engine are obtained.

2.5   Using Ant Colony Algorithm to Determine Digital Resource Retrieval Path

In the storage space of educational information digital resources in the smart library, the retrieval path of digital resources is obtained through ant colony algorithm using the established educational information digital resources search engine. Figure 4 shows the basic operation principle of the ant colony algorithm.

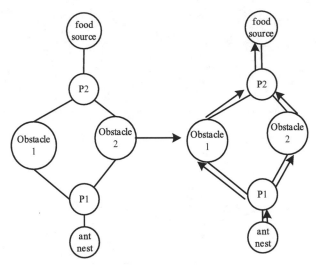

Fig. 4. Schematic diagram of ant colony algorithm

Node in Fig. 4 P_1 and P_2 Choose the way forward for Ma Nu.Ant colony algorithm mimics the process of ants searching for food through pheromones in nature. Its essence is to use the positive feedback characteristics of pheromones to make ant colony algorithm continuously self correct until it finds the optimal solution to the problem. The logic structure of the ant colony algorithm is shown in Fig. 5.

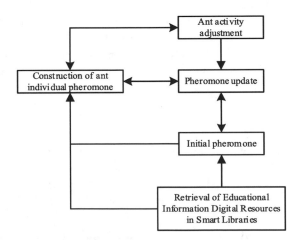

Fig. 5. Logic structure of ant colony algorithm

In the process of determining the path of digital resource retrieval, the path calculation process of ant colony algorithm can be divided into three steps: initialization, path selection, pheromone update. Nodes are set in the initialization phase i And nodes j The pheromone concentration between is 0, that is:

$$\lambda_{ij}(0) = \beta \tag{9}$$

In Formula 9 β Is a random number, that is, the path pheromone concentration between nodes at the initial time is the same, both are β. The content of path selection is to choose a path for ants, that is, to choose ants k The node to be reached in the next step. Ants k From node i Move to node j The calculation result of probability is:

$$P_{ij}^k = \begin{cases} \left[\dfrac{[\lambda_{ij}(t)]^\rho [f_{ij}(t)]^\sigma}{\sum\limits_{s \in G}[\lambda_{is}(t)]^\rho [f_{is}(t)]^\sigma}\right] & ,j \in G \\ 0 & , \text{otherwise} \end{cases} \tag{10}$$

among G It is a collection of educational information digital resources for smart libraries, $f_{ij}(t)$ For heuristic function, ρ and σ Is pheromone concentration factor and heuristic function factor, ρ It reflects the importance of accumulated information in the process of path selection, σ It reflects the importance of heuristic information in the process of selecting the path. Where heuristic function $f_{ij}(t)$ The expression of is:

$$f_{ij}(t) = \frac{1}{w_{ij}} \tag{11}$$

among w_{ij} Is the weight element in the data weight set, representing the node i And nodes j Distance between. In the path selection stage, according to the probability result calculated by Formula 10, from the collection of educational information digital resources of the smart library G Search for accessible nodes in the. Every time one is found, the node to which you want to search will be changed from G The process continues $n - 1$ Until all the educational information digital resources of the smart library have been traversed. In the pheromone update phase, when ants k When nodes in the path are discovered, pheromones will be left on the path between nodes. At the same time, with the passage of time, the pheromone left by ants on the path will also volatilize. When all ants have completed the traversal of nodes, the pheromone concentration on the path between nodes needs to be updated. Assume that the traversal completion time is $t + m$, the pheromone update result is:

$$\lambda_{ij}(t + m) = \kappa_{\text{residue}} \lambda_{ij}(t) + \Delta\lambda_{ij}(m) \tag{12}$$

among κ_{residue} Is the pheromone residual coefficient, $\Delta\lambda_{ij}(m)$ Is the change of pheromone, and the calculation formula of this variable is as follows:

$$\begin{cases} \Delta\lambda_{ij}(m) = \sum\limits_{k=1}^{m} \Delta\lambda_{ij}^k(i,j) \\ \Delta\lambda_{ij}^k(i,j) = \begin{cases} \dfrac{Y}{L_k}, k \in l_{ij} \\ 0, else \end{cases} \end{cases} \tag{13}$$

Variables in Eq. 13 Y and L_k They are the total number of pheromones released by ants in one traversal, L_k For ants k The total length passed in one traversal, l_{ij} Is an edge ij. After the above process, the initial selection result of digital resource retrieval path can be obtained. In ant colony algorithm, only globally optimal ants are allowed to release pheromones [11]. The purpose of this selection, as well as the use of pseudo-random proportional rules, is to make the search process more instructive: ants' search mainly focuses on the field of the best path found by the current cycle. The global update is executed after all ants have completed their paths. The update result of the initially generated search path is:

$$\lambda(i, j) \leftarrow (1 - \kappa_{\text{volatilize}})\lambda(i, j) + \kappa_{\text{volatilize}} \cdot \Delta\lambda(i, j) \qquad (14)$$

Among $\kappa_{\text{volatilize}}$ It is a pheromone volatilization parameter. Only pheromones on the edge of the global optimal path will be enhanced.The optimal selection result of digital resource retrieval path is obtained.

2.6 Realize the Retrieval of Educational Information Digital Resources in Smart Libraries

In the process of intelligent library education information digital resource retrieval, first judge the credibility of the search results by the ant colony algorithm. The formula for measuring the credibility of the search path is as follows:

$$\psi = \frac{U}{\sqrt{D}} * \rho \qquad (15)$$

where U and D They are path and path distance.For dependability above threshold ψ_0 To calculate the similarity between digital resources and input keywords. The calculation result is:

$$\phi = \frac{\tau_{con}(i) \cdot \tau_{\text{keyword}}}{\|\tau_{con}(i)\| \cdot \|\tau_{\text{keyword}}\|} \qquad (16)$$

Variables in Eq. 16 τ_{keyword} If the calculation result of formula 16 is higher than the threshold value ϕ_0, Certification $\tau_{con}(i)$ The corresponding digital resource belongs to the resource retrieval result, otherwise it is considered that the resource does not belong to the retrieval result. Finally, the search results that meet the similarity conditions will be arranged in order from large to small, and the search results of the educational information digital resources of the smart library will be output in a visual form.

3   Experimental Analysis of Retrieval Performance Test

To test and optimize the retrieval performance of the intelligent library educational information digital resources retrieval method based on ant colony algorithm, design a performance test experiment. The basic idea of this experiment is to prepare the intelligent library educational information digital resources and retrieval keywords, and determine the number of similar resources corresponding to the retrieval keywords according to the preparation of the resources. This is used as the comparison standard to judge the retrieval performance of resources. By calculating the gap between the output retrieval results and the expected retrieval results, the quantitative test results reflecting the retrieval performance of the optimized design method are obtained.

3.1 Building an Experimental Test Environment

This experiment selects OverSim platform as the development environment. OverSim is an open source overlay framework based on OMNeT+ +.OverSim is highly modular, and its structure can be divided into three levels, namely underlay, overlay, and application from bottom to top. Underlay layer mainly constructs the underlying basic network, overlay layer mainly contains structured and unstructured running protocols, and application layer is the running program based on the application layer. Because OverSim has a strong graphical interface and real-time display capability, it can find errors in the running process of the retrieval program in the early stage of running. In addition to the development platform of intelligent library educational information digital resource retrieval method based on ant colony algorithm, it is also necessary to select hardware equipment. The optimization design method uses Dell server as the hardware support, Acer EX214 as the main computer of the retrieval method, and SQLServer2005 as the storage environment of intelligent library educational information digital resource samples.

3.2 Prepare Samples of Educational Information Digital Resources of Smart Library

This experiment selects all text and image education information in the smart library as digital resource samples. Text information includes the basic introduction of each book, such as book name, book author, publishing house, preface, catalog, book content, etc., while image education information mainly includes book cover images and electronic books. Through data statistics, the educational text information resources prepared for this performance test experiment are 65 GB in total, and the image resources are 55 GB in total.

3.3 Generating Digital Resource Retrieval Cases

According to the preparation of educational information digital resource samples of the smart library, search cases of digital resources are generated. The results of some search cases are shown in Table 1.

In the actual retrieval process, the retrieval keywords in the generated retrieval cases are input into the retrieval program. In order to ensure the credibility of the experimental results, the experiment generated a total of 80 search cases, and clearly marked the search keywords and the expected number of searches.

Table 1. Digital resource retrieval use case table

Case number	Search keywords	Search Type	Expected retrieval quantity/GB
1	Journey to the West	Text retrieval	16.4
2	Journey to the West	image retrieval	9.8
3	The Dream of Red Mansion	Text retrieval	12.6
4	The Dream of Red Mansion	image retrieval	5.2
5	Five thousand years in China	Text retrieval	13.7
6	Five thousand years in China	image retrieval	6.8
7	the Imperial Palace	Text retrieval	11.7
8	the Imperial Palace	image retrieval	7.1

3.4 Input Operation Parameters of Ant Colony Algorithm

Because the optimization design method uses the ant colony algorithm as the support, it is necessary to set the operation parameters of related algorithms. The details are shown in Table 2.

Table 2. Running parameters of Ant colony optimization algorithms

Parameter	Value
Volatility coefficient of Pheromone	0.5
Initial value of distance heuristic factor	1
Initial value of Pheromone heuristic factor	5
Total Pheromone	200
Ant number	20
Maximum survival cycle of ants	12
Path selection probability threshold	0.5

Input the setting results of the above ant colony algorithm running parameters into the running program corresponding to the intelligent library's educational information digital resource retrieval method.

3.5 Describe the Performance Test Experiment Process

Through the file configuration parameters, load the corresponding modules of the system to realize the steps of initial resource sample import, optimization design method coding, retrieval method switching, etc. Input the keywords in the generated search cases one by one into the running program of the intelligent library education information digital resource retrieval method based on the ant colony algorithm, and output the corresponding search results, including the output results of the No. 1 and No. 2 search cases, as shown in Fig. 6.

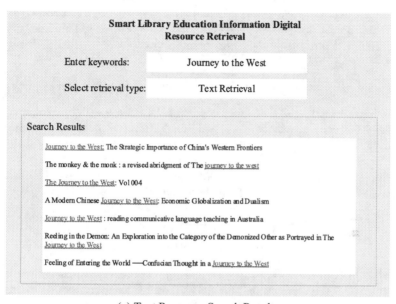

(a) Text Resource Search Results

(b) Image resource retrieval results

Fig. 6. Retrieval Results of Library Educational Information Digital Resources

The output results of all search cases can be obtained according to the above method. In order to reflect the advantages of the optimization design method in the resource retrieval performance, the traditional intelligent library resource retrieval method based on deep learning and the resource retrieval method based on the Internet of Things technology are set as the experimental comparison method, and the above process is repeated to realize the development and operation of the comparison method, and output the resource retrieval results of the comparison method.

3.6 Setting Experimental Indexes for Retrieval Performance Test

The precision ratio and recall ratio of the educational information digital resources of the smart library are set as the quantitative test indicators to verify the retrieval performance. The precision ratio is the percentage of the number of relevant documents detected and the total number of documents detected. Recall ratio refers to the amount of relevant documents detected and the total amount of relevant documents in the retrieval system percentage. The numerical results of the above indicators are as follows:

$$\begin{cases} \eta_{accurate} = \frac{N_{correct}}{N_{out}} \times 100\% \\ \eta_{all} = \frac{N_{correct}}{N_{all}} \times 100\% \end{cases} \tag{17}$$

Variables in Eq. 17 $N_{correct}$, N_{out} and N_{all} They respectively represent the amount of resource data retrieved correctly, the amount of resource data successfully output by the retrieval method, and the total amount of resource data prepared for the experiment. By comparing the output results of the retrieval method with the expected results, we can get $N_{correct}$ The specific value of.In addition, in order to verify the operation performance of the retrieval method of educational information digital resources of the smart library, the response time is set as a quantitative test indicator, and the test results of this indicator can be expressed as:

$$T = t_{out} - t_{retrieval} \tag{18}$$

among $t_{retrieval}$ and t_{out} It should be the start time of the retrieval program and the output time of the retrieval results. Finally calculate the precision $\eta_{accurate}$ And recall η_{all} The larger the value is, the better the retrieval performance of the corresponding method is, and the response time is T The smaller the size, the better the performance of the corresponding retrieval method.

3.7 Retrieval Performance Test Results

Through the statistics of relevant data, the test results reflecting the retrieval performance are obtained, as shown in Table 3.

By substituting the data in Table 3 into Formula 17, it is calculated that the average precision and recall of the two comparison methods are 89.5% and 92.6%, 84.7% and 89.0%, respectively, while the average precision and recall of the optimized design method are 98.4% and 97.4%, respectively. In addition, the response time test results of

Table 3. Results of Information Digital Resource Retrieval Performance Test

Case number	Intelligent Library Resource Retrieval Method Based on Deep Learning		Resource retrieval method based on Internet of Things technology		Retrieval Method of Educational Information Digital Resources in Smart Library Based on Ant Colony Algorithm	
	Retrieve the correct amount of resource data/GB	Amount of resource data successfully output by retrieval method/GB	Retrieve the correct amount of resource data/GB	Amount of resource data successfully output by retrieval method/GB	Retrieve the correct amount of resource data/GB	Amount of resource data successfully output by retrieval method/GB
1	15.1	15.8	15.5	16.0	16.2	16.3
2	8.7	9.3	9.0	9.5	9.6	9.8
3	11.1	11.7	11.2	12.1	12.5	12.6
4	4.1	5.0	4.4	5.0	5.0	5.1
5	12.7	13.3	13.1	13.4	13.4	13.6
6	5.2	6.2	5.4	6.3	6.4	6.7
7	10.4	11.1	10.5	11.2	11.3	11.5
8	5.1	6.6	6.2	6.8	7.0	7.0

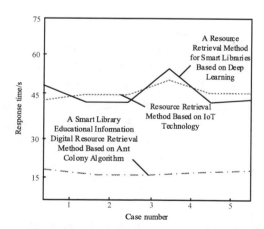

Fig. 7. Response Time Test Comparison Results of Resource Retrieval Methods

three retrieval methods are obtained through the calculation of Formula 18, as shown in Fig. 7.

From Fig. 7, it can be intuitively seen that the response time of the three algorithms is different. Among them, the response time of the comparison method is around 45 s, while the response time of the optimization design method is shorter and always below 20 s, indicating that the optimization design method has obvious advantages in terms of operational performance.

4 Conclusion

Smart library is an intelligent building formed by applying intelligent technology to library construction. Intelligent building with highly automated management Digital Library Organic combination and innovation. Smart library is a concept that is not limited by space, but can be actually perceived at the same time. Someone once said that the smart library will realize intelligent service and management through the Internet of Things. In fact, it also includes some cloud computing and intelligent equipment, through which to transform the traditional library. In order to improve the service quality of the smart library, this paper puts forward a retrieval method of educational information digital resources of the smart library based on ant colony algorithm, and applies it to the actual library environment. Collect educational information digital resources using a smart library as a platform, integrate the collected resources, extract resource features using information augmentation technology, establish a search engine to analyze the extracted information features, and use ant colony algorithm to achieve efficient resource retrieval. From the experimental results, we can see that compared with traditional methods, the retrieval quality and retrieval speed of the optimized design method have been significantly improved, which is of positive significance for the operation of smart libraries.

With the continuous updating of educational resources, the retrieval of educational information digital resources in smart libraries may involve multiple objectives or constraints. In future research, Ant colony optimization algorithms can be extended to multi-objective optimization and multi constraint problems to achieve more flexible and comprehensive resource retrieval and provide appropriate resources for teachers and students.

Aknowledgement. "Theory and Practice Research on the Integration and Aggregation Ability of Digital Resources in Higher Vocational Colleges under the Background of Smart Campus", Tianjin Higher Vocational Education Research Association 2022 (project approval number: 2022-H-093).

References

1. Li, H.: Internet tourism resource retrieval using pagerank search ranking algorithm. Complexity **2021**(1), 1–11 (2021)
2. Liu, S., He, T., Dai, J.: A survey of CRF algorithm based knowledge extraction of elementary mathematics in Chinese. Mobile Netw. Appl. **26**(5), 1891–1903 (2021)
3. Mothe, J.: Analytics methods to understand information retrieval effectiveness—a survey. Mathematics **10**(2135), 2022 (2022)
4. Lechtenberg, F., Farreres, J., Galvan-Cara, A.L., et al.: Information retrieval from scientific abstract and citation databases: a query-by-documents approach based on Monte-Carlo sampling. Expert Syst. Appl. **199**, 116967 (2022)

 5. Verberne, S., Zwenne, G.J., Loon, W.V., et al.: Exploration of domain relevance by legal professionals in information retrieval systems. Leg. Inf. Manag. **22**(1), 49–67 (2022)
 6. Dobreski, B., Zhu, X., Ridenour, L., et al.: Information organization and information retrieval in the LIS curriculum: an analysis of course syllabi. J. Educ. Libr. Inf. Sci. **3**, 63 (2022)
 7. Nguyen, H.D., Tran, T.V., Pham, X.T., et al.: Design intelligent educational Chatbot for information retrieval based on integrated knowledge bases. IAENG Int. J. Comput. Sci. **49**(2), 531–541 (2022)
 8. Yang, G., Liu, K., Chen, X., et al.: CCGIR: information retrieval-based code comment generation method for smart contracts. Knowl.-Based Syst. **237**, 107858 (2022)
 9. Pennington, D.R., Haynes, D.: Metadata for information management and retrieval: understanding metadata and its use. J. Librarianship Inf. Sci. (4), 53 (2021)
10. Zhang, J.: Application of Data mining based on computer algorithm in personalized recommendation service of university smart library. In: Journal of Physics: Conference Series, vol. 1955, no. 1, 012008 (2021)
11. Shu, Z., Jiang, Y., Liu, J., et al.: Analysis of mobile push service model of smart library based on big data. In: Journal of Physics Conference Series, vol. 1883, no. 1, p. 012055 (2021)

Mobile Education, Mobile Monitoring, Behavior Understanding and Object Tracking

A Prediction Method of Students' Output and Achievement in Higher Vocational English Online Teaching Based on Xueyin Online Platform

Dan Wang[✉] and Lihua Sun

Liaoning Petrochemical College, Jinzhou 121000, China
wdd6162023@126.com

Abstract. By predicting students' achievements in advance, teachers and administrators can make and improve teaching plans, optimize teaching resources and improve teaching results in advance. Aiming at the problem of insufficient accuracy of traditional prediction methods, this paper studies a prediction method of students' output scores in online English teaching in higher vocational colleges based on the online platform of Xueyin. Collect and sort out student behavior data on the online platform of Xueyin, and divide learning behavior into two categories, namely, student basic data and online learning behavior data. Implement data cleaning, data transformation and missing value filling for student behavior data. The attention mechanism is introduced into LSTM to build an Att-LSTM prediction model. The attention mechanism helps LSTM quickly filter out important information from a large number of feature data, focus LSTM on the information that is most helpful for completing the current task, and improve the prediction accuracy of the model by filtering out unimportant data. The results show that the average absolute error and root mean square error are smaller and the coefficient of determination is larger under the application of the research method, which shows that the research prediction method has good effect and higher accuracy in predicting student performance.

Keywords: Xueyin Online Platform · Vocational English · Student Output Performance · Att-LSTM Prediction Method

1 Introduction

With the popularization of the Internet and the progress of information technology, more and more groups accept the teaching and sharing of knowledge through digital information. The new coronal epidemic in early 2020 further pushed online teaching to the track of mainstream teaching mode. Online teaching is no longer only the main way for primary and secondary students to cram lessons, but also the inevitable choice for colleges and universities to change the teaching mode under the epidemic. At present, there are

L. Yun et al. (Eds.): ADHIP 2023, LNICST 547, pp. 275–290, 2024.
https://doi.org/10.1007/978-3-031-50543-0_19

many network teaching systems for teachers to teach courses through the network. However, online teaching is not as intuitive as classroom teaching, which leads to teachers' less accurate judgment of students' learning than classroom teaching. Therefore, how to effectively analyze the effect of online teaching has always been a hot issue in teaching research. Many researches on online teaching show that the data related to recording students' learning paths and learning performance can be used as an important basis for teachers to evaluate students' learning effectiveness and diagnose students' learning difficulties. These data have valuable research value [1]. Therefore, the effective evaluation of students' academic performance is an urgent problem to be studied. This research proposes an integrated analysis algorithm, which processes these heterogeneous large-scale learning records and integrates multiple perspectives to analyze these learning record information, so as to identify students' learning behavior, and predict students' possible learning effects according to their current learning situation. So that teachers can provide auxiliary teaching strategies to students who may have learning difficulties according to these predicted information.

At present, relevant literature has conducted research on methods for predicting student grades. Reference [2] applies machine learning technology to educational systems, constructs a prediction and analysis model for student academic performance data, and develops a data analyzer to predict students' future academic performance by analyzing their academic performance data. However, the information mined by this method is not accurate and comprehensive enough, and the prediction accuracy needs to be further improved. Reference [3] uses deep learning techniques, CNN models are used to extract local features, and LSTM models consider the advantages of global text order. By classifying educational texts on online learning platforms and analyzing fine-grained emotional tendencies, factors that affect academic performance are explored to achieve prediction of students' online learning performance. But this model requires more parameters, which means that deep learning requires more training data, so it is not suitable as a universal algorithm.

This research proposes a prediction method of students' output performance in online English teaching in higher vocational colleges based on the online Xueyin platform. Through students' learning behavior data on the online Xueyin platform, we can observe students' various behavior performances, and let students' learning behavior data speak, so as to find out students' learning behavior that has greater relevance to students' learning effect after learning. Finally, a performance prediction model is constructed to study the behavioral characteristics that affect students' performance.

2   A Study on the Prediction of Students' Output in Online English Teaching in Higher Vocational Colleges

2.1   Collection of Student Behavior Data on the Online Platform of Xueyin

Student achievement prediction is a process of discovering the potential effective information in teaching data through mining and analysis based on the data of curriculum, students' historical achievements, students' behavior, etc., and then evaluating and inferring students' performance in the future learning stage. Based on this, data collection is the first step of performance prediction.

"Xueyin Online" platform is developed and operated by Beijing Xueyin Online Education Technology Co., Ltd., a wholly-owned subsidiary of Superstar Group Co., Ltd. It is a new generation of open learning platform based on the concept of credit bank jointly launched by Superstar Group and National Open University, and a public platform for higher education, vocational education and lifelong education, It is also one of the selection and operation platforms of national high-quality online open courses. "Xueyin Online" provides a new educational structure and learning mode characterized by "lifelong learning", "ubiquitous learning" and "future learning" for social learners, builds personal end learning files based on credit banks that realize the storage, certification, accumulation and conversion of learning achievements, and helps build a learning society. "Xueyin Online" provides learners with diversified and personalized learning services by integrating a large number of high-quality digital learning resources and courses from various schools and educational institutions, and authenticates, accumulates and converts in accordance with a unified learning achievement framework and standard to achieve the goal of "learning without boundaries, credits can be accumulated, achievements can be converted, and quality can be trusted". The online Xueyin platform provides support for online English teaching in higher vocational colleges. Its teaching scheme covers three teaching links before, during and after class. Pre class preparation, student preview, online teaching and software practice operation in class, review, assessment and teaching evaluation after class, through this whole process, students can realize the integration of knowledge. At the same time, the online platform's real-time data collection, cloud processing analysis and real-time feedback of results are conducive to the innovation and reform of teachers' teaching mode and teaching organization form, thus forming a complete online platform curriculum framework (see Fig. 1).

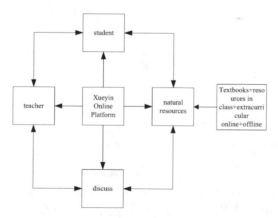

Fig. 1. Course framework of online platform of Xueyin

It can be seen from this that the online platform of Xueyin has completely saved the whole learning process of students, including their basic information, student number, name, social occupation, initial education background, learning terminal, learning duration, number and score of physical examination, credit, number of students' posts and replies, etc.; The database of the system records the interaction between students and the platform server in detail, which can truly and completely record the online learning process of students, and can obtain a large amount of student data at any time.

This paper collects students' learning behavior data through our school's online teaching platform, sorts out and classifies the factors that may affect students' performance according to the collected data, and then divides the learning behavior into two categories. The first category is students' basic data: student student number; The second category is the online learning behavior data of students: task point completion rate, video task point completion rate, video viewing duration (minutes), chapter test completion rate, chapter test average score, chapter learning times, homework completion rate, homework average score, course interaction volume, sign in completion rate, and score, of which the score is the score prediction category.

In order to facilitate the subsequent performance prediction and analysis, the online learning behavior data of students are described, as shown in Table 1.

Table 1. Characteristics and Meaning of English Online Learning Behavior

Characteristics of learning behavior	meaning
Completion amount of video task points	Number of completed watching videos
Video viewing duration	Total time spent watching videos
Chapter quiz completion count	Number of completed chapter tests
Number of completed assignments	Number of assignments completed
Chapter learning frequency	Number of learning times for each chapter
Number of completed check-in	Number of times students attend classes on time
Course interaction volume	Number of interactions between teachers and students
Chapter Test Average Score	Average score of all chapter tests
Average homework score	Average score of all assignments
Final exam	Final Exam Total Score

The student performance data of higher vocational colleges is saved in the school's educational administration system, which stores the scores of each student and each course as well as their comprehensive rankings. With the consent of relevant departments, this paper obtained the final scores of students in 7 different majors through the school educational administration system of the school. Student performance data includes the examination results, classroom results and credits of each course obtained by students in different courses. The fields in the obtained performance data include academic year, semester, student student number, course code, examination results, classroom results and credits. For the score information, this paper obtained the final exam scores of all students from the educational administration system with the approval of the relevant departments of the school. The score information generally includes the student's ID, name, course name, course hours, course credits, assessment methods, assessment scores and other information [4]. In this paper, the test scores are summarized for the purpose of studying the relationship between students' online behavior on campus and their scores.

Next, in order to build a more effective performance prediction model, we must further refine and process these data, and analyze whether each learning behavior feature is related to the performance and the correlation between each behavior feature. For the applicability of data types and correlation coefficients involved in this study, three major statistical correlations are selected.

The Pearson correlation coefficient, one of the coefficients, is shown in Formula (1) to analyze the relevant attributes of this article and verify the analysis results.

$$s = \frac{n \sum x_i C_i - \sum x_i \sum C_i}{\sqrt{n \sum x_i^2 - \left(\sum x_i\right)^2} \sqrt{n \sum C_i^2 - \left(\sum C_i\right)^2}} \tag{1}$$

where, x_i Represent sample data, such as age, learning duration, etc.; C_i Represent the learning effect, i.e. the end result; n Number of representative samples; s Represents the correlation coefficient between the sample data and the final grade. correlation coefficient s The size of represents the correlation between the sample data and the final grade, which is linear and most commonly used [5].

2.2 Data Pre-Processing

The obtained original data has many types, a wide range of sources and a large amount of data, which makes it inevitable that there are some problems in the data, such as data missing, data exceptions, etc. Data pre-processing refers to the process of filtering, reviewing and integrating the acquired data before conducting research and operation. In order to ensure the reliability of data mining results, the data is preprocessed to eliminate the redundancy, inconsistency and other problems in the original data. Data cleaning, data transformation, missing value filling, etc. are common methods of data preprocessing [6].

(1) Data cleaning

In the process of data collection, the preliminary work is only based on the teaching business of the school and the needs of teachers and students.

Combine and improve, integrate and process the data according to the actual meaning of the data fields of the three party data sources, and the data will be.

The integrated data is cleaned and improved with the research of this paper. Due to the existence of null value, data noise, data redundancy and other data format problems, it is not suitable for data mining [7]. Due to the particularity of online academic education, the courses chosen by students are also very different. Some students fail to take the final exam due to reasons such as dropping out, repeating a grade, or not registering, or some students fail to take the final exam due to other reasons such as work. Although they have student information in the information database, their scores are all empty, so students who repeat a grade or drop out only consider their scores, All absent students fill in the score as default, which is used to smooth the noise data. The general principles of data cleaning in this paper are as follows:

When selecting data mining objects, because there are many courses offered by schools and most people choose different courses, the selected data are the public courses selected by most people and the corresponding student and learning information.

The students who lack the final grade of the course are meaningless for analyzing learning behavior and constructing the grade prediction model, so they are deleted directly. The data set in this paper excludes invalid data of more than 10000 people,

Due to the special reasons of online academic education, the scattered places where students sign up, and the different branch campuses, these data that have nothing to do with their scores are directly deleted. This is based on the results of educational psychology and research on the factors affecting their scores in the previous research, such as credits, semesters, branch campuses, etc. The combination of manual and script procedures is mainly used to reduce the data dimension.

(2) Data transformation

Most of the data in the original dataset belong to different unit magnitudes or have different dimensions, such as gender, region, pre school education and other attributes. Or because the data types are different, these data cannot be calculated. In view of these problems, we use numerical methods to conduct numerical operations on character data. The main work is to conduct numerical processing on character data, standardize the data, and summarize and overview the data,A part of data standardization and summary has been done, so data normalization [8] is mainly carried out in this section. After comparing a large number of literature, we choose the minimum maximum normalization method in this paper, so that the normalized data fall within the 0–1 interval, which is convenient for subsequent data analysis and mining. For example, students' age, online learning duration, learning times, physical examination completion, number of posts and replies and other relevant data, see formula (2) for conversion.

$$x_i = \frac{\dot{x}_i - \min \dot{x}}{\max \dot{x} - \min \dot{x}} \tag{2}$$

where, x_i is the normalized data value, $\dot{x}_i$ Is the original data value of this type, $\min \dot{x}$ is the maximum value in the sample, $\max \dot{x}$ is the maximum value in the sample.

(3) Missing value filling

Due to the particularity of distance education, adult education and school running history, it is inevitable that there is no data or missing data in the integrated data, such as the students' score ratio, course completion ratio, learning times, online days, number of posts and replies, and learning duration. For individual students, when there are multiple missing learning data, based on the method of ignoring tuples, all relevant information of the student is directly eliminated without analysis. For students with few missing attributes, multiple interpolation method is used to fill the missing values [9]. For some attributes, the overall missing value has exceeded 50%. According to previous research, it is considered that they are not worth studying in this dataset.

2.3 Performance Prediction Model

The research on the prediction model of student output performance should first clarify the purpose, then carry out correlation analysis from the perspective of students, extract the independent variables of the correlation function, and finally establish the relationship between students and characteristic behaviors. Under the guidance of the previous theoretical research, based on the learning behavior selected in Table 1, as the input of the performance prediction model, students' final scores are taken as the output results of the prediction model. Similarly, the input-output relationship of the final score prediction model is as follows (3).

$$Y(x) = \{x_i | i = 1, 2, ..., 9\} \tag{3}$$

This paper takes students' achievement data and curriculum knowledge points as the main research object, and analyzes the collected student achievement data by constructing a student achievement prediction model. First, preprocess the test score data accumulated by students, and calculate all the change indicators [10] of students in this test with each test as the basic unit. In this paper, the problem of student achievement prediction is regarded as a time series prediction problem, and a short-term memory artificial neural network achievement prediction model integrating attention mechanism is proposed. By integrating the attention mechanism into the long-term and short-term memory artificial neural network time series model, the model can screen key information from a large number of feature data, so as to focus on the feature information that is helpful for students' performance prediction tasks.

Similar to the commonly used neural network, attention mechanism is a method of using neural network to process input data. Attention mechanism is derived from the inspiration generated by the observation process of things. People will focus on a specific position and ignore other relatively unimportant parts. Similarly, the attention mechanism will focus on the key areas after obtaining the overall overview, which is a model similar to human focus. The characteristics of the input model will first be assigned a different importance value, indicating that the importance of each feature is different, and then find the feature that has the greatest impact on the results from all the features, so that the model can make better decisions [11]. Generally, for the same model, increasing the number of input parameters of the model can improve the prediction ability of the model. Correspondingly, this model needs to store more intermediate data. Sometimes, due to too much intermediate data, it is easy to put a lot of pressure on the model, resulting in reduced model effect. By introducing the attention mechanism, we can focus on the feature information that can improve the prediction ability of the current model, capture a large amount of feature information, and filter out the feature information that has little influence. This reduces the pressure on the model to a certain extent, thus improving the overall processing capacity of the model. The use steps of attention mechanism can be roughly divided into three stages:

In the first stage, according to Query and available methods such as vector dot product, Key selects an appropriate method to calculate the correlation coefficient between Query and Key. Vector similarity method and MLP neural network method. The formula of vector similarity method is as follows:

$$B_{(Q,K_i)} = \sqrt{\frac{Q^T K_i}{\|Q^T\| * \|K_i\|}} \qquad (4)$$

Among them, $B_{(Q,K_i)}$ Represents the value of the correlation coefficient, $*$ Denotes modular operations, K_i Indicates different keywords, Q Represents a query. about K_i Correlation coefficient value of $B_{(Q,K_i)}$ Can be used D_i Means that $B_{(Q,K_i)} = D_i$.

In the second stage, the Softmax method is used to process the previous correlation coefficient to obtain the weight coefficient. This processing has two advantages. One is to process the weight of all keywords as a probability distribution with a sum of 1. The other is to increase the importance of important keywords through its own functions. The Softmax formula is as follows:

$$d_i = soft \max(D_i) = \frac{\ln D_i}{\sum\limits_{j=1}^{L} \ln D_i} \qquad (5)$$

Among them, d_i express K_i Corresponding weight coefficient; L Indicates the total length of the input feature.

In the third stage, the weight coefficient d_i and the attention weight of each position W_i Calculate to get the final attention value, and the formula is as follows:

$$Att_{(Q,K,W)} = \sum\limits_{i=1}^{L} d_i W_i \qquad (6)$$

Long term and short term memory network (LSTM) is a special recurrent neural network (RNN) used to process long time series. It is designed to solve the problem of long-term dependence. LSTM adds screening of past states on the basis of RNN, so that it can effectively select more influential states, and extract long-term dependence information from long series data. The LSTM neural network has effectively avoided the problems of gradient disappearance and explosion, and has achieved certain success in speech modeling, translation, recognition and picture description. LSTM model is mainly used to obtain information in input feature data. The attention mechanism helps the model quickly filter out important information from a large number of feature data, and focuses the model on the information that is most helpful for completing the current task [12]. Improve the training efficiency of the model by filtering out unimportant data. The Att-LSTM model is designed and implemented based on LSTM model and attention mechanism. This model has a good effect on predicting students' achievements. The Att-LSTM performance prediction model consists of five layers:

Input layer: Input students' test data into the model. In Chapter 3, various data related to students and courses have been screened and constructed through feature engineering [13]. According to the historical information of examination papers, knowledge points and achievements, the feature vectors that can be recognized are constructed. By setting the exam times window to m, get continuous m Score training sample of the second exam Y_m.

$$Y_m = \{y_1, y_2, ..., y_m\} \qquad (7)$$

$$y_m = \{x_1, x_2, ..., x_9\} \tag{8}$$

Among them,y_j Indicates the student's j Times of examination results, x_i It is the student behavior characteristics on the online platform of the Xueyin corresponding to this exam. See Table 1.

LSTM layer: It is controlled by three gates, which are called forgetting gate, input gate and output gate respectively. LSTM needs to filter out the input that needs to be discarded from the cell state first[14]. This operation is carried out through an activation function of forgetting the door. It passed the last time $t - 1$ Hidden state of E_{t-1} And current time t Input information for x_t, to calculate a vector in which the value range of each value is between [0, 1]. Indicates the retention degree of corresponding data in the cell state. 0 means that the information is not reserved at all, and 1 means that it is reserved at all. The specific calculation of this process is as follows:

$$g_t = f\left(w_g x_t + R_g E_{t-1} + \delta_g\right) \tag{9}$$

$$f(\cdot) = \frac{1}{1 + e^{-\cdot}} \tag{10}$$

Among them, g_t Indicates the forgotten door, x_t Represents the input of the current time, w_g, R_g Indicates the importance parameter, δ_g Indicates the correction offset parameter. E_{t-1} Indicates the last operation $t-1$ the hidden state of LSTM cells. Formula (10) is a sigmoid expression.

After the forgetting gate calculation is completed, the next step is to calculate what new data should be input for the cell state. First, by calculating the last time $t - 1$ Hidden state of E_{t-1} and x_t At the input door p_t To determine the data to be updated. Then use E_{t-1} and x_t Obtain new candidate memory cells through tanh function [15] q_t. The calculation is as follows:

$$p_t = f(w_i x_t + R_i E_{t-1} + \delta_i) \tag{11}$$

$$q_t = \tanh\left(w_q x_t + R_q E_{t-1} + \delta_q\right) \tag{12}$$

Among them, p_t Indicates the current operation t Input door of, w_i, R_i, w_q, R_q Represents the importance parameter,δ_q, δ_i Indicates the correction offset parameter, x_t Indicates the current operation t Input data, E_{t-1} Indicates the last operation $t-1$ Hidden state of LSTM cells,q_t Indicates that the current operation may be selected to remember the cell status, tanh() Represents a hyperbolic tangent function.

After calculating the memory cells that may be selected q_t After that, the original memory cells need to be updated $\dot{q}_{t-1}$, determining memory cells $\dot{q}_t$ Subsequent values. The process is when x_t get into g_t When,g_t Will pass $f()$ discard E_{t-1} Some data of When x_t adopt p_t When, p_t Will use $f()$ hold q_t Some data of $\dot{q}_t$ Medium. The calculation is as follows:

$$\dot{q}_t = g_t \otimes \dot{q}_{t-1} + p_t \otimes q_t \tag{13}$$

Among them, $\dot{q}_t$ Indicates the updated memory cell status, q_t Indicates the status of candidate memory cells, g_t Stands for Forgotten Gate, $\dot{q}_{t-1}$ Indicates the last operation $t - 1$ The state of the memory cell when.

Calculate the cell state $\dot{q}_t$ After, according to E_{t-1} and x_t To calculate the hidden layer information that needs to be sent to the next cell by E_{t-1} and x_t stay r_t And then calculate the vector through tanh layer, which is the same as r_t The final result is obtained by calculating the obtained result. This step is as follows:

$$r_t = f(w_r x_t + R_r E_{t-1} + \delta_r) \tag{14}$$

$$E_t = r_t \otimes \tanh(\dot{q}_t) \tag{15}$$

Among them, r_t Represents an output gate. w_r, R_r Indicates the importance parameter, δ_r Indicates the correction offset parameter, $_r x_t$ Indicates the current operation t's data source, E_{t-1} Indicates the last operation $t - 1$ Hidden layer state of, E_t Indicates the current operation t The hidden state of Attention mechanism layer: capture input feature information and filter out feature information with little influence.

Full connection layer: aggregate the results of attention mechanism layer into final prediction data $\hat{y}_j$.

Full connection layer output prediction results $\hat{y}_j$ Then it is necessary to calculate the MAPE value between the actual score and the training sample. When the MAPE value is less than the set error threshold, the optimal Att-LSTM model is obtained and saved. Otherwise, the Att-LSTM model parameters need to be adjusted until the training meets the end requirements.

3   Method Test

3.1   Data Set Preparation and Processing

Att-LSTM is used to establish a prediction model for experimental training, which is used to predict the final scores of online academic education students and guide a vocational college to improve teaching strategies in a timely manner. This paper selects 6000 students' relevant data for model training. In the stage of experimental results display, this paper randomly selects the original behavior data of 100 students on the "Xueyin Online" platform to show the actual prediction effect, and analyzes the relevant data. Next, on the basis of the basic data set in Table 1, the original behavior data of 100 students is processed into a data set conforming to P-MIML. The data set contains information about students' learning behavior and performance in all semesters. It should be noted that the data set contains the information of students' required courses and optional courses. When forecasting, only the same courses learned by students in the same data set are predicted, because the students in the same major of online academic education in this school have the same required courses, but some optional courses are different. If an optional course is given for forecasting,There will be some students who do not choose this course, and their grades are missing, which will lead to the problem of missing marks in the subsequent prediction, and will also affect the prediction results. Therefore, this paper only forecasts public courses in the same major.

3.2 Evaluation Index

Obtaining a model with strong generalization ability is a continuous goal of machine learning tasks. In the process of continuously improving the model, the model evaluation method guides the direction of model improvement. Choosing appropriate model evaluation methods and indicators can judge the performance of the model more objectively and accurately. This paper uses three evaluation indicators to evaluate the student achievement prediction model, as follows.

(1) Mean absolute error (ς)This indicator can reflect the true state of the difference between the predicted value and the actual value, and can also deal with the problem of the offset between the positive and negative errors. Generally, the smaller the MAE value, the better the model fitting effect. The formula is as follows:

$$\varsigma = \frac{\sum\limits_{j=1}^{m} |\hat{y}_j - y_j|}{m} \tag{16}$$

Among them, $\hat{y}_j$ Is the predicted value of the sample, y_j Is the true value of the sample, m Is the number of predicted samples.

(2) Root mean square error (ζ), used to measure the difference between the predicted value and the true value, which represents the sample standard deviation between the predicted value and the true value. Its value is equal to the square root of the mean square error. The formula is expressed as follows:

$$\zeta = \sqrt{\frac{\sum\limits_{j=1}^{m} (\hat{y}_j - y_j)^2}{m}} \tag{17}$$

(3) Coefficient of determination (ξ) The formula is as follows. Its molecule is similar to the mean square error, that is, the total error generated by the prediction model. At the same time, its denominator can also be understood as a prediction model, and all the predicted values of the model are considered as the average value of the sample. If $\xi = 0$ means that the forecast model is equal to the benchmark model; If $\xi = 1$, it means that the prediction model will not have any error and the prediction result is perfect. ξ The value range is between [0, 1]. In general, the coefficient of determination ξ The closer the value of is to 1, the better the prediction model is.

$$\xi = 1 - \frac{\sum\limits_{j=1}^{m} (\hat{y}_j - y_j)^2}{\sum\limits_{j=1}^{m} (\bar{y}_j - y_j)^2} \tag{18}$$

Among them,$\bar{y}_j$ Represents the average of the true values of all samples.

3.3   Prediction Results

The prediction adopts a cross validation method of 50 times, which involves five cycles to obtain five test results. Finally, these five results are used as the average of the model's prediction results. To verify the feasibility and performance of each module, attention mechanism and LSTM model were removed from the complete model, and only LSTM model and traditional machine learning methods were used for prediction. Compare the prediction results after removing the LSTM model with the prediction results after removing the attention mechanism and the complete model, and evaluate the contribution of the LSTM model and attention mechanism to prediction accuracy. And compare the method studied in this article with traditional machine learning based performance prediction methods and deep learning based performance prediction methods to verify the prediction effect of this method. The specific results are shown in Figs. 2, 3 and 4 below.

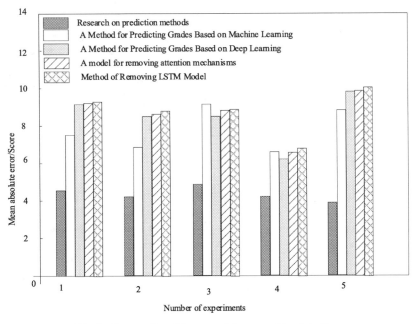

Fig. 2. Comparison Diagram of Average Absolute Error

From Figs. 2, 3 and 4, we can see that compared with LSTM model, traditional machine learning method, traditional machine learning based performance prediction method and deep learning based performance prediction method, the Mean absolute error of the research method is smaller, which indicates that the prediction method in this paper has higher accuracy in predicting student scores, and can more accurately predict student scores; The Root-mean-square deviation is small, which indicates that the difference between the prediction results of this method and the true value is small, and the prediction accuracy is higher; The larger Coefficient of determination indicates that the method in this paper can better explain the variance of student performance changes

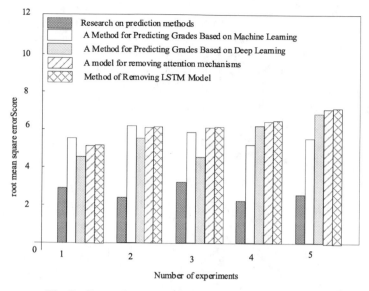

Fig. 3. Comparison Diagram of Root Mean Square Error

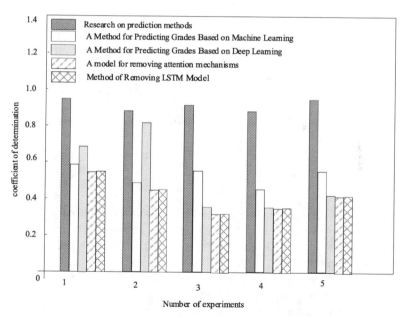

Fig. 4. Comparison Diagram of Determination Coefficient

and has a better fitting effect. This is because attention mechanisms can help models pay more attention to important information and improve the accuracy of predictions; The LSTM model can capture long-term dependencies in time series data, thereby improving the accuracy of prediction.

In order to verify the applicability and effectiveness of the method proposed in this article, three additional groups of 100 students' raw behavior data were randomly selected, labeled as dataset 1, dataset 2, and dataset 3 as the research objects. After processing the data, the prediction under the method proposed in this article was completed. The number of experiments is set to 5 cycles, and the predicted results are shown in Fig. 5.

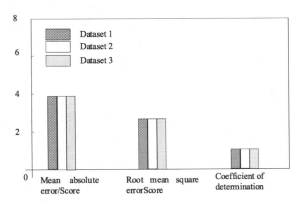

Fig. 5. Comparison of Prediction Results under Different Datasets

It can be seen from Fig. 5 that the results of Mean absolute error, Root-mean-square deviation and Coefficient of determination of experimental data sets 1, 2 and 3 are consistent after five cycles of testing. This indicates that the proposed method has consistent prediction performance and stability across different datasets. By verifying multiple datasets, the generalization ability and applicability of the method can be more reliably evaluated.

4   Conclusion

Student achievement prediction is one of the important research topics in the field of education. Accurate prediction results can help improve learners' academic performance and help managers make scientific decisions. It has important research significance and application value in the field of personalized service and adaptive learning. Through analysis, it is found that there are still some deficiencies in the existing performance prediction models in the use of various attribute characteristics and other factors to predict student performance. In view of this, this paper first proposes a data preprocessing method, and from the perspective that students are affected by different factors to different degrees, proposes a student performance prediction method based on the self attention mechanism to achieve effective prediction of student performance;Secondly, it analyzes the relationship between students' final grades and historical scores, and proposes a prediction method of students' output scores in online English teaching in higher vocational colleges based on the online platform of Xueyin, which makes more comprehensive and accurate use of various attribute features, thus further improving the prediction ability of the model;Finally, through the visualization of the attention results of each attribute feature on the final grade, the personalized analysis and guidance of students can be realized. This research also provides a new idea for students' performance prediction. The main work of the paper includes the following two aspects:

(1) In order to overcome the shortcomings of existing machine learning based performance prediction research methods under the online learning environment, this paper introduces an efficient performance prediction model. Compared with other traditional machine learning algorithms, this algorithm can be efficiently and accurately applied to the online learning behavior dataset of students with large amount of data, sparse features and high dimensions. Secondly, based on the experimental results of the constructed performance prediction model, combined with Pearson's phase.

The relationship number together analyzes the impact of different learning behavior characteristics on students' online learning performance, so as to provide meaningful guidance for the construction and optimization of online learning.
(2) The self attention mechanism is introduced into LSTM to fully explore the relationship between attribute characteristics and their importance to the final grade, so as to give different attention weights to different attribute characteristics, and solve the problem that different factors have different degrees of influence on the same student and different students have different degrees of influence by the same factor.

This research not only solves the problem of individual differences ignored in the current research on the prediction of students' performance using various attribute characteristics, improves the accuracy of performance prediction and the interpretability of the model, but also realizes the personalized analysis and guidance for students, which also provides a new idea for solving the problem of performance prediction. However, there are still some deficiencies in the research, which need to be further improved:

(1) Only the relationship between the attribute characteristics and the final grade has been preliminarily excavated, and the combined characteristics may also be related to the final grade to some extent. Next, more consideration and design can be given to the combination of different characteristics or the influence of higher-order characteristics on the performance prediction results to improve the prediction accuracy of the model.
(2) The selected data set is relatively mature and the data volume is not large enough. In the next step, we can consider to carry out experimental verification on the constructed model in the data set with larger data volume. It is better to conduct further thinking and research after cleaning the actually collected data.
(3) When the system or indicators of the platform change, it is necessary to re collect and reorganize student behavior data, and conduct corresponding data processing and Feature engineering. In addition, it is necessary to adjust the parameter configuration of the prediction model based on new systems or indicators, and may even require retraining the model. Only after adapting to new changes in data and model adjustments can the accuracy and precision of the prediction method be ensured. Therefore, in practical applications, it is necessary to combine the collected data for research, establish a flexible process and mechanism, update data and models in a

timely manner, and analyze individual differences of students more accurately to maintain the effectiveness of prediction methods.

References

1. Yu, J.: Academic performance prediction method of online education using random forest algorithm and artificial intelligence methods. Int. J. Emerg. Technol. Learn. (iJET) **16**(5), 45 (2021)

2. Li, G.: Decision system for performance prediction based on machine learning technology in the context of intelligent education. J. Ningde Normal Univ. (Nat. Sci.) **33**(1), 36–41 (2021)

3. Zheng, A., Wang, Y., Hao, C.: Research on online academic achievement prediction with deep learning. Comput. Era (12), 69–72, 75 (2021)

4. Yekun, E.A., Haile, A.T.: Student performance prediction with optimum multilabel ensemble model. J. Intell. Syst. **30**(1), 511–523 (2021)

5. Maurya, L.S., Hussain, M.S., Singh, S.: Developing classifiers through machine learning algorithms for student placement prediction based on academic performance. Appl. Artif. Intell. **35**(3), 1–18 (2021)

6. Liang, K., Liu, J., Zhang, Y.: The effects of non-directional online behavior on students' learning performance: a user profile based analysis method. Future Internet **13**(8), 199 (2021)

7. Klingner, M., Fingscheidt, T.: Online performance prediction of perception DNNs by multi-task learning with depth estimation. IEEE Trans. Intell. Transp. Syst. **22**(7), 4670–4683 (2021)

8. Yong, H.E., Chung, Y., Min, S.P., et al.: Deep learning-based prediction method on performance change of air source heat pump system under frosting conditions. Energy **228**(1), 120542 (2021)

9. Lee, I., Sawant, A., Shin, J., et al.: Accurate gyrotron performance prediction based on full 3-d magnetic field and electron beam information altered by nonideal factors. IEEE Trans. Electron Devices **68**(3), 1276–1283 (2021)

10. Hao, M., Xiaoling, D., Shulong, W., Shijie, L., Jincheng, Z., Yue, H.: GaN JBS diode device performance prediction method based on neural network. Micromachines **14**(1), 188 (2023)

11. Chen, W., He, W., Zhu, H., Zhou, G., Mu, Q., Han, P.: A processor performance prediction method based on interpretable hierarchical belief rule base and sensitivity analysis. Comput. Mater. Continua **74**(3), 6119–6143 (2022)

12. Rossopoulos Georgios, N., Papadopoulos, C.I.: A journal bearing performance prediction method utilizing a machine learning technique. Proc. Inst. Mech. Engineers Part J: J. Eng. Tribol. **236**(10), 1993–2003 (2022)

13. Luo, Y., Han, X.: A model for predicting student performance in hybrid courses based on incremental learning algorithm. E-educ. Res. **42**(7), 83–90 (2021)

14. Wei, X.: The evaluation and prediction of academic performance based on artificial intelligence and LSTM. Chin. J. ICT Educ. **28**(4), 123–128 (2022)

15. Zhang, B., Chen, G., Chen, X., Lu, H.: Hardware implementation of tanh function based on second-order approximation and error compensation. Microelectronics **51**(6), 905–909 (2021)

Classification Algorithm of Sports Teaching Video Based on Wireless Sensor Network

Zhipeng Chen[✉]

Chongqing Vocational Institute of Engineering, Jiangjin 402260, China
chenzhipeng1993@126.com

Abstract. In view of the problems of low recall and accuracy caused by the huge amount of physical education teaching videos, a physical education teaching video classification algorithm based on wireless sensor network is proposed. The classification framework of physical education teaching videos based on wireless sensor networks is constructed, and the video node coordinates are located according to the structural relationship of physical education teaching videos. Initialize the histogram index, calculate the similarity of any two frames of the video, and set the clustering index of key frames of the physical education teaching video based on the distance between the two frames. Retrieve the video to be classified, find the sensor node with the largest weight, calculate the distance between the target and the detection sensor node, design the video classification steps of physical education teaching, and realize video classification. The experimental results show that the minimum recall rate of this algorithm is 87%, and the maximum classification accuracy rate is ninety-seven percent, which has the advantages of high classification recall and accuracy.

Keywords: Wireless Sensor Network · Physical Education · Video Classification · Accuracy Rate · Recall Rate

1 Introduction

Sports teaching video is an important video resource with a wide range of users. Due to the rapid development of information technology, the research on the classification of sports teaching videos has gradually become a hot issue concerned by relevant personnel. At present, China's sports cause is in a rapid development stage, and sports teaching video data also shows a massive growth. Efficient sports teaching video classification has important application value for high-quality sports teaching video browsing and retrieval. There is a large amount of sports teaching video information in the network. If there is no efficient sports teaching video retrieval system, the sports teaching video in the network will be disordered. Therefore, how to organize sports teaching video resources with high efficiency and precision, and realize the classification and sorting of sports teaching videos is conducive to helping users to obtain their own sports teaching video content efficiently.

L. Yun et al. (Eds.): ADHIP 2023, LNICST 547, pp. 291–305, 2024.
https://doi.org/10.1007/978-3-031-50543-0_20

In the past, sports teaching videos were managed and classified in the form of manual annotation, which not only wasted more human resources, but also had low classification accuracy due to a large number of subjective human factors. Therefore, the research and development of a method that can accurately and reasonably classify sports teaching videos has great scientific research value. Reference [1] proposed a key frame extraction method based on inter frame difference, which is to extract key frames based on the change value of image information (color, texture, brightness, etc.). When the change value is less than the preset threshold, it can be selected as a key frame. The key frame extraction method based on the difference between frames is simple and fast, but the calculation is large and the results are easily affected by the selection of starting frame and threshold setting. The key frame extraction method based on clustering clusters the frames with similar image characteristics, and takes the intermediate frame of each class as the key frame. Reference [2] proposed a method to obtain specific motion frames based on human posture estimation and clustering. Each frame of the motion video is classified according to similarity, and one frame is selected as the key frame for each category according to specific criteria; Reference [3] proposed a dynamic clustering method. First, the predicted number of classifications is calculated according to the similarity of adjacent video sequences and the characteristics of motion parameters to complete the video decomposition; Then, the ISODATA algorithm is used to adaptively calculate different class thresholds, and the video sequence clustering and key frame extraction are completed by combining and splitting operations. It can be seen that the above algorithm does not consider the motion expression ability of key frames, which easily leads to the distortion of motion sequence analysis.

From the existing academic achievements, due to the influence of action diversity, background complexity, etc., there is no universal method of action key frame extraction that can fully achieve the above goals. As far as the sports video review of physical education teaching is concerned, because it needs to judge whether the sports action is correct or not according to the frame sequence of interest in the aerobics video, the requirement for sports expression ability is higher. Therefore, a classification algorithm of sports teaching video based on wireless sensor network is proposed.

2   Video Key Frame Clustering Based on Wireless Sensor Networks

The classification framework of sports teaching video based on wireless sensor network is shown in Fig. 1.

According to the characteristics of video data, video content information is extracted from two aspects. First, data irrelevant to content information, including file path, playing date and time length; The second is data related to content information, including personal entities, semantic events and feature data [4]. The retrieval module is mainly responsible for extracting features according to the user's retrieval information, matching them with the information in the database, and returning the retrieval results. For the segmentation of the original video stream, video boundary detection and audio detection are combined to improve the efficiency of video segmentation. In fact, due to the characteristics of sports teaching videos, most of the other shot changes are abrupt, except for gradual change detection of the replay scene shot. The general algorithm based on color histogram threshold segmentation can achieve better results.

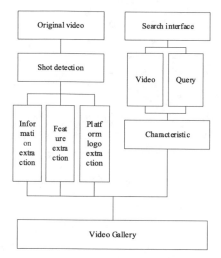

Fig. 1. Classification framework of sports teaching videos

2.1 Coordinate Positioning of Video Nodes in Wireless Sensor Networks

In addition to the general video data structure, the structure of sports teaching video content also has its own characteristics. First of all, the semantic information in the sports teaching video is relatively clear, reducing the fuzziness and subjectivity between the underlying semantic and high-level semantic. Secondly, different events have specific structures and rules. There are obvious differences between the number of people, groups, time, equipment and other factors in the event. These differences are conducive to the analysis and understanding of the video, it can help users quickly find the content they need from a large number of videos, and directly and accurately locate and browse [5]. Video is an unstructured data stream. For video structure, it can be divided into four levels from small to large: frame, shot, scene and video.

Frame represents a still image, which is the smallest unit of video. Shot is a video sequence composed of adjacent consecutive frames, and shot is the basic unit of video retrieval. In the retrieval process, one or more key frames are usually extracted to describe a shot, and then a video is retrieved through the matching similarity of key frames. A scene is composed of a set of semantically related shots that express a complete event, usually with the same background and character information.

In conclusion, the number of hops between nodes is used as the eigenvector to calculate the similarity. Based on the location information of beacon nodes, the positioning algorithm of wireless sensor networks classifies and determines the coordinate interval of unknown nodes for many times. Therefore, the process of determining the node location by the location algorithm is divided into three stages: hop count calculation stage, beacon broadcast stage, and node self location stage.

Hop count calculation stage: beacon node broadcasts its own location information packet to neighbor node, including hop count field and node number. The hop count field is initialized to 0. The receiving node records the minimum hops to each beacon node, ignoring packets with larger hops from the same beacon node. Then add 1 to the hop value and forward it to the neighbor node. With this method, all nodes in the network can record the minimum hops of each beacon node.

Beacon broadcasting stage: each beacon node forms an eigenvector according to the hops from other beacon nodes recorded in the first stage. Accordingly, the unknown node forms the eigenvector according to the hops away from the beacon node. Then, beacon nodes broadcast packets containing node numbers and eigenvectors to the network, and unknown nodes are only responsible for forwarding packets from beacon nodes to ensure that nodes in the network receive packets from each beacon node [6].

Based on the positioning algorithm of wireless sensor network, the X axis and Y axis of unknown nodes are classified respectively to determine the range of coordinates: After the X axis is classified, its space is $X \in \left[x\frac{L}{2^n}, (x+1)\frac{L}{2^n}\right]$ after classification, the space of Y axis is $X \in \left[x\frac{L}{2^n}, (x+1)\frac{L}{2^n}\right]$, $Y \in \left[y\frac{L}{2^n}, (y+1)\frac{L}{2^n}\right]$, where L indicates the length of 2D space, n indicates the number of classifications. According to the above classification process, if the location algorithm of wireless sensor network is correct after 2^n classification, the space where the unknown node is located can be expressed as:

$$O_n \in \left[x\frac{L}{2^n}, (x+1)\frac{L}{2^n}\right] \times \left[y\frac{L}{2^n}, (y+1)\frac{L}{2^n}\right] \tag{1}$$

It can be seen from the formula that the unknown node is in the square area.

Node self localization stage: the unknown node starts the self localization stage after receiving the packets of all beacon nodes in the beacon broadcast stage. According to the feature vector, calculate the similarity with the beacon node, and the wireless sensor network localization algorithm classifies the X axis and Y axis of the unknown node respectively to determine the node coordinates.

2.2   Key Frame Clustering of Sports Teaching Videos

Features represent a certain target and some quantifiable attributes. For sports teaching videos, they mainly include general features and specific field features. Considering the efficiency of key frame extraction of sports teaching video, the image features of sports teaching video are set as color histogram and color distribution descriptor [7]. In general, the description of the color of the sports teaching video image belongs to the color space problem. The key frame extraction algorithm based on the similarity coefficient of the shot boundary is used. After the image is in the HSV color space and the color histogram is derived, the histogram index is initialized first. The similarity of the two frames of the sports teaching video histogram can be seen as:

$$sim_k(a, b) = \sum_{k=1}^{3} \min\{H_a(l_a, d_a, h_a), H_b(l_b, d_b, h_b)\} \tag{2}$$

In formula (2), l_a, d_a, h_a represent histograms respectively H_a length, width and height of; $H_b(l_b, d_b, h_b)$ represent histograms respectively H_b length, width and height of; k represents the number of calculations; a, b represents two frames of the histogram. In the formula, 0 describes that the color histogram difference between the two graphs is very large, and 1 describes that the color histogram difference between the two graphs is the same [8]. For the color segment descriptor, which is used to represent the spatial part of the color in the sports teaching video image, the feature extraction process is: sports teaching video image segmentation, dominant color selection, discrete cosine conversion of 64 pixels (brightness signal Y, blue difference b, red difference r) components, obtaining three groups of coefficients, and finally Zigzag scanning of the obtained discrete cosine coefficients, A new distribution descriptor [9] is established by selecting a few low frequency coefficients. At this time, the distance d_k between frames a, b is set as:

$$d_k = \sqrt{\omega \times \lambda(Y)^2} + \sqrt{\omega \times \lambda(b)^2} + \sqrt{\omega \times \lambda(r)^2} \tag{3}$$

In formula (3), ω represents weight; $\lambda(Y)$, $\lambda(b)$ and $\lambda(r)$ respectively represent the discrete cosine coefficients of Y, b and r describing the frame.

Because the same shot will appear repeatedly in the same sports teaching video, resulting in repeated key frame sequences obtained. In order to reduce the repeatability of the last key frame sequence obtained, cluster the key frame sequences: K average clustering, and finally set according to the clustering effectiveness method K value side. Clustering performance indicators are:

$$\eta = s_a \times V(a) + s_b \times V(b) \tag{4}$$

In formula (4), $V(a)$, $V(b)$ respectively a, b two frame key frame sequence class; s_a, s_b represent and describe the distance between two frames. Because the value ranges of these two items are very different, set a weight factor, and the maximum number of preset clusters obtained when this value is at the minimum value is the optimal number of clusters.

3   Video Classification Algorithm Based on Wireless Sensor Networks

3.1   Video Retrieval to Be Classified Based on Multi View Cooperative Operation

Correctly describing the relationship model between video data and establishing an efficient index can effectively help the system filter out a lot of irrelevant information [10]. In this paper, a hierarchical classification structure is given, which is illustrated by taking the project event type stratification as an example. The hierarchical structure is shown in Fig. 2.

Figure 2 shows the hierarchical relationship between project event entities. For example, sports events are first divided into ball games, field events and gymnastics, and then ball games are subdivided into football, basketball and volleyball, and gymnastics is subdivided into rings, vault and horizontal bar. Similarly, hierarchical relationships are also established for the extracted video information such as individuals and icons. When video retrieval is carried out, a multi view collaborative retrieval method is used, and the steps are as follows:

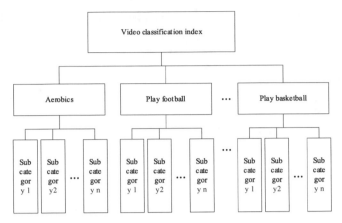

Fig. 2. Structure of Classified Events

Step 1: The information related to the retrieval of video content in the video library can be expressed by the following formula:

$$W = G \cap H \tag{5}$$

In formula (5), G is the video segment of the search video related to frame a; H is the video segment of the retrieved video related to the frame b. For multiple relationships, the intersection of multiple relationship sets is obtained according to the above method, and the content of the set is a candidate video clip.

Step 2: Let the total matching similarity between a video segment in W set and the retrieved video for the relationship between frame a and frame b be:

$$Sim' = \frac{Sim_a + Sim_b}{2} \tag{6}$$

In formula (6), Sim_a and Sim_b represent the similarity of two frames respectively. By analogy, the total similarity of multiple relationships can be expressed as:

$$Sim'' = \frac{\sum_{k=1}^{N} Sim_N}{N} \tag{7}$$

In formula (7), N represents the total number of mapping relationships.

Step 3: During video retrieval, query the video or auxiliary text information according to a segment given by the user, and get the video information with high similarity and return it to the user according to the extracted classification information after steps 1 and 2.

Because step 1 combines the relational semantics of multiple information, the information that is obviously irrelevant to the problem can be filtered out after the operation, which greatly reduces the amount of videos to be searched when retrieving videos. The comprehensive similarity is obtained through the multi view collaborative operation in step 2, and the retrieval efficiency and detection rate will be greatly improved.

3.2 Retrieval Video Tracking Based on Wireless Sensor Network

The master node sends request information to all other neighbor nodes that are one hop away from the master node, and takes them as slave nodes and forms a cluster directly. Each adjacent sensor node sends a join request message to the master node, which contains location information, and informs the master node to use it as a slave node. After the cluster is formed, the master node establishes a time division multiple access plan and sends it to its slave nodes accordingly. This mechanism avoids the conflict of sending messages. After all slave nodes know the TDMA scheduling, the slave node of each detection target will find its residual energy and use the TOA method to calculate its distance from the target. This data is sent to the master node through the slave node data message in the allocated TDMA timeslot. On the other hand, the master node obtains data from the node message and aggregates its own data to calculate the distance between the detection sensor node and the base station. The master protocol basically maintains three lists: the first list contains the distance from the target to the detection sensor node; The second type includes detecting the residual energy of sensor nodes; The third includes the distance between the detection sensor node and the receiving node. The master protocol uses these lists to find the sensor node with the highest weight according to the following formula:

$$w_{\max}(i) = \left(\frac{E(i)}{d_1 \times d_2} \right) \qquad (8)$$

In formula (8), $E(i)$ represents the weight of detecting the sensor node to be selected as the master node i; d_1 and d_2 represent the distance from the detection sensor node to the base station and the distance from the detection sensor node to the target, respectively. If the current master node still has the highest weight value, it will still be the master node in the next round and reassign the same slave node. Otherwise, you must select a new master node. In addition, the first list maintained in the master node is used to estimate the target location in the location algorithm, which will be discussed later. After the target location is estimated, the main node records the estimated target location in the target archive, and then sends it to the base station through a cluster node message.

In order to calculate the distance between the moving target and the sensor node, TOA method is used. In the proposed algorithm, TOA represents the time from the sensor node to the moving target. Therefore, considering the signal speed, the distance between the target and the detection sensor node can be directly calculated according to the following formula:

$$d' = \Delta t \times v \qquad (9)$$

In formula (9), Δt indicates signal delay; v indicates the signal speed. In some applications, obstacles may prevent the direct line of sight between the sensor node and the target. Therefore, TOA readings may be affected. However, the location algorithm used in IAH is the same as the cluster head adaptive clustering algorithm, which mainly depends on the circle exchange principle. The positioning algorithm uses the results of the following formula, combined with the position of the sensor nodes, to obtain a set of circular equations:

$$(x - i_k)^2 + (y - j_k)^2 = r_k^2 \qquad (10)$$

Stay (i, j) each sensor node knows its position when the network is deployed; r_k It is the distance between the sensor node and the moving target. The position of the target tracking result is obtained by solving these equations.

3.3 Design of Video Classification Steps for Physical Education Teaching

On the basis of obtaining the key frame features of sports teaching video images, the sports teaching video image classification method based on the deep learning coding model is adopted to realize the sports teaching video classification. The detailed process is as follows:

The multi-level restricted Boltzmann machine is used to encode and learn the key frame feature library of sports teaching video images, and become a model visual dictionary. According to the spatial information of the key frame feature library of sports teaching video images, set the features of the adjacent key frame feature library of sports teaching video images as the input of RBM, train RBM with wireless sensor network positioning algorithm, and obtain the hidden layer features; Then the adjacent hidden layer features are regarded as the input of the lower layer RBM, and the output dictionary is obtained. When any weight belongs to the connection weight of RBM, RBM has an explicit layer and a hidden layer, while neurons based on the same layer in RBM do not have connection relationship. During network training, the hidden layer and the visible layer of RBM are connected according to the conditional probability distribution.

By setting the weight matrix ω_h and the hidden layer bias vector ε, the input layer features β can be encoded into the corresponding visual dictionary φ, and by setting the ω_h and the explicit layer bias matrix ϕ, the physical education teaching video features can be reconstructed through the visual dictionary. For a set of input layers and coding layers in RBM, its energy function is:

$$f(E) = -\sum_{k=1}^{i}\sum_{k=1}^{j} \beta_{ij}\omega_h\varphi_{ij} - \sum_{k=1}^{i}\sum_{k=1}^{j} \phi_{ij}\beta_{ij} \tag{11}$$

The calculation of the energy function can obtain the joint probability distribution characteristics of the energy function and the edge distribution of the joint distribution of the sports teaching video - the probability of the feature input node. RBM network training is mainly to maximize the probability of the input node. In general, the Monte Carlo Romankov chain method can be used to obtain the key frame feature vector of the sports teaching video.

The CD algorithm is used to implement rapid learning of RBM to improve the convergence efficiency of parameters. The updated amount of weight obtained is:

$$\Delta\omega_h = v_h(\langle\beta\varphi\rangle - \langle\beta\varphi\rangle') \tag{12}$$

In formula (12), v_h describe learning speed; $\langle\beta\varphi\rangle$ and $\langle\beta\varphi\rangle'$ represent the actual weight and threshold of the visible layer and hidden layer respectively.

If all layers of sports teaching video features are trained at the same time, the time complexity will increase; If only one layer is trained at a time, the transmission of each layer will aggravate the degree of underfitting. Therefore, when using deep learning to encode the characteristics of sports teaching video, first obtain the visual dictionary through bottom-up unsupervised RBM hierarchical training.

According to the position (x_t, y_t) of the visual dictionary and the previous position (x_{t-1}, y_{t-1}), the current speed is obtained, and the calculation is as follows through message sending:

$$v_t = \frac{\sqrt{(x_t - x_{t-1})^2 + (y_t - y_{t-1})^2}}{t - (t - 1)} \tag{13}$$

In formula (11), t represents time.

Predict the next position. Based on the current speed, assume that the speed and direction of the target movement remain the same as the original. Therefore, after a given time, the predicted position of the target represents:

$$\begin{cases} x'_{t+1} = x_t + v_t \Delta t \cos \theta \\ y'_{t+1} = y_t + v_t \Delta t \sin \theta \end{cases} \tag{14}$$

In formula (14), θ indicates the current direction; Δt represents a given time.

Step 4: When selecting the master node, use the predicted location as the classification parameter. Therefore, the new primary node selection criteria are defined as:

$$w'(i) = \left(\frac{E(i)}{d_1 \times d_2 \times d_3} \right) \tag{15}$$

In formula (15), d_3 indicates the distance between the predicted position and the detection node.

The number of active nodes in the formed cluster is determined according to the angle value obtained in the previous step. In other words, if the angle value is less than a certain error threshold, that is, the prediction error is relatively small, the number of active nodes in the formed cluster will be reduced to half, based on which the teaching video classification will be realized.

4 Experiment

The experiment adopts the sports teaching video submitted by students on the hybrid teaching platform of a science and technology college, with the sampling frequency of 24 frames/s, the total length of the video of 2.03 min, and a total of 2953 frames. The simulation experiment is carried out in the Matlab2015 environment.

4.1 Experimental Objects

Through the extraction of action video clips based on beat, each video clip obtained represents a sports action clip. Because the key frame extraction only focuses on the moving human body in each sports action video clip. So first, with the video segment unit and the previous frame as the reference, the L-K method is used to calculate the optical flow, and it is detected that the moving target in each frame may be the background rather than the human body image at some times due to the background jitter and other noise effects in the complex scene, as shown in Fig. 3 (a). For this reason, after obtaining the moving target, HoG classifier is used to complete the human detection, and the bounding box containing the human body is obtained, as shown in Fig. 3 (b).

aerobics   play football

(a) L-K optical flow diagram with the previous frame as reference

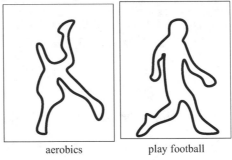

aerobics   play football

(b) Human body bounding box obtained by HoG classifier

Fig. 3. Actual moving target detection

In order to describe the motion characteristics in the human body bounding box, after the human body bounding box is normalized, the previous frame is the reference frame. According to the cell unit (unit: pixel), the L-IC method is used to estimate the motion direction of each cell unit, and the motion characteristics of the human body bounding box in each frame are obtained.

4.2 Experimental Environment

In order to further verify the algorithm performance, CC2510 node is used to build a wireless sensor network experiment test platform in the open area, as shown in Fig. 4.

121 nodes in the laboratory (beacon nodes evenly distributed among them) cover 30m × 30 m area, the distance between nodes is 3 m, forming 10 × 10 Grid distribution. After testing, beacon nodes can "reach each other". The experimental parameters are shown in Table 1.

Floor mounted
camera

Fig. 4. Experimental Test Platform

Table 1. Experimental Parameter Setting Table

Serial number	Parameter	Describe the content
1	Network type	Wireless sensor network
2	network topology	Network composed of multiple wireless sensor nodes
3	data acquisition	Wireless sensor nodes collect video data
4	data transmission	Wireless sensor networks transmit video data to algorithm processing nodes
5	Experimental evaluation index	Classification node energy consumption, classification recall and classification accuracy

4.3   Experimental Results and Analysis

The key frame extraction method based on inter frame difference, the specific motion frame acquisition method based on human posture estimation and clustering, the dynamic clustering method and the classification algorithm based on wireless sensor network are respectively used to compare and analyze the energy consumption of classification nodes. The comparison results are shown in Fig. 5.

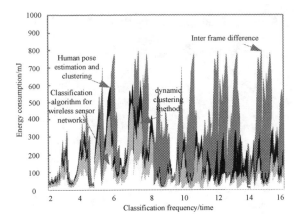

Fig. 5. Comparison and analysis of node energy consumption by different methods

As shown in Fig. 5, the key frame extraction method based on inter frame difference has the largest energy consumption of nodes, the reason is that this method needs to compare each frame and calculate the difference between frames, so as to extract key frames. And followed by the specific motion frame acquisition method based on human posture estimation and clustering, and the dynamic clustering method. Both methods need to estimate human posture or cluster analysis, and then extract specific motion frames. The wireless sensor network classification algorithm has the smallest energy consumption of nodes, with the maximum energy consumption value of only 240 mJ. The energy consumption of nodes should be considered when designing the video classification algorithm for physical education teaching. Choosing a low energy consumption method can reduce the energy consumption of nodes and prolong the network life.

In order to further verify the effectiveness of the algorithm studied, the classification recall rates of the three methods are compared, and the comparison results are shown in Fig. 6.

It can be seen from Fig. 6 (a) that the recall rate of classification using the three traditional methods is less than 70%, because it is difficult for traditional methods to accurately classify complex sports actions, some key frames or specific sports frames are missed, which affects the recall rate. And the minimum recall rate using the algorithm studied can also reach 87%, compared with the traditional method, it has been significantly improved.

It can be seen from Fig. 6 (b) that the recall rate of specific motion frame acquisition method and dynamic clustering method based on human posture estimation and clustering exceeds 70%, it shows that these methods can better capture the key information of sports movements. The recall rate of key frame extraction method based on inter frame difference is less than 60%, because this method can't accurately capture the important changes of actions when extracting key frames. And the minimum recall rate of classification algorithm based on wireless sensor network can also reach 94%, it shows that the algorithm has a high recall rate in video classification of physical education teaching.

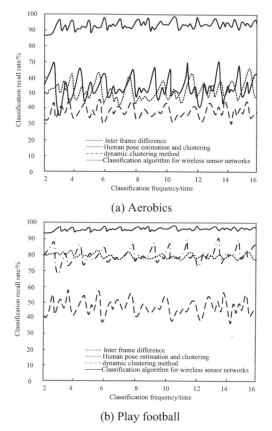

(a) Aerobics

(b) Play football

Fig. 6. Comparative Analysis of Recall Rates by Different Methods

The classification accuracy of the three methods is compared as shown in Fig. 7.

It can be seen from Fig. 7 (a) that the maximum classification accuracy of the three traditional methods is 71%, there is a big classification error when dealing with complex sports movements, which leads to low accuracy of classification results. While the maximum classification accuracy of the algorithm studied is 96%, compared with the traditional method, it has been significantly improved.

It can be seen from Fig. 7 (b) that the maximum classification accuracy of specific motion frame acquisition method and dynamic clustering method based on human posture estimation and clustering is 87%, this shows that these methods can better capture the key information of sports movements and accurately distinguish different types of movements in the classification process. The maximum classification accuracy of key frame extraction method based on inter frame difference is 59%, because this method can not accurately extract representative key frames. And the maximum classification accuracy of the algorithm studied is 97%, it shows that the algorithm has high classification accuracy in video classification of physical education teaching.

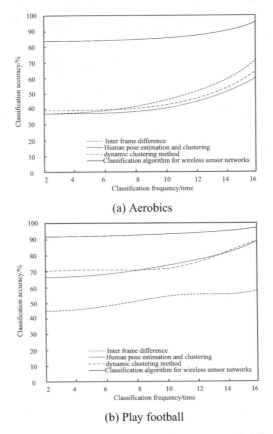

(a) Aerobics

(b) Play football

Fig. 7. Comparison and analysis of classification accuracy of different methods

5   Conclusion

Aiming at the problem of poor classification effect of sports teaching videos using traditional methods, this paper proposes a classification algorithm of sports teaching videos based on wireless sensor networks. Experimental results show that, compared with existing algorithms, this algorithm can accurately track targets, and has lower energy consumption and higher classification accuracy, which shows the effectiveness and feasibility of using the proposed algorithm. The algorithm studied in this paper has a high recall rate in the classification of sports teaching videos, and can better capture the key information of sports actions. At the same time, the algorithm studied in this paper has high classification accuracy in video classification of physical education teaching, and can identify and classify different sports actions more accurately.

References

1. Cao, C.P., Yuan, K.G.: Key frame extraction algorithm of reinforcement learning based on multi-channel feature and attention mechanism. Appl. Res. Comput. **39**(4), 1274–1280 (2022)

2. Cai, M.M., Huang, J.F., Lin, X., et al.: Acquisition method of specific motion frame based on human attitude estimation and clustering. J. Graph. **43**(1), 44–52 (2022)
3. Huang, W., Wang, Y., Zhang, L., et al.: Key frame extraction algorithm for theodolite image sequence. J. Appl. Opt. **43**(3), 430–435 (2022)
4. Zhang, M.Q., Li, W.P.: An automatic classification method of sports teaching video using support vector machine. Sci. Program. **2021**(9), 1–8 (2021)
5. Ass, A., Sh, B., Jp, B., et al.: Classification of educational videos by using a semi-supervised learning method on transcripts and keywords. Neurocomputing **2021**(7), 637–647 (2021)
6. Lin, Y., Liu, H., Chen, Z., et al.: Machine learning-based classification of academic performance via imaging sensors. IEEE Sens. J. **21**(22), 24952–24958 (2021)
7. Zheng, Y., Shi, G.: Research on data retrieval algorithm of English microlearning teaching based on wireless network information classification. J. Sens. **2021**(7), 1–10 (2021)
8. Hong, S., Kim, J., Yang, E.: Automated text classification of maintenance data of higher education buildings using text mining and machine learning techniques. J. Archit. Eng. **28**(1), 1–10 (2022)
9. Liang, X., Yin, J.: Recommendation algorithm for equilibrium of teaching resources in physical education network based on trust relationship. J. Internet Technol. **23**(1), 133–141 (2022)
10. Li, J.: Application of mobile information system based on internet in college physical education classroom teaching. Mob. Inf. Syst. **2021**(9), 1–10 (2021)

Monitoring Method of Students' Achievement of Curriculum Objectives in Higher Vocational English Online Teaching Based on Xueyin Online Platform

Lihua Sun[✉] and Dan Wang

Liaoning Petrochemical College, Jinzhou 121000, China
lnpcsunlihua@126.com

Abstract. Due to the common influence of various factors on the achievement of curriculum goals, the final results are significantly different from the actual difficulties in the process of monitoring them. Therefore, this paper puts forward a research on the monitoring method for the achievement of curriculum goals of online English teaching students in higher vocational colleges based on the online platform of Xueyin. Combined with the characteristics of the online platform of Xueyin, the monitoring indicators for the achievement of online teaching students' curriculum goals are designed from the three perspectives of teaching elements, learning elements and management elements, which are taken as the basis for the analysis of the achievement of curriculum goals. In the specific analysis process, the impact of different curriculum goal assessment methods and assessment difficulty differences on the achievement evaluation is fully considered, The weighted piecewise function is used to calculate the quantitative score of each monitoring indicator to accurately monitor the achievement of students' curriculum goals. In the test results, there is no significant difference between the monitoring results of online listening and oral English teaching course goal attainment and the actual situation, and the design method has good application value.

Keywords: Xueyin Online Platform · Online English Teaching · Degree of Achievement of Curriculum Objectives · Achievement Monitoring Indicators · Weighted Piecewise Function · Quantitative Score

1 Introduction

Through comparative analysis, research and reference to the domestic and foreign methods [1] for monitoring learners' learning quality and evaluating learning effects in the online education process, this paper analyzes and summarizes the defects in the two aspects of learning behavior monitoring and learning effect evaluation in the existing online teaching system. On this basis, the main research contents at this stage are mainly divided into the following aspects: The first is online teaching element monitoring [2].

© ICST Institute for Computer Sciences, Social Informatics and Telecommunications Engineering 2024
Published by Springer Nature Switzerland AG 2024. All Rights Reserved
L. Yun et al. (Eds.): ADHIP 2023, LNICST 547, pp. 306–320, 2024.
https://doi.org/10.1007/978-3-031-50543-0_21

Online teaching elements are important indicators of online education information monitoring. The objects of element monitoring include learners, teachers, and administrators. Different monitoring elements are constructed from three aspects: teaching, learning, and management. Different approaches are adopted to monitor different individuals, with learners as the main focus [3]. Through analyzing and comparing learners' learning elements, big data analysis technology is utilized to identify behavior elements that effectively reflect learners' online teaching situation. These behavior elements are then monitored, analyzed [4], and presented to the teaching platform using data visualization technology.T he second is online teaching effect evaluation [5]. Online teaching effect evaluation is an evaluation method based on the analysis of learners' participation in course learning on the learning platform, including learning effect evaluation system, evaluation method, technical scheme, evaluation effect, etc. [6]. During learners' participation in learning, the system analyzes and summarizes the learning behavior logs generated by learners in the database. It constructs an effective evaluation system that supports learners' learning effects, compares and analyzes learners' learning behavior logs, and evaluates the rationality of learners' curriculum performance evaluation methods and results based on the database behavior logs [7]. Furthermore, an online teaching warning prompt is implemented. With regard to improper learning behaviors exhibited by learners in online learning, a recognition mode is established to automatically identify learners engaging in such behaviors in the system. These learners are then graded accordingly [8]. Finally, the system automatically provides learners with reminders, warnings, or deducts behavior points as appropriate. The other is the optimization of online teaching integral strategy. In the online teaching system, learners adjust and improve the setting of course learning progress and learner scoring rules [9], enhance the integrity of the learning effect evaluation system, and improve the rationality of the learning effect evaluation system. Compared with other online platforms, the online platform of ICBC has the following results [10]. First, while learning from MOOC's experience [11], the online platform of Xueyin integrates the credit banking system, provides MOOC with a dynamic mechanism [12], and emphasizes the learning mechanism of learning achievement transformation and small deposit and lump sum withdrawal. Second, the online platform of Xueyin introduced the concept of educational Taobao. First of all, Xueyin Online is a third-party platform, a public platform for the whole society, and does not belong to any institution. Secondly, all kinds of educational institutions are the main body of the platform, and can independently develop the scope of course sharing, select partners, determine business models, set market prices, and target audience groups on the online platform of the Xueyin. Moreover, for learners, ubiquitous learning can be realized on the online Xueyin, including self selected majors, self selected courses, credit deposit and withdrawal, certificate application, etc. Once the online platform of Xueyin is registered, it is valid for life. For learning, whether there is a learning environment and quality is crucial. Without quality, there is no credibility. Therefore, an integrated learning environment has been created on the platform. The so-called integrated learning environment includes three aspects: first, it can involve multiple terminals. National Open University, New Education Research Institute and Superstar Group have jointly created an integrated learning environment - Learning Connect, which provides a wide range of communication channels for institutions participating in Xueyin Online. TV, cloud classrooms, mobile terminals, etc. can all be involved. Second, Xueyin Online can not only provide learners with a large number of high-quality online learning courses, but also establish lifelong learning achievement files for learners to authenticate, accumulate and transform their learning achievements, truly serving the career and life development of individuals. Third, Xueyin Online contains high-quality learning content both in and out of class. At present, the platform gathers a large number of high-quality courses at home and abroad, including various video and audio learning resources. In order to ensure the quality, the online platform of ICBC has also set up an expert committee. In the future, the expert committee will be divided into various discipline groups to check the key links of all quality processes, including credit banks.

2   Design of Monitoring Method for Students' Achievement of Curriculum Objectives in Online English Teaching in Higher Vocational Colleges

The process of monitoring the implementation of online English teaching course objectives in vocational colleges is shown in Fig. 1.

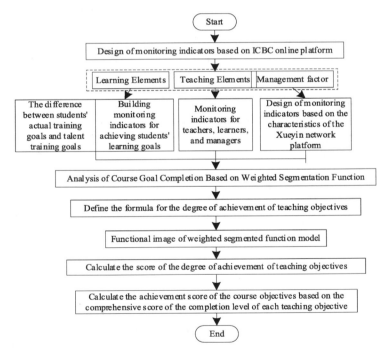

Fig. 1. Monitoring Method Flow for the Implementation of Online English Teaching Curriculum Objectives in Vocational Colleges

2.1 Monitoring Index Design Based on the Online Platform of ICBC

The degree of achievement of curriculum objectives refers to the extent to which talent training objectives are fulfilled based on the established talent curriculum objectives of the responsible colleges and universities [13]. It involves cultivating a solid understanding and mastery of professional theories and skills in accordance with the relevant curriculum objectives and professional systems formulated by these institutions. The ultimate goal of the school is to cultivate talents with high quality and solid professional ability. In order to achieve this ultimate goal, it must be assisted by the purpose of education, positioning and formulation of relevant professional training goals [14]. However, from a practical standpoint, it is crucial to implement the formulated training objectives and subsequently cultivate students who can meet these objectives. In light of this, this paper synthesizes various aspects of students' assessments and examines the disparities between the formulated talent training objectives and the actual outcomes in student development. The analysis of online teaching elements primarily focuses on the significant factors and activities that contribute to effective teaching when online teaching participants engage in instructional activities. The elements of online teaching mainly come from three aspects: teachers [15], learners and administrators. Therefore, based on the historical data in the online platform of the Xueyin, this paper constructs the monitoring indicators for the achievement of students' curriculum goals from three aspects: teachers, learners and managers. The analysis of the online teaching elements of the online platform of Xueyin mainly includes the following three aspects: teaching elements, learning elements and management elements. Teaching elements mainly include online tutoring and question answering, teaching resource construction, teaching task management, etc. Among them, online tutoring and Q&A includes initiating discussion, forum Q&A, etc.; The construction of teaching resources includes adding test questions, editing test papers, etc.; Teaching task management includes assigning and reviewing assignments. Learning elements mainly include login, video learning, PPT learning, posting comments, deleting comments, submitting homework, chapter exercises, and online homework, among others. The elements of management mainly include teaching plan maintenance, online course maintenance, online test question maintenance, teaching resources maintenance, assignment layout maintenance, assignment review maintenance, forum management maintenance, learning early warning control, etc. Among them, learning early warning control includes short message early warning, pop-up early warning, improper learning behavior monitoring, etc. On this basis, this paper combines the elements involved in the learning process of students using the online Xueyin platform to achieve the design of monitoring indicators for students' achievement of curriculum goals.

(1) Learning Elements

The composition of the monitoring indicators for the achievement of students' curriculum objectives with the learning elements in the online platform of the Xueyin as the core is shown in Fig. 2.

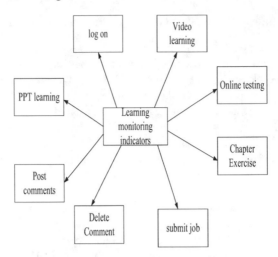

Fig. 2. Learning monitoring indicators

The specific information of the monitoring indicators in Fig. 2 is detailed.

Login: learners must log in to the system before starting online teaching using the online platform of Xueyin. In the process of learners' participation in online learning, the number and time of login to the learning platform can indirectly reflect the enthusiasm of learners to participate in online learning.

Video learning: course video is an important content of online course teaching resources on the online platform of Xueyin. After opening the course, learners can learn the course by watching the course video. The learning behavior log of the platform database can automatically record learners' learning of the selected course video, including the time when learners start learning the video, the time when they end video learning, learning content, learning progress, etc.

PPT learning: course PPT is an important resource in the online courses of the online platform of the Xueyin. After opening the course, learners can learn by browsing the course PPT. The learning behavior log of the platform database can automatically record the learners' learning of the course PPT they selected.

Comment: The online forum serves as the main hub for online communication between learners and teachers. In the process of online teaching, forum posting is the main form of interaction between learners and teachers. When learners encounter doubts in the learning process or have questions about teaching methods, forms and teaching resources, they can interact in the form of forum posts. The content and number of posts posted by learners reflect the enthusiasm of learners to participate in online teaching of online courses.

Delete comments: Posting on the forum can reflect the enthusiasm and initiative of learners to participate in the teaching platform, which is usually a factor to be considered when assessing learning achievements. This will also lead some learners to publish a large number of invalid comments in the forum, deliberately show their enthusiasm to participate in online teaching, and cheat learning points. In this paper, these comments are called "invalid learning comments". For invalid learning comments of learners, managers or teachers will regularly delete them and remind, warn or punish learners.

Submit homework: homework is a way for teachers to test learners' learning. When learners finish learning each video or PPT of the course, they can enter the corresponding homework module, and evaluate the learners' mastery of the course content by detecting the completion of homework, exercises and other ways.

Chapter exercise: similar to homework, it corresponds to each chapter of online courses. After learning each chapter of the course, learners can enter the chapter exercise module in the platform. Through the chapter exercise, learners can check their mastery of the learned chapter knowledge, further clarify their learning objectives, and guide their next learning direction.

Online test: it is a kind of online simulation test for learners, which can be used as one of the evaluation factors for learners on the learning effect of courses. Before starting the online test, learners can conduct a simulation test on the learning content in the course to consolidate their knowledge learning.

(2) Teaching Elements

The composition of the monitoring indicators for the achievement of students' curriculum goals with the teaching elements in the online platform of the Xueyin as the core is shown in Fig. 3.

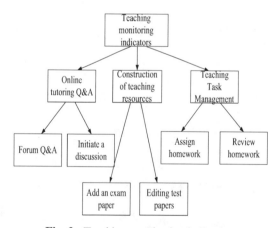

Fig. 3. Teaching monitoring indicators

The specific information of the monitoring indicators in Fig. 3 is detailed.

Online tutoring Q&A: online tutoring Q&A corresponds to the interaction of learners' forums. Learners will feed back the problems encountered in the learning process to teachers through the learning platform. Teachers will answer or summarize the questions and suggestions raised by learners through the platform, and strive to form a set of interactive online teaching guidance mechanism.

Construction of teaching resources: the construction of teaching resources is closely related to the quality and effect of learners' curriculum learning, and is the premise and foundation of learners' curriculum learning. In the online course learning platform, teachers and administrators should put the construction of course resources in the first place, and build rich and high-quality online teaching resources for learners.

Teaching task management: the content of teaching quality supervision should also include the supervision and management of various behaviors of teachers in the process of online learning assistance, such as teacher editing test questions, assigning homework, marking homework, editing test papers, submitting grades, adding face-to-face teaching schedule, adding test questions, forming test papers, etc., as well as teacher monitoring and management of learners' learning behavior in the process of learners' learning, so as to standardize the learning behavior of learners and ensure the learning quality and efficiency of learners.

(3) Management Elements

The composition of the monitoring indicators for the achievement of students' curriculum objectives with the management elements in the online platform of the Xueyin as the core is shown in Fig. 4.

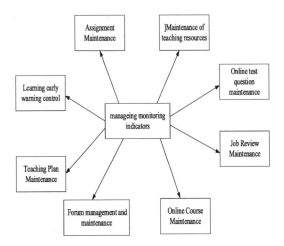

Fig. 4. Management Monitoring Indicators

The specific information of the monitoring indicators in Fig. 4 is detailed.

Teaching plan maintenance: it is mainly responsible for compiling the course name, assessment method, course credit, course score composition ratio, course class hours and course semester in the teaching plan of each grade, and then publishing the teaching plan information.

Online course maintenance: mainly responsible for editing and maintaining the name, course number, course type, course classification, course class hours, course web page style, course construction year and course construction completion of online courses.

Online test question maintenance: mainly responsible for the maintenance of online course test question types (multiple choice questions, blank filling questions, judgment questions, short answer questions, calculation questions, etc.) and test categories (chapter exercises, comprehensive tests, questionnaires, etc.).

Maintenance of teaching resources: mainly responsible for the maintenance of chapter information (chapter code, chapter name, etc.), resource categories (syllabus, handouts, teaching videos, PPT, experiments, etc.), resource synchronization, etc. of course resources.

Job layout maintenance: mainly maintain the job category, job release, job submission time and other information.

Assignment review maintenance: maintain the assignment review of each course.

Forum maintenance: maintain the posting term, posting course, posting time, posting section (comprehensive discussion, course assignments, teaching video reference resources, course experiments, etc.), posting title, posting content, etc. of the online forum.

Pop up alert: it is an effective means to intervene and urge learners to actively participate in learning. For learners in the platform, the system will automatically calculate the average learning progress information of all learners for each course through algorithms. For learners whose learning progress is lower than a specific proportion of the average learning progress (such as 50%, which can be set), the system will automatically give an alert prompt when the learners log on to the platform to urge learners to improve the course learning progress. SMS alert is similar to pop-up alert. Similarly, the average learning progress of each course is calculated first, and then the system will send the relevant information to the mobile phone terminal in the form of SMS through the mobile phone number reserved by the learners in the teaching platform for the learners who are lower than the specific proportion of the average learning progress, so that the learners can accept supervision without logging into the learning platform. The monitoring of improper learning behavior refers to screening out the factors that can be used to judge whether learners have improper learning behavior according to the learning behavior logs generated by learners in the database, and constructing a monitoring mechanism to monitor and identify the improper learning behavior of learners and ensure the learning quality of learners.

In the way shown above, the construction of monitoring indicators for students' achievement of curriculum goals on the online platform of Xueyin will be achieved, providing a reliable basis for subsequent monitoring and analysis.

2.2 Analysis of Achievement Degree of Curriculum Objectives Based on Weighted Piecewise Function

For the courses to be evaluated, due to the differences in assessment methods and assessment difficulty, the qualified scores of different teaching goals in the same year or the same teaching goal in different years usually have some differences. Therefore, the degree of achievement of different teaching goals is not comparable, and the degree of achievement of curriculum goals in different years is also not comparable. To solve this problem, this research will achieve an objective analysis of the degree of achievement of curriculum objectives based on the weighted piecewise function model.

There are m teaching objectives for the courses to be evaluated, and i The total assessment score of the teaching objectives s_i, the average score is recorded as x_i, qualified points are recorded as c_i, because the total score and qualified score of different teaching objectives are different, this study records the degree of achievement of teaching objectives as d_i Defined as:

$$d_i = \begin{cases} 0.6 + 0.4\dfrac{(x_i - c_i)}{(s_i - c_i)}, x_i \geq c_i \\ 0.6\dfrac{x_i}{c_i}, x_i < c_i \end{cases} \qquad (1)$$

Among them, s_i Can be expressed as

$$s_i = \sum l_i c_{li} + \sum t_i c_{ti} + \sum m_i c_{mi} \qquad (2)$$

Among them, l_i, t_i and m_i They respectively represent the specific data information of learning monitoring indicators, teaching monitoring indicators and management monitoring indicators in the online platform of Xueyin, c_{li}, c_{ti} and c_{mi} They respectively represent the quantitative scores of learning monitoring indicators, teaching monitoring indicators and management monitoring indicators. And there are

$$c_{li} + c_{ti} + c_{mi} = 1 \qquad (3)$$

According to formula (1), d_i It's about x_i Its function image is shown in Fig. 5.

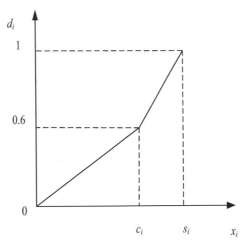

Fig. 5. Function image of course goal attainment

It is easy to find the degree of achievement of teaching objectives defined by Formula (1) in combination with Fig. 5d_i It has the followingcharacteristics:

(1) d_i Is a dimensionless number with a range of [0,1]; And d_i The higher the value of indicates the teaching goal i The better the achievement of.
(2) When d_i When $\in$ [0,0.6), it indicates that the average score of teaching goal i is lower than the qualified threshold c_i.
(3) When d_i When $\in$ [0.6,1], it indicates that the average score of teaching goal i reaches or exceeds the qualification threshold c_i.

Therefore, this study takes 0.6 as the degree of achievement of teaching objectives d_i Eligibility threshold of; Define [0, 0.6) as the degree of achievement of teaching objectives d_i Non conformity range of; Define [0.6, 1] as the degree of achievement of teaching objectives d_i The qualified range of. Achieving degree of course objectives D The weighted comprehensive value of the achievement degree of each teaching goal is used for calculation, that is

$$D = \sum (w_i d_i) \tag{4}$$

Among them, w_i For i Weight of teaching objectives. It is easy to find. After weighted calculation, D It has the following characteristics:

(1) D Is a dimensionless number with a range of [0,1]; And D The higher the value of, the better the achievement of the course to be evaluated.
(2) When each teaching goal is achieved d_i When the threshold value is 0.6, the value after weighted comprehensive calculation is also 0.6.

Therefore, this study takes 0.6 as the degree of achieving the curriculum objectives D Eligibility threshold of; Define [0, 0.6) as the degree of achievement of curriculum objectives D Non conformity range of; Define [0.6, 1] as the degree of achievement of curriculum objectives D The qualified range of.

In this way, we can effectively monitor students' achievement of curriculum objectives in online English teaching in higher vocational colleges.

3   Test Experiment Analysis

3.1   Test Parameter Setting

In order to ensure the reliability and persuasiveness of the research, this study first conducted a four month English listening test on 50 first-year students in a certain vocational college. Based on this, combined with data from listening tests, oral tests, and questionnaire surveys, the reliability of the monitoring methods used was compared, statistically analyzed, and analyzed.

In this study, three tools were used in the experiment, namely, the online platform APP of Xueyin, English listening and speaking test papers and questionnaires. In order to better meet the new requirements of modern English assessment, accurately assess students' English listening and speaking abilities, establish a connection between the assessment format of the senior high school entrance examination and the college entrance examination, and enhance examinees' adaptability, this study aims to test students' listening and speaking abilities using the online platform APP of Xueyin, which serves as the primary data source for this paper. The question type design of the pre-test and post-test papers for oral and listening sections used in the experiment adheres to the requirements of the curriculum standards. It covers the fundamental elements of English listening and speaking skills, encompassing micro skills at different levels of proficiency. The application of intelligent speech technology for scoring is feasible, ensuring the scientificity and flexibility of the test items. Additionally, it is suitable for developing an intelligent scoring question bank, making it beneficial for classroom use. Among them, the listening test questions involve listening to dialogues, sentences, and monologues. The oral test primarily focuses on reading aloud, including question types such as word reading, sentence reading, paragraph reading, and topic expression.

3.2 Test Standards

(1) Hearing test standard

After interpreting the requirements of the listening part of the curriculum standards, the listening test covers words, sentences and paragraphs. The English teaching and research group of the school provides listening papers and answers. The main question types of the listening test paper are to listen to the dialogue and choose the pictures that match each dialogue, listen to the sentences and choose the answers, listen to the monologue and choose the answers. All the question types meet the test scope of this experiment. The answers to the listening questions are fixed, and the standard answers are given by the test group. Through the analysis of listening texts and answers, the listening standard is effective. The listening scoring criteria are shown in Table 1.

Table 1. Scoring Criteria for Listening Test

Test dimension	Question type	Score per question	Number of questions	Total score
word	Listen to sentences (choice question)	1	5	15
sentence	listen to the dialogue (choice question)	1	5	15
Paragraph	Listening to a monologue (choice question)	1	5	15

According to the criteria shown in Table 1, there are 15 listening test questions in total, each of which has a score of 1 point, and the full score is 15 points. One point will be deducted for errors and one point for correctness. Students' final listening scores will be counted according to the standard.

(2) Oral Test Standards

There are some differences in subjectivity and judgment in oral English tests. Therefore, in order to reduce errors, the dimensions of oral English tests at home and abroad are collected and sorted out in the test process. For example, the standard of college English speaking test evaluates learners' oral proficiency from six dimensions: accuracy, sentence length, topic flexibility, coherence and relevance. The scoring criteria for oral test are shown in Table 2.

Table 2. Scoring Criteria for Oral Test

Test dimension	Question type	Score per question	Number of questions	Total score
word	Read to sentences	1	5	15
sentence	Read to the dialogue	1	5	15
Paragraph	Read to a monologue	1	5	15

In the specific evaluation process, four raters are set, and the final score of students is the average score of four raters.

In this way, the reliability, accuracy and differentiation of scoring can be more advantageous.

3.3 Test Results and Analysis

After the test is completed, the collected data is processed with Microsoft Excel. SPSS19.0 and Microsoft Excel are used to carry out independent sample T test for the test data results. The t value, P value and standard deviation in statistics are used to explore the relationship between the achievement of the designed course objectives and the actual learning situation of students in a quantitative way.

(1) Oral Test Results

The independent sample t-test is used to study the monitoring effect of the achievement of oral teaching curriculum goals. The specific results of the significant differences between word reading, sentence reading, short passage reading, topic expression, and total scores are shown in Table 3.

According to Table 3, the scores of actual results and monitoring results in word reading are 2.2805 and 2.1829 respectively,

And the t value is 0.729, which does not reach a significant level at the 0.05 level, so there is no significant difference in word reading between the actual results and the monitoring results.

In sentence reading, the scores of actual results and monitoring results are 2.3171 and 2.2683 respectively, and t value is 0.396, did not reach a significant level at the 0.05 level, so there was no significant difference in sentence reading between the actual results and the monitoring results.

Table 3. Difference Analysis of Monitoring Results of Oral English Teaching Course Goal Achievement

index	option	M	SD	t	p
Word recitation	actual value	2.2805	.5814 7	.729	.468
	Monitor Value	2.1829	.63004		
Sentence recitation	actual value	2.3171	.56741	.396	.693
	Monitor Value	2.2683	.54883		
Short passage reading aloud	actual value	3.5854	1.16137	−.096	.924
	Monitor Value	3.6098	1.13750		
Topic expression	actual value	1.8537	1.31455	−.173	.863
	Monitor Value	1.9024	124107		
Total score	actual value	10.0610	3.09676	.108	0914
	Monitor Value	9.9878	3.04238		

In short text reading, the scores of actual results and monitoring results were 3.5854 and 3.6098 respectively, and the t-value was −0.096, which did not reach a significant level at the 0.05 level, so there was no significant difference between the actual results and monitoring results in short text reading.

In topic expression, the scores of actual results and monitoring results are 1.8537 and 1.9024 respectively, and the t-value is −0.173, which does not reach a significant level at the 0.05 level, so there is no significant difference in topic expression between the actual results and monitoring results.

In the total score, the actual result and the monitoring result are 10.0610 and 9.9878 respectively, and the t value is 0.108,

There is no significant difference in the total score between the actual results and the monitoring results at the 0.05 level.

The results show that there is no significant difference between the actual results and the monitoring results in the overall level of oral English before the experiment.

(2) Hearing test results

The independent sample t is used to test the monitoring effect of the achievement of teaching curriculum objectives. The specific results of the significant differences between the measured word reading, sentence reading, short passage reading, topic expression, and total scores are shown in Table 4.

It can be seen from Table 4 that in listening dialogue, the scores of actual results and monitoring results are 4.5366 and 4.5122 respectively, and the t value is 0.139, which does not reach a significant level at the 0.05 level, so there is no significant difference between the actual results and monitoring results in listening dialogue.

Table 4. Difference analysis of monitoring results of the achievement of listening teaching curriculum objectives

index	option	M	SD	t	p
Listen to the conversation	actual value	4.5366	.80925	.139	.890
	Monitor Value	4.5122	.77852		
Listen to sentences	actual value	3.9268	1.03417	.000	1.000
	Monitor Value	3.9268	1.10432		
Listening to a monologue	actual value	2.1951	1.74956	.063	.950
	Monitor Value	2.1707	1.75930		
Total score	actual value	10.5610	3.11487	−0.71	.944
	Monitor Value	10.6098	3.14546		

In listening sentences, the scores of the actual results and the monitoring results are 3.9268 and 3.9268 respectively, and the t value is 0.000, which does not reach a significant level at the 0.05 level, so there is no significant difference between the actual results and the monitoring results in listening sentences.

In listening to monologue, the scores of actual results and monitoring results were 2.1951 and 2.1707 respectively, and the t value was 0.063, which did not reach a significant level at the 0.05 level, so there was no significant difference in listening to monologue between the actual results and monitoring results.

In the total score, the scores of the actual results and the monitoring results are 10.5610 and 10.6098 respectively, and the t-value is −0.071, which does not reach a significant level at the 0.05 level, so there is no significant difference in the total scores of the actual results and the monitoring results.

The results show that there is no significant difference between the actual results and the monitoring results in the overall level of English listening before the experiment.

Based on the above test results, it can be concluded that the monitoring method designed in this paper, which is based on the online Xueyin platform, can achieve accurate analysis of the achievement of curriculum goals for students in online English teaching in higher vocational colleges. The corresponding monitoring results are highly reliable and have certain guiding value for the actual online teaching work.

4 Conclusion

This article studies the monitoring methods for achieving the goals of online English teaching courses in vocational colleges using the Learning Citation Network Platform as the platform. Based on the test results, the following conclusions can be drawn:

(1) By fully considering the three perspectives of teaching elements, learning elements, and management elements, combined with the characteristics of the Xueyin network platform, it is possible to comprehensively cover different influencing factors and improve the reliability and accuracy of monitoring results.

(2) The use of weighted segmented functions to calculate the quantitative scores of monitoring indicators effectively solves the problem of deviation caused by different evaluation methods and evaluation difficulties.

(3) Be able to adapt to the actual teaching environment and student needs, and improve the consistency between monitoring results and actual situations.

Therefore, the monitoring method based on the learning citation network platform designed in this article can accurately evaluate the achievement of students' curriculum goals, and has good application value in online English teaching in vocational colleges. The application of this method helps to improve teaching quality, promote students' learning outcomes, and provide strong decision-making basis for teachers and managers.

References

1. Li, Z.W., Wang, Z.K., Xia, W.W.: Research on online and offline public opinion linkage of college students based on system dynamics. Comput. Simul. **40**(2), 6 (2023)
2. Co, M., Chu, K.M.: A prospective case-control study on online teaching of ultrasonography skills to medical students during COVID-19 pandemic. Heliyon **8**(1), e08744 (2022)
3. Lahon, J., Shukla, P., Singh, A., et al.: Coronavirus disease 2019 pandemic and medical students' knowledge, attitude, and practice on online teaching-learning process in a medical institute in North India. Natl. J. Physiol. Pharmacy Pharmacol. **11**(7), 1 (2021)
4. Chinnasamy G . A study on the instant switch to online teaching and learning in private schools during the Covid-19 pandemic. Int. J. Anal. Exp. Mod. Anal. **VII**(VIII), 7 (2021)
5. Chen, L.: Application of artificial intelligence technology in personalized online teaching under the background of big data. J. Phys. Conf. Ser. **1744**(4), 042208 (2021)
6. Luo, W.W.: Building a neural network model to analyze teachers' satisfaction with online teaching during the COVID-19 ravages. J. Comput. Commun. **10**(1), 91–114 (2022)
7. Zhang, N.: A brief analysis of how online teaching can develop steadily and be far-reaching in the new era-a case study of practical course for college music programs. Psychol. Res. **12**(7), 525–529 (2022)
8. Ironsi, C.S.: A switch from flipped classrooms to emergency remote online teaching (EROT): misconceptions, instructors and preservice teachers perceptions. Int. J. Inf. Learn. Technol. **39**(1), 13–28 (2022)
9. Siegel, V., Moore, G., Siegel, L.: Improving nursing students' knowledge and assessment skills regarding skin cancer using online teaching resources. J. Dermatol. Nurses' Assoc. **13**(6), 305–308 (2021)
10. Huang, Y.Y.: Integrating online teaching into public physical education - taking vocational colleges in Chongqing as examples. J. Contemp. Educ. Res. **6**(4), 102–107 (2022)
11. Zhuang, H.: Practice of blended learning activities in higher vocational education under MOOC environment. Office Inf. **26**(21), 22–23 (2021)
12. Lian, Y., Wang, M., Wang, Z., Sun, L.: MOOC recommendation model based on visual analysis and graph convolutional network. Exp. Technol. Manag. **39**(6), 34–42 (2022)
13. Chen, Q., Zhang, Y., Wang, S., Xia, L.: The importance and experience of curriculum objective design. Chin. Med. Mod. Dist. Educ. Chin. **19**(16), 177–179 (2021)
14. Lin, X., Nan, G., Yang, D., Zhang, C., Wang, L., Zhu, Q.: Development and practice of college-hospital cooperated, two-stage module-based vocational nursing curriculum based on modern apprenticeship. J. Nurs. Sci. **36**(10), 67–70 (2021)
15. Lai, L., Peng, L.: Online teaching service quality evaluation index system construction in colleges and universities. Heilongjiang Res. Higher Educ. **39**(6), 147–154 (2021)

Research on the Push of Online Teaching Resources for Innovation and Entrepreneurship Based on User Characteristics

Meiling Ou[✉]

Chongqing Vocational Institute of Engineering, Jiangjin 402260, China
m19912412058@163.com

Abstract. One of the main reasons for the low extraction rate of online teaching resources is that push resources do not meet user needs. In order to improve the resource extraction rate of online teaching resource push, optimize the efficiency and quality of push, and achieve the ideal effect of teaching resource push, the principle of user characteristics analysis is introduced. Taking innovation and entrepreneurship courses as an example, the research on innovation and entrepreneurship online teaching resource push based on user characteristics is carried out. First, select online teaching resource filter encoder to obtain the characteristics of innovation and entrepreneurship online teaching resources. Secondly, design the frequency of online teaching resources push cycle and push presentation mode to lay a good foundation for online teaching resources push. In order to ensure the security of the storage of massive online teaching resources for innovation and entrepreneurship, a database of teaching resources is established. On this basis, using the user characteristics analysis method, this paper makes an objective analysis of the specific characteristics of the users who push online teaching resources for innovation and entrepreneurship. According to the user characteristics, it uses targeted personalized teaching resources push methods to achieve the goal of resource oriented push. The experimental analysis results show that after the application of the new push method, the resource extraction rate of the resource package has reached more than 97%, and the push effect has significant advantages.

Keywords: User characteristics · Innovation and entrepreneurship · Online teaching · Resources · Push

1 Introduction

In recent years, with the continuous development of "Internet + education" construction, China's education has been constantly changing and innovating. With the development of the education industry in China, new online learning platforms are emerging on the network platform, which provides great convenience for learners to learn [1]. Due to the existence of massive learning resources on the online learning platform, learners are prone to learning disorientation, cognitive overload and fragmented learning in the

© ICST Institute for Computer Sciences, Social Informatics and Telecommunications Engineering 2024
Published by Springer Nature Switzerland AG 2024. All Rights Reserved
L. Yun et al. (Eds.): ADHIP 2023, LNICST 547, pp. 321–336, 2024.
https://doi.org/10.1007/978-3-031-50543-0_22

process of learning, resulting in the inability to fully master the learning curriculum. In this context, the concept of personalized learning has brought a new direction for the development of online learning platform [2].

The concept of push service first appeared in an article of American Information Weekly. This service comes from information recommendation research. Its main goal is to actively recommend resources to demanders. Knowledge push is defined as a network knowledge service mode. This technology has received extensive attention since it was proposed, and many organizations and individuals have invested in relevant research [3]. In the short term, it has achieved successful commercial applications, such as Dangdang, Amazon, eBay, Taobao, etc., which we are very familiar with. The successful application of the recommendation system in the above online business activities has promoted the continuous improvement of the recommendation algorithm, but so far there is no mature recommendation service for educational resources [4].

Push service provides a new mode and method for the acquisition of educational resources and the establishment of scientific research knowledge system under the network environment, and fundamentally changes people's means of information acquisition and learning [5]. Starting from the sharing level of educational resources, promote new ways for services to provide education, learning and scientific research. The active information service based on user interest changes the passive state of looking for information into the active learning mode of looking for information. The main forms of active information service include personalized recommendation, personalized retrieval, and personalized customization. The educational resources involved in this article are push services, which mainly involve personalized recommendation services. The former passive query of online education resources has been transformed into the active recommendation service of education resources [6].

In recent years, the state has attached great importance to innovation and entrepreneurship education, and application-oriented undergraduate universities have become the main force in cultivating innovative talents. The upsurge of innovation and entrepreneurship education is sweeping through application-oriented undergraduate colleges. Innovation and entrepreneurship courses play an important role in the process of cultivating innovative and entrepreneurial talents in colleges and universities. However, the current curriculum construction is not yet mature, the innovation and entrepreneurship curriculum system is not yet perfect, and the online teaching resource push method is not mature enough to achieve the best effect of innovation and entrepreneurship teaching. User feature analysis can improve the above problems, which means that it can be used in the user groups facing our product design to analyze the typical user characteristics, understand the situation of target user groups, describe the typical target users, understand their group behavior, clearly know the user demands, and provide the core effect basis for the design of function points. Based on this, this paper puts forward the research on online teaching resource push of innovation and entrepreneurship based on user characteristics.

2 Research on the Push of Online Teaching Resources for Innovation and Entrepreneurship

2.1 Select Online Teaching Resources to Filter Encoder

In the online teaching resource push method of innovation and entrepreneurship based on user characteristics designed in this paper, first, simulate the real teaching environment of innovation and entrepreneurship courses to generate a teaching resource dataset. It is stored in the data warehouse in the form of JavaSript Object Notation (JSON). Secondly, the curriculum relationship chain is generated according to the students' learning behavior to further describe the characteristics of the existence of resources. Finally, open the online teaching platform and use the crawler technology to process the resource data. If the data filtered out at this time is missing, you can use the user data file to fill in, so as to obtain the correlation between the data [7]. Combining the above steps, online course teaching resource filter encoder can be generated, as shown in Fig. 1.

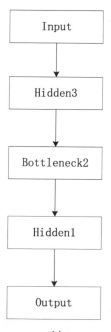

Fig. 1. Online course teaching resource filter encoder

It can be seen from Fig. 1 that the online course resource filtering encoder can use the hidden function to carry out vector mapping, activate the internal coding space, and obtain the final resource collection result. The calculation formula of the descriptive formula y is:

$$y = f(qX + c) \tag{1}$$

Among them, f Represents the activation function of online teaching resources for innovation and entrepreneurship; q Represents the encoder weight; X Represents the description offset; c Represents hiding input vectors. It can be decoded by combining the above calculated values x_i The calculation formula of is:

$$x_i = f(q_i X + c_2) \tag{2}$$

Among them, q_i Represents decoding weight; c_2 Represents the decoding offset. At this point, you can generate a loss function that meets the resource push requirements Q, the calculation formula is:

$$Q = \sum_{i=1}^{M} \|x_i - y\|^2 \tag{3}$$

where, M Represents the number of learners. If the loss value of the resource collection encoder is too high, it needs to perform linear interactive processing to generate the final resource representation vector k, the calculation formula is:

$$k = soft \max(wq_i + c_2) \tag{4}$$

Among them, w Indicates the resource attention parameter. Using the above resource representation vector, we can effectively obtain the characteristics of online teaching resources for innovation and entrepreneurship, and then achieve effective filtering of online teaching resources.

2.2 Frequency Design of Online Teaching Resources Push Cycle

In the implementation process of push, researchers and teachers need to design and control the cycle frequency and presentation mode of personalized online teaching resources to provide basic support for subsequent resource push.

The frequency of the push cycle of personalized online teaching resources is determined by the teacher according to the teaching needs. This paper suggests that the teacher provide the prepared subject test after each new lesson. Input the test data into the push system, and the push system will automatically provide personalized online teaching resources for learners. For other courses, such as review lessons and exercise lessons, teachers should decide whether to provide personalized online teaching resources to students according to teaching needs [8]. Although it is possible to push personalized learning resources supported by computer technology, in the field of basic education, the popularity of computers and mobile devices is still relatively low. Most schools and parents do not support students to use electronic devices for learning, which has caused some obstacles to our personalized online teaching resources push [9]. Therefore, in the actual teaching process, the push cycle frequency of personalized online teaching resources is determined by the cycle frequency of student evaluation data collection. According to the teaching needs, the teacher prepares the subject test in the system before class, prints it into paper version, and distributes the test to students after class. After retrieving the student evaluation data, input the data into the push system, and provide students with personalized online teaching resources output by the system within 12 h.

2.3   Design of Online Teaching Resources Push Presentation Mode

After completing the design of the frequency of online teaching resources push cycle, the next step is to comprehensively design the presentation mode of teaching resources push.

In order to ensure that learners can obtain personalized online teaching resources, this paper provides learners with two presentation methods, electronic version and paper version. If students have electronic equipment, they can directly access electronic personalized online teaching resources on the equipment, and can quickly obtain feedback results of consolidation exercises. If students are boarding schools and cannot obtain electronic online teaching resources, teachers are required to print the personalized online teaching resources of each learner in the system into a paper version and distribute them to students. After they finish, they will consolidate the collection of evaluation data in the exercise part and update the learner model iteratively.

2.4   Establish Online Teaching Resource Database for Innovation and Entrepreneurship

After the design and control of the cycle frequency and presentation mode of personalized online teaching resources are completed, in order to ensure the safety of the storage of massive innovation and entrepreneurship online teaching resources, a teaching resources database is established.

First, build a knowledge map of online teaching resources for innovation and entrepreneurship, connect the dependencies between curriculum knowledge, and plan the path for resource data storage. The construction process of online teaching resource knowledge map is shown in Fig. 2.

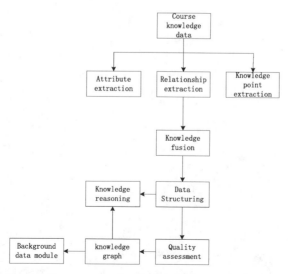

Fig. 2. Construction process of online teaching resource knowledge map

As shown in Fig. 2, Step 1: determine the content outline of online innovation and entrepreneurship teaching courses that need to build a knowledge map. Step 2: analyze the curriculum content outline, complete the extraction of all knowledge points contained in the curriculum content outline, analyze the extracted knowledge points, determine the association relationship between different knowledge points according to the curriculum outline, and form the knowledge point data set and association relationship set. Step 3: Clean and integrate the course content knowledge point data set extracted in step 2 and the association relationship set between different knowledge points to remove errors and redundant data. Step 4: Add the processed knowledge entity to the knowledge entity database, complete the description of the concept hierarchy system, and build a preliminary model of the curriculum knowledge map. Step 5: Design the storage structure [10] for the constructed curriculum knowledge map in the background data module.

On this basis, the database is established. The teaching resource data is mainly uploaded and managed by the system administrator. At the same time, the teacher user can upload and manage the resources through the teacher end application. The resource types are text, picture, sound, and video. The data storage structure of innovation and entrepreneurship online teaching resource database is shown in Table 1.

Table 1. Online Teaching Resources Database

field	explain	type
ZY-ID	Resource id	Int
ZY-MC	Resource name	String
ZY-LX	Resource Type	Int
ZY-YT	Resource use	Int
ZY-TGZ	Resource provider id	Int
ZY-SCSJ	Resource upload time	String
ZY-BCLJ	Resource saving path	String
SSYHFL	User classification	Int
ZY-SM	Resource description	Int
WJGS	file format	Int
FLJB	Classification level	Int
FLQZ	Classification weight	Int
LLKSSJ	Browse Start Time	String
LLJSSJ	Browse End Time	String

Store and manage innovation and entrepreneurship online teaching resources according to the database table shown in Table 1.

2.5 Personalized Teaching Resources Push Based on User Characteristics

After the establishment of online teaching resource database based on the above innovation and entrepreneurship, the goal of massive online teaching resource storage has been achieved. On this basis, the user characteristics analysis method is used to make an objective analysis of the specific characteristics of the users of innovative and entrepreneurial online teaching resource push. According to the user characteristics, targeted personalized teaching resource push methods are adopted to achieve the goal of resource oriented push and achieve the best effect of teaching resource push.

The user feature analysis structure is shown in Fig. 3.

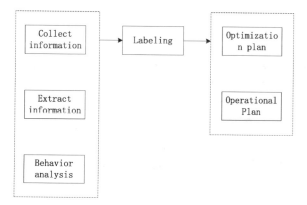

Fig. 3. Structure diagram of user characteristics analysis

As shown in Fig. 3, first, collect user related information to extract useful information data. According to the extracted information data, the user behavior characteristics are analyzed based on the operation logic of user feature analysis. According to the analysis results of user behavior characteristics, users are labeled accordingly. The analysis of user characteristics is the key to understand the users' demands. Most of the users' demands and some of the more hidden demands cannot be mined through questionnaires and user interviews, which are relatively superficial methods. By analyzing users' behaviors and characteristics, we can have a deeper understanding of users' demands. When analyzing the characteristics, the method of mapping groups from individuals is adopted. Individual analysis - > key characteristics - > recall verification (technology returns the characteristics to the user database to find the corresponding users) - > solutions.

On this basis, design personalized teaching resources push system. The online teaching resources push system designed in this paper mainly includes three important modules, namely user model, push object model and push algorithm. User model is the key module to obtain user information and determine user needs; The push object model is a content model pushed for users. The push algorithm is a principle algorithm that matches user needs with the pushed content. The general model of innovation and entrepreneurship online teaching resource push system based on user characteristics designed in this paper is shown in Fig. 4.

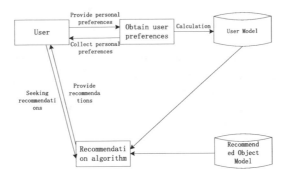

Fig. 4. General Model of Online Teaching Resources Push System

As shown in Fig. 4, the construction of the user model is the basis of the push system. Before building the user module, you need to think about the objects to be modeled and the data to be collected. The general process of user modeling includes six steps: obtaining user information, establishing user model, using the model to push, providing push results, pushing results feedback or demand changes, and updating the user model. There are two ways to obtain user data: explicit input and implicit acquisition. The output content of the user model is mostly the user's interests or needs. How to model the content of the push object is also a key factor affecting the push system. Before modeling, you need to consider which features of the push object to describe. How to establish the relationship between the feature description of the push object and the user model. Whether the properties of the push object need to be updated automatically, etc. There are two ways to build the push object model, namely, the manager's own tag and the user generated data. The appropriate way can be selected according to the needs of this study. Push algorithm is the core module of the push system, which mainly includes four push algorithms: content-based push, collaborative filtering based push, association rule based push, knowledge-based push and hybrid push.

First of all, determine the learners' nearest development zone, locate their learning needs, and push learning resources suitable for their learning level to promote their learning effect, knowledge structure integrity and understanding level. After combining the advantages and disadvantages of several common recommendation algorithms, combined with the characteristics of this study, a hybrid push strategy suitable for this situation is designed, including automatic recommendation of the system and manual recommendation of teachers.

The online teaching resource push method designed in this paper adopts a hybrid recommendation algorithm, including knowledge-based push and association rule based push algorithms. Knowledge based recommendation can be seen as a reasoning technology, which is different from other recommendation algorithms and is not based on user preferences and needs. Functional knowledge is a kind of knowledge about how a project meets a specific user, so it can explain the relationship between needs and push. Therefore, user data can be any knowledge structure that supports reasoning, can be a standardized query by users, or can be a more detailed representation of user needs. The push algorithm based on association rules allows managers to manually set push rules according to the push requirements to provide users with push content. Teacher recommendation means that when students have reached their learning goals and the system has been unable to automatically associate related resources, teachers need to manually push learning resources according to students' learning conditions.

In order to improve the usability of the recommendation model in online teaching practice and enable students to effectively obtain personalized learning resources, this study designed a personalized learning resource push process based on the analysis of evaluation data, and established a push mechanism and specification for specific push cycle frequency and push presentation mode. The push process in online teaching of push model is shown in Fig. 5.

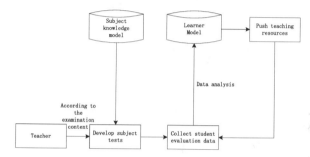

Fig. 5. Schematic diagram of online teaching resources push process

As shown in Fig. 5, first of all, teachers divide the examination content into corresponding knowledge points, call the corresponding questions in the question resource database, and prepare subject tests according to the test preparation requirements proposed by this research; Secondly, the subject test is distributed to students to complete, and the intelligent evaluation data collection system is used to obtain the student evaluation data, and the student evaluation data is analyzed to obtain the knowledge structure and understanding level of learners; Then, a personalized learning resource push model based on evaluation data analysis is adopted to push personalized learning resources for learners; Finally, the evaluation data of students are collected iteratively, analyzed, processed and pushed in a new round. On this basis, the intelligent customized push strategy of teaching resources is designed from two aspects: teacher push and learner push.

Teachers' push strategies are divided into two types: one is resource push for students in the class, and the other is resource push for a student. In the existing education environment, for a certain course, the teaching materials used by teachers are usually unified and customized, and the teaching content and objectives are relatively fixed. The advantage of this kind of teaching materials is that the quality of teaching resources is high, suitable for most learners, and can make teachers stand at a higher level to teach, but also make teachers' teaching less flexible. In actual teaching, students in different classes have different mastery of learning content due to factors such as learning preferences and teachers' teaching methods. However, due to limited classroom time, it is difficult for teachers to have time to provide personalized teaching to students after completing the teaching tasks required by the syllabus. Therefore, the personalized learning resource push system needs to assist teachers to carry out personalized teaching. Through the use of the system, teachers can provide targeted learning resources to students in their own class according to their learning characteristics, teaching plans and teaching objectives, combined with their own teaching experience, and students can access the resources provided by teachers through the system. The diagram of online teaching resources pushed by teachers is shown in Fig. 6.

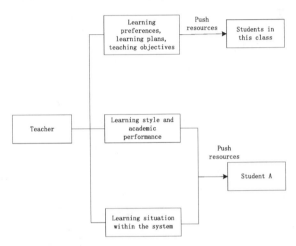

Fig. 6. Schematic diagram of online teaching resources pushed by teachers

As shown in Fig. 6, the resources uploaded by teachers can be the existing learning resources in the system, and teachers can provide them to students through screening; It can also be the learning resources made by teachers themselves. After teachers upload the resources to the server, the system will push the resources uploaded by teachers to the designated class or a student.

The diagram of customized push of online teaching resources by learners is shown in Fig. 7.

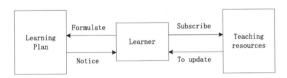

Fig. 7. Schematic diagram of customized push of online teaching resources for learners

As shown in Fig. 7, there are two kinds of customized push for learners: learner subscription and learner plan. Learner subscription refers to that if a learner is interested in a certain type of resource during the learning process, he or she can subscribe to this type of resource. When this type of resource in the system is updated, the system will immediately push it to the learner, and the learner can also cancel the subscription. Then the system will not push this type of resource to the learner. Learner plan means that learners can make a learning plan for themselves according to their own learning situation. If learners do not complete the corresponding learning plan within the specified time, the system will remind users to learn. If learners complete the learning plan, they will not be reminded. The learning plan can be a learning plan for a certain day or a certain period of time. The customized push of learners can help learners to complete the learning plan.

Through the above process steps, we can achieve the goal of pushing online teaching resources for innovation and entrepreneurship.

3 Experimental Analysis

3.1 Experiment Preparation

Experimental testing is an important step in the process of method design. After the development of the proposed online teaching resource push method is completed, we can know whether the method has realized the functional requirements through experimental testing, and run according to the designed workflow. Before the proposed push method is opened to users, through testing, we can find out whether there are functional defects in the operation of the method and abnormal problems in the development process, and repair them in time to provide users with a better experience. This chapter mainly introduces the testing process of the online teaching resource push method designed and implemented in this paper. Firstly, the mobile devices used for testing are introduced, and the methods are tested and analyzed from the aspects of method function and performance. First, build the test environment for this experiment. The innovation and entrepreneurship online teaching resource push method designed and implemented in this paper provides users with mobile applications including teacher end applications and student end applications, teacher end applications for teacher users, and mobile end applications for student users. Since the application end of the system supports the Android operating system and iOS operating system, in order to verify that the application end can operate normally on these two operating systems, this paper selects the Huawei tablet with the Android operating system and the iPhone 13 with the iOS operating system as the test platform. Among them, the software and hardware information of tablet devices equipped with Android operating system is shown in Table 2.

Table 2. Android operating system tablet device configuration

number	project	to configure
1	operating system	Android 7.0
2	processor	Snapdragon 425
3	Running memory	3.0GB
4	storage space	32.00GB
5	resolving power	1280 * 800 pixels

Table 3 shows the software and hardware information of iPhone13 of ios operating system.

Table 3. Configuration of iOS operating system iPhone13

number	project	to configure
1	operating system	iOS 11.2.6(15D100)
2	processor	AppleA10 + M10 coprocessor
3	Running memory	2.0GB
4	storage space	32.00GB
5	resolving power	1334 * 750 pixels

According to the configuration parameters shown in Table 2 and Table 3, after building the test environment, test the function of the push method to see if the system meets the functional requirements, find the functional deficiencies of the method through the test, and repair the functional deficiencies of the method in a timely manner. The operation method used in the test is to list the expected effect of the teacher end application and personal end application functions respectively, and then test each function of the application according to the user process. If the effect is consistent with the expectation, then the push method meets the functional requirements. Table 4 shows the test results of online teaching resource push methods for innovation and entrepreneurship.

Table 4. Test Results of Teaching Resource Push Function

Test content	Expected results	test result
Resource upload	Teacher users can select local resources to upload to the system database	success
resource management	Teacher users can view and use authorized resources on the application end, and can delete uploaded resources	success
Resource push	Teacher users can select resources to push to class students for learning	success
Receive teaching resources	Student users can normally receive the learning resources pushed by the system and the class teachers	success
Complete resource learning	Student users can select resources to complete learning. During the learning process, the system can complete the collection, upload and storage of student user learning behavior data	success
Information prompt during use	The application can correctly use the pop-up frame to complete the information prompt according to the process	success

As shown in Table 4, the resource upload module provides a convenient way for teachers to easily upload local resources to the system database, facilitating the sharing and management of teaching resources with students. The resource management module enables teachers to easily view, use, and manage authorized resources, while also providing the function of deleting uploaded resources, which is beneficial for teachers to better organize and control teaching resources. The resource push module allows teachers to choose suitable resources to push to class students, providing flexible teaching tools for teachers to effectively convey learning resources to students. The receiving teaching resource module ensures that students can receive the learning resources pushed by the system and teachers normally, providing convenient learning pathways and resource acquisition channels for students. The completion of the resource learning module allows students to choose resources for learning and ensures that the system can collect, upload, and store students' learning behavior data, providing teachers with an understanding and evaluation of students' learning situation. The information prompt module during user use ensures that the application can accurately use pop-up boxes for information prompts according to the process, providing a good user experience and guidance. The functions of the innovation and entrepreneurship online teaching resource push method proposed in this paper based on user characteristics meet the requirements and pass the functional test. The online teaching resource push method not only needs to meet the functional requirements of the design, but also should have efficient execution efficiency, bringing smooth operation experience to users. Next, test the performance of the push method, and then determine the feasibility of the proposed method.

3.2 Result Analysis

In order to verify the actual push effect of different online teaching resource push methods, the appropriate course resource package was selected, and the innovation and entrepreneurship online teaching resource push method designed in this paper based on user characteristics was set as the experimental group, the traditional teaching resource push method was set as the control group, and the two online teaching resource push methods were compared. In order to facilitate the acquisition of experimental data, an effective Scrapy grabbing framework is generated using Python. This grabbing framework can effectively extract important teaching resources in the interface and generate resource packages required for experiments. The resource fetching framework is shown in Fig. 8.

It can be seen from Fig. 8 that in the process of running the Scrapy framework, resource scheduling instructions need to be obtained through the crawler first for sorting and classification processing. Next, transfer the intermediate file to the downloader to save the download data. Finally, stop the resource extraction program, generate 10 resource packages, and give them numbers according to the total amount of their internal resources. The larger the number, the higher the total amount of internal resources. In order to improve the experiment efficiency and reduce the experiment time, the multi process technology is used to analyze and allocate data, as shown in Fig. 9.

It can be seen from Fig. 9 that the experimental curriculum data after the above data analysis and distribution process meets the needs of this experiment, and then experimental indicators can be selected. Resource extraction rate of resource package P. As an experimental index, its calculation formula is:

$$P = \frac{L_R}{F} \times 100\% \tag{5}$$

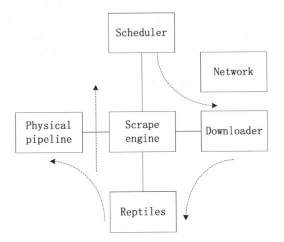

Fig. 8. Resource Grabbing Framework

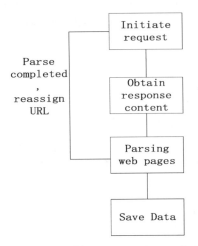

Fig. 9. Flow chart of online teaching resource data analysis and distribution

Among them, L_R Indicates the amount of online teaching resources targeted for innovation and entrepreneurship; F It represents the total amount of resources in the online teaching resource package of innovation and entrepreneurship. The higher the resource extraction rate is, the better the actual extraction effect is. On the contrary, the lower the resource extraction rate is, the worse the extraction effect is. In order to minimize the impact of the differences in experimental indicators on the final experimental results, the curriculum data was normalized. The above two online teaching resource push methods were used to carry out experiments, and SPSS statistical analysis software was used to measure and calculate the resource extraction rate of the resource package and make a comparison. The comparison results are shown in Table 5.

Table 5. Comparison Results of Resource Extraction Rate of Resource Packages

Resource package	experience group	control group
ZYB01	98.62%	78.47%
ZYB02	97.05%	62.16%
ZYB03	98.13%	80.26%
ZYB04	98.06%	77.61%
ZYB05	98.49%	60.52%
ZYB06	97.65%	78.96%
ZYB07	97.88%	79.46%
ZYB08	98.63%	77.43%
ZYB09	98.47%	76.59%
ZYB10	97.49%	77.26%
ZYB11	99.02%	76.43%
ZYB12	98.57%	81.29%
ZYB13	98.66%	83.49%
ZYB14	97.05%	83.64%
ZYB15	98.63%	80.54%

The comparison results in Table 5 show that the two online teaching resource push methods show different performance effects after application. Among them, after the application of the online teaching resource push method of innovation and entrepreneurship based on user characteristics proposed in this paper, the resource extraction rate of the resource package is always higher than that of the traditional method, reaching more than 97%, while the resource extraction rate of the traditional method is the highest, not more than 83.64%. It is not difficult to see from the comparison results that the push method proposed in this paper has high feasibility, high push effect and certain application value.

4   Conclusion

With the reform of educational informatization, online and offline hybrid teaching mode has gradually replaced the traditional teaching mode as the leading teaching mode, and various colleges and universities have used various.

The teaching is carried out on an information-based teaching platform. These teaching platforms are often poorly targeted in the process of pushing teaching resources, and cannot be guaranteed the actual resource extraction rate. Therefore, Research on the push of online teaching resources for innovation and entrepreneurship based on user characteristics. This method first selects an online teaching resource filtering encoder to obtain the characteristics of innovation and entrepreneurship online teaching resources. Secondly, design the frequency and presentation of online teaching resource push cycles, laying a solid foundation for online teaching resource push. To ensure the security of the storage of massive online teaching resources for innovation and entrepreneurship, a teaching resource database is established. On this basis, the user feature analysis method is used to objectively analyze the specific characteristics of users who push online teaching resources for innovation and entrepreneurship. Based on user characteristics, targeted personalized teaching resource push methods are adopted to achieve the goal of resource targeted push. The experimental results show that the method has higher resource extraction rate, better push effect, and is conducive to improving teaching efficiency, broadening students' knowledge, presenting online teaching resources in a variety of ways. The learning form is more flexible, highly targeted, and has certain application value.

References

1. Lu, J., Gao, H.: Online teaching wireless video stream resource dynamic allocation method considering node ability. Sci. Program. **2022**, 1–8 (2022)
2. Dong, X., Chen, X.: Research on online teaching of college teachers under the background of education informatization. MATEC Web Conf. **336**, 05005 (2021)
3. Turk, M., Heddy, B.C., Danielson, R.W.: Teaching and social presences supporting basic needs satisfaction in online learning environments: How can presences and basic needs happily meet online? Comput. Educ. **180**, 104432 (2022)
4. Zheng, H.Y., Ran, X.C.: Application of QR Code Online Testing Technology in Nursing Teaching in Colleges and Universities. Sci. Program. **2021**(Pt. 13) (2021)
5. He, Y.: Design of online and offline integration teaching system for body sense dance based on cloud computing. J. Interconnect. Networks **22**(Supp05) (2022)
6. Alsuwaida, N.: Online courses in art and design during the coronavirus (COVID-19) pandemic: teaching reflections from a first-time online instructor. SAGE Open **12**(1), 59–66 (2022)
7. Nurtayeva, A., Abdirassilova, G., Karbozova, M., et al.: Teaching online enantiosemia and word diffusion aspects during the coronavirus pandemics. XLinguae **14**(1), 289–298 (2021)
8. Chen, H., Huang, J.: Research and application of the interactive English online teaching system based on the internet of things. Sci. Program. **2021**(S1), 1–10 (2021)
9. Joycilin, S.A., Shakeel, Z., Sylvia, V.: Classroom teaching to remote teaching: preparedness of teachers of HEIs in India. J. Crit. Rev. **7**(9), 3303–3309 (2021)
10. Cheng, P., Ming, D., Man, X., et al.: Optimized allocation of tennis teaching resources based on big data. J. Phys. Conf. Ser. **1744**(4), 042138 (2021)
11. Liu, T., Yang, Z.: Personalized recommendation method of sports online video teaching resources based on multiuser characteristics. Math. Probl. Eng. 2022 (2022)
12. Feng, J., Zhang, W., Tsai, S.B.: Construction of a multimedia-based university ideological and political big data cloud service teaching resource sharing model. Mathematical Problems in Engineering: Theory, Methods and Applications, 2021 (2021-Pt.52) (2021)
13. Darmoni, S., Thirion, B., Pourchez, B., et al.: Use of information and communication technologies to retrieve French pre-residency examination program teaching resources on the Internet (2022)
14. Zheng, W., Muthu, B.A., Kadry, S.N.: Research on the design of analytical communication and information model for teaching resources with cloud: haring platform. Comput. Appl. Eng. Educ. (1) (2021)
15. Han, Y.: The construction of Chinese online resource database and teaching implementation in higher vocational education (2021)

A Method for Digital Resource Allocation in Mobile Online Education Based on Ant Colony Algorithm

Yan Huang[1(✉)] and Xiaotang Geng[2]

[1] School of Tourism and Humanities, Heilongjiang Polytechnic, Harbin 150086, China
hy8161@126.com
[2] Heilongjiang Polytechnic, Harbin 150086, China

Abstract. In response to the problems of low accuracy and long time consumption in traditional methods for mobile online education digital resource allocation, a mobile online education digital resource allocation method based on ant colony algorithm is proposed. Firstly, the characteristics of mobile online education digital resources were extracted and the key nodes for resource allocation were determined. Based on this, we can accurately grasp the core issues of resource allocation and improve the accuracy and efficiency of resource allocation. On this basis, the Ant colony optimization algorithms is used as the basis of resource allocation. By initializing the mobile online education digital resource allocation Pheromone, and according to the resource allocation path selection rules, the dynamic Pheromone update is carried out. By simulating the behavior of ant colonies in searching for food, the allocation of digital resources for mobile online education is achieved. The resource allocation method based on Ant colony optimization algorithms has better global search ability and adaptability, can better guide the process of resource allocation, and improve the accuracy and efficiency of resource allocation. The experimental results show that this method has a shorter extraction time for educational digital resource information, better integration and output of educational digital resources, better accuracy in resource allocation, and higher allocation efficiency. The experimental results show that the method in this paper takes less time to extract the information of educational digital resources, the integration and output of educational digital resources are better, the accuracy of resource allocation is better, and the allocative efficiency is better.

Keywords: Ant Colony Algorithm · Mobile Online Education · Characteristics of Teaching Resources · Digital Resource Allocation

1 Introduction

Mobile online education is a rapidly emerging form of education with the rapid development of the mobile internet. It advocates autonomy, interactivity, and diversity, allowing learners to access personalized learning services anytime and anywhere without time and

L. Yun et al. (Eds.): ADHIP 2023, LNICST 547, pp. 337–349, 2024.
https://doi.org/10.1007/978-3-031-50543-0_23

spatial distance constraints. It has become a new direction and trend in current educational development. At the same time, mobile online education has also brought a large amount of digital resources, such as video, audio, graphic and other forms of course materials, which require efficient allocation and management to meet the rapidly growing needs of learners. In this trend, the allocation of digital resources for mobile online education is becoming increasingly important [1]. The problems of digital resource allocation include resource waste, unfairness, and low efficiency, which pose challenges to the development of mobile online education. How to effectively allocate digital resources, fully utilize the benefits of resources, and improve user experience has become an urgent problem to be solved. The allocation of digital resources involves various factors, such as network bandwidth, user scenario requirements, resource types, number of users, and so on. Currently, statistical methods are mainly used for the allocation of digital resources, but their shortcomings lie in their inability to address the differentiated needs of learners. With the continuous development of internet technology and artificial intelligence technology, researchers have begun to consider how to apply internet technology and artificial intelligence technology to achieve accurate and efficient allocation of digital resources for mobile online education, and continuously optimize and adjust configuration methods to improve the quality and effectiveness of resource allocation [2].

Reference [3] proposed an online ideological and political education resource allocation method based on the decision tree algorithm. It uses the mutual information method to extract the characteristics of online ideological and political education resources, uses the decision tree algorithm to build a decision tree classifier, and inputs the extracted characteristics of ideological and political education resources into the classifier to complete resource allocation. This method has a high accuracy of educational digital resource allocation, but poor allocative efficiency. Reference [4] proposes a classification method for college English teaching resources based on density clustering algorithm. Firstly, fully consider the correlation between resources between adjacent grids and construct a weighted grid for each resource partition; Secondly, the corresponding weights are set based on resource correlation. After calculating the density parameters of the grid cells, the COMCORE-MR algorithm is used to determine the range of Key value parameter values; Finally, when the Key value parameter value is within the given density threshold parameter range of the grid unit, the corresponding educational resources and the central target grid object are divided into similar resources. This method has higher allocative efficiency of educational digital resources, but poor accuracy.

In response to the problems of the above methods, this article proposes a mobile online education digital resource allocation method based on ant colony algorithm. The Ant colony optimization algorithms is used as the basis of resource allocation, and the allocation of digital resources for mobile online education is realized by simulating the behavior of ant colony in the process of searching for food. The dynamic update mechanism can adjust the weight of Pheromone according to the search results of Ant colony optimization algorithms and the changes of the objective function, so as to better guide the process of resource allocation and improve the accuracy and efficiency of resource allocation. The resource allocation method based on Ant colony optimization algorithms has better global search ability and adaptability, and can effectively solve the problems of low precision and long time consumption in traditional methods.

2 Mobile Online Education Digital Resource Preprocessing

In mobile online education, the preprocessing of digital resources plays an important role in improving the availability and effectiveness of resources, providing better support and assistance for the teaching process.

For digital resource preprocessing, multiple aspects need to be considered. Firstly, it is necessary to process educational resources, including cleaning, integrating, optimizing, etc., to ensure the quality and availability of resources, and to improve the access and browsing speed of resources. Secondly, it is necessary to consider the characteristics and limitations of mobile devices and adapt or tailor resources to ensure smooth playback and display effects on mobile devices. In addition, it is also necessary to consider the combination with teaching scenarios to better serve teaching and enable it to be better integrated into the teaching process.

Digital resource preprocessing is an indispensable part of mobile online education, which can improve the quality and effectiveness of digital resources and provide better learning support and assistance for students. At the same time, digital resource preprocessing is also a continuous process of updating and optimizing, requiring continuous exploration and experimentation with new technologies and strategies to meet the constantly changing educational and student needs.

2.1 Feature Extraction of Digital Resources in Mobile Online Education

In mobile online education, feature extraction of digital resources is an important link in achieving efficient allocation and management of digital resources. By extracting features from digital resources, they can be transformed into numerical data that machine learning algorithms can process for better analysis and application. The features of digital resources also include but are not limited to various types such as visual features, speech features, text features, etc. These features can provide a foundation and basis for the classification, recommendation, and search of digital resources. In the application process of mobile online education digital resources, with the increase of teaching course hours, resource data shows an incremental development trend. In the process of extracting the features of mobile online education digital resources, both new and historical data should be considered, and feature extraction should be implemented based on a global perspective to avoid ignoring the hidden information contained in the resources [5].

The adaptive sliding window mutual information method is used to process the historical data and incremental data of mobile online education digital resources, and realize the feature extraction of mobile online education digital resources.

Use matrix $X_1=[x_1, x_2, \cdots, x_m]$ to represent the original window data, and matrix $X_2=[x_{m+1}, x_{m+2}, \cdots, x_{m+r}]$ to represent the incremental window data; All data contained in mobile online education digital resources is represented by $X = [X_1, X_2]$; Z_1 and Z_2 represent the mutual information matrix of original window data and new window data of mobile online education digital resources respectively; Z represents the mutual information matrix of all mobile online education digital resource samples.

According to the definition of mutual information, the expression of mutual information matrix can be obtained as follows:

$$Z = \frac{1}{m+r}(Z_1 + Z_2) \tag{1}$$

The eigen decomposition formula of diagonalization using identity matrix to represent Z_1 is as follows:

$$I = G_1^T Z_1 G_1 \tag{2}$$

Using the space formed by G_1 to receive the projection of Z_2, the formula can be obtained as follows:

$$\overline{Z_2} = G_1^T Z_2 G_1 \tag{3}$$

Summing formula (1) and formula (2) yields:

$$G_1^T (Z_1 + Z_2) G_1 = I + \overline{Z} \tag{4}$$

The formula for feature decomposition $\overline{Z_2}$ is as follows:

$$\overline{Z_2} = P_2 \Lambda_2 P_2^T \tag{5}$$

Substitute formula (5) into formula (4) to obtain the following expression:

$$P_2^T G_1^T (Z_1 + Z_2) G_1 P_2 = I + \Lambda_2 \tag{6}$$

Through the above process, the feature decomposition results of all mobile online education digital resources can be obtained.

According to formula (2):

$$G_1 = B_1 \Lambda_1^{-\frac{1}{2}} \tag{7}$$

In formula (7), $\Lambda_1 \in R^{m \times k}$ and $B_1 \in R^{n \times k}$ respectively represent the matrix composed of the first k eigenvalues and the principal component decision matrix of the original mobile online education digital resources.

The eigenvalue Λ_2 and eigenvector P_2 of the mutual information matrix of the newly added window data, and $\Lambda_2 = [\mu_1, \mu_2, \cdots, \mu_n]$ and $P_2 = [\beta_1, \beta_2, \cdots, \beta_n]$, are obtained through the above process.

The formula for obtaining the eigenvalues of all mobile online education digital resources based on feature vectors and eigenvalues is as follows:

$$\Lambda = \frac{1}{m+r}(I + \mu_i) \tag{8}$$

In formula (8), m represents historical mobile online education digital resource data; r represents the addition of mobile online education digital resource data.

The feature vector formula for mobile online education digital resources is as follows:

$$P = G_1\beta_i \tag{9}$$

Using the obtained feature vectors to establish a principal component decision matrix, mapping mobile online education digital resources to the established principal component decision matrix can achieve data dimensionality reduction [6]. Repeat the above process in the subsequent window to achieve feature extraction of all mobile online education digital resource samples.

2.2 Determine Key Nodes for Digital Resource Allocation in Mobile Online Education

Before allocating complex mobile online education digital resources, determine the key nodes in the allocation process. In the current mobile online education digital resources, with continuous reform and practice, there are more and more applied education digital resources. Mobile online education digital resources no longer only refer to theoretical knowledge that may be used in teaching classrooms, but also include many practical content. In this case, this article divides mobile online education digital resources into theoretical education digital resources and applied practical education digital resources, and organically combines and allocates them [7–9]. This article uses genetic method theory to determine the allocation nodes. According to genetic methods, the classification of digital resource types for mobile online education is completed through multiple iterations. Subdivide mobile online education digital resources into N resource category area, and set the dispersion set of education digital resources as F, with the expression:

$$F = \{f_1, f_2, \cdots f_N\}_{N+1} \tag{10}$$

In the formula, $f_1, f_2, \cdots f_N$ represents the dispersion of educational digital resources in the 1, 2, ..., n region. Calculate the demand for educational digital resource allocation channels, set the coverage of allocation nodes as D, and the calculation matrix is:

$$D = \{d_1, d_2, \cdots d_N\}_{N+1} \tag{11}$$

In the formula, $d_1, d_2, \cdots d_N$ represents the coverage of educational digital resource allocation nodes in the 1, 2, ..., n region. Set interference constraints based on the calculation results of coverage, calculate the minimum value of non interference between different allocation nodes, and the maximum distance between allocation nodes in different regions. Set the distance of the allocation node to D. The calculation formula is:

$$\begin{cases} C_{\max} = (d_N + f_N)_{m \times n} \\ C_{\min} = (d_N - f_N)_{m \times n} \end{cases} \tag{12}$$

In the formula, m represents the allocation frequency of educational digital resource allocation nodes, and n represents the distance that generates allocation interference. If the key node of educational digital resource allocation is A, its calculation formula is:

$$A = \{d_N | d_N \in \{0, 1\}\}_{m \times n} \tag{13}$$

According to the above formula, calculate the key nodes for the allocation of digital resources in mobile online education, complete the division of theoretical education digital resources and application practical education digital resource allocation channels in mobile online education, and achieve the preprocessing of mobile online education digital resources.

By pre processing mobile online education digital resources, educational resources can be processed, including cleaning, optimization, and integration, to ensure the quality and availability of resources, improve resource access and browsing speed, and thereby improve the effectiveness and availability of digital resources. It can also better serve teaching and integrate resources into the teaching process, becoming an important auxiliary tool for teaching.

3   Educational Digital Resource Allocation Based on Ant Colony Algorithm

Educational digital resource allocation based on ant colony algorithm is a method that solves the problem of educational digital resource allocation by simulating the behavior of ant colonies. In this algorithm, educational digital resources are seen as a resource, establishing a weight relationship between learners and educational resources. Educational digital resources also include multiple resources, each with different characteristics in terms of availability, quality, and other aspects. The ant colony algorithm simulates the behavior of the ant colony. When searching for resource allocation schemes, the ant colony's search ability and route selection are taken as the basic methods, and the priority of each resource is determined through pheromone, distance and other factors, so as to finally provide learners with the best resource allocation scheme.

The advantage of ant colony algorithm lies in its ability to effectively handle multi-objective, complexity, and change issues in the allocation of educational digital resources. It simulates the behavior of ants in the process of searching for resources, taking into account the priority, availability, and other characteristics of resources, and can better meet the needs and requirements of learners in the learning process.

Based on the above construction and allocation model, the dynamic optimization of mobile online education digital resources using ant colony algorithm is carried out. The specific optimization process is described as follows:

(1) Assign pheromone initialization

Release all resource information in the current construction model with equal amount of pheromone, and complete initialization of all pheromone according to the following formula.

$$\tau_{ij}(0) = T0 \qquad (14)$$

where, $\tau_{ij}(t)$ represents the total amount of pheromone of mobile online education digital resources remaining on the corresponding resource allocation path at t time points, and c represents the vector coefficient.

(2) Resource allocation path selection rules

In order to ensure the integrity of the resource allocation network, when an ant individual in the ant colony is assigned a corresponding resource, its state transition rule is set to the next moving position as the state point, thereby ensuring that the entire ant colony can simultaneously receive adjacent allocation resource information [10–12]. $allow_k$ represents the allocation status point that Ant k has not passed through. The probability of ant k transferring from position i to position j at time point $\tau_{ij}(t)$ represents t is:

$$P_{ij}^k(t) = \begin{cases} \dfrac{[\tau_{ij}(t)]^\alpha \cdot [\mu_{ij}(t)]^\beta}{\sum_{j \in allow_k} [\tau_{ij}(t)]^\alpha \cdot [\mu_{ij}(t)]^\beta} & j \in allow_k \\ 0 & other \end{cases} \tag{15}$$

In the formula, $\mu_{ij}(t)$ represents the activation function, and represents the information concentration of the allocated resources transferred from status point i to status point j, generally $\mu_{ij}(t) = 1/d_{ij}$; $\mu_{ij}(t)$ represents the residual pheromone concentration on the corresponding distribution path at t time points; α represents the pheromone activation factor, and represents the importance of the resource allocation track. The larger the coefficient value, the greater the decisive role of the pheromone accumulated in the process of ant movement; β represents the activation factor, representing the range of influence of assigned visibility. The larger the coefficient value, the higher the weighting of the activation information of ants on the selection of assigned paths during their movement [13–15].

(3) Dynamic allocation pheromone update

Considering that in the process of dynamic allocation, the information parameters have the attribute of short-term change, and the pheromone on the corresponding allocation resource information track will speed up the volatilization speed, which is not conducive to the ant information search at the next state point [16, 17]. In order to better extend the residual time of the dynamic allocation resource information, and stimulate the pheromone concentration propagation ability, the correlation between the pheromone concentration and the time variable is conducted for all ants who have completed a cycle, Keep its pheromone update frequency consistent with the time variable of resource allocation [18–20], and obtain the dynamic allocation and update rules of mobile online education digital resources based on ant colony algorithm as follows:

$$\tau_{ij}(t+n) = (1-\rho)\tau_{ij}(t) + \Delta\tau_{ij}, 0 < \rho < 1 \tag{16}$$

$$\Delta\tau_{ij} = \sum_{k=1} \Delta\tau_{ij}^k, 0 < \rho < 1 \tag{17}$$

In the equation, parameter ρ represents the volatility coefficient of the information concentration element; $1 - \rho$ represents the concentration of residual pheromone of ants at the previous time point; $\Delta\tau_{ij}$ represents the total concentration of pheromone released at status point i and status point j in the current resource allocation cycle; $\Delta\tau_{ij}^k$ represents the total concentration of pheromone released by the seventh ant at status point i and status point j, and the relationship function is $\Delta\tau_{ij}^k = W/L_k$, where W represents the residual concentration of pheromone, and L_k represents the iteration cost after ant k completes a dynamic allocation of pheromone updates.

4   Experimental Analysis

The experiment takes a school in a certain region as the experimental object, and randomly selects one class with a population of 30 people. The selected mobile online education digital resources are C language education digital resources, consisting of pre class, in class, and post class teaching activity information. The original information and newly added C language teaching information before, during, and after class are shown in Table 1.

Table 1. C Information set of Language Education Digital Resources

C Language Education Digital Resources Information set Name	Original information quantity/GB	New information volume/GB
Pre class C language teaching information	30.6	5.2
C language teaching information in class	25.4	8.8
C language teaching information after class	40.5	3.6

Using the raw information of C language educational digital resources before, during, and after class in Table 1 as the experimental object, the effect of different window sizes on the feature extraction of educational digital resources before, during, and after class was tested. The experimental setting was an adaptive sliding window with a width of 50–300 bytes. The time for extracting C language teaching features was observed, and the average value was calculated as the experimental result. The experimental results are shown in Fig. 1.

As shown in Fig. 1, the time for extracting information from educational digital resources before, during, and after class decreases with the increase of window width. When the window width is greater than 200 bytes, the time for extracting information from educational digital resources gradually increases. This is because when the window width is small, information is often extracted from the buffer, which takes up more time. When the window width is too large, the time for decomposing information features of educational digital resources increases, The temporal trend of feature extraction of educational digital resources is consistent under different information quantities. The experimental results show that when the window width is 200 bytes, the time for feature extraction of educational digital resource information before, during, and after class is the shortest, and the speed of feature extraction is the fastest.

In order to test the performance of this method in the information integration of C language education digital resources, the experiment takes the information of three kinds of education digital resources in Table 1 of the C language education digital resources Information set as the test sample, and uses this method to integrate the historical information and new information of the C language teaching world before, during and after

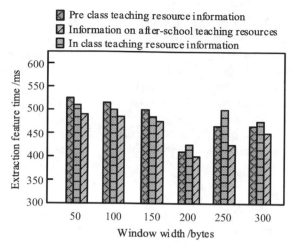

Fig. 1. Information extraction results of digital resources for pre class, in class, and post class education

class. Blue represents the historical information before, during and after class of C language teaching, The red color indicates the pre class, in class, and post class information of the newly added C language teaching history. Integrate the method of this article into pre class, in class, and post class educational digital resource information, and output it in the educational digital resource information space. The output results are shown in Fig. 2.

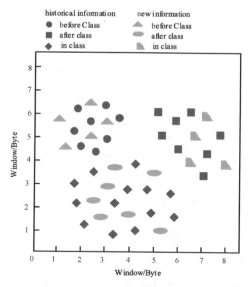

Fig. 2. Integrated Output of Digital Resources for Pre class, In class, and Post class Education

As shown in Fig. 2, it can be seen from the integration of historical and newly added information of educational digital resources before, during, and after class that after the integration of the method in this article, the historical and newly added educational digital resource information of C language teaching before, during, and after class is evenly distributed in the educational digital resource information space, with no irregular distribution of educational digital resource information. The method in this article has a good balance in the integration of C language educational digital resource information, And the integration effect is stable, and the balance of integration is consistent across different educational digital resource information. The experimental results indicate that the method proposed in this paper has a high balance in integrating C language resource information, and the integration effect of C language education digital resource information is better.

On this basis, the methods of Reference [3] and Reference [4] were used as experimental comparison methods. Analyzed and compared the accuracy of digital resource allocation for mobile online education corresponding to different methods, and its calculation method can be expressed as:

$$P_r = \frac{x_r}{X} * 100\% \tag{18}$$

In the formula, P_r represents the accuracy of the corresponding method, x_r represents the total amount of correctly allocated educational resources, and X represents the total amount of educational resources.

According to the above calculation method, the accuracy of allocation results for different methods is shown in Fig. 3.

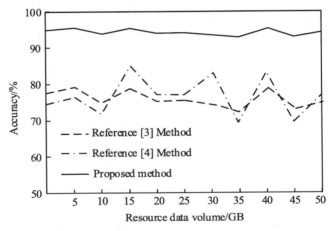

Fig. 3. Comparison of accuracy of allocation results using different methods

By analyzing the data information in Fig. 3, it can be seen that among the three methods, the accuracy of allocation of educational resources shows a certain degree of fluctuation, but there are significant differences in the specific allocation results. Among them, the allocation results of the method in Reference [3] showed relatively high stability overall, with an accuracy rate of 72% −80% for the allocation of educational resources. In the test results of the method in Reference [4], there is a significant fluctuation in the accuracy of allocation of educational resources, with an accuracy rate ranging from 68% to 85%. In the test results of this method, the accuracy of allocation of educational resources is relatively stable, ranging from 92% to 96%. The test results indicate that this method can achieve accurate allocation of digital resources for mobile school education.

When calculating the allocative efficiency of the three methods for mobile online education digital resources with different window sizes, the comparison results are shown in Fig. 4.

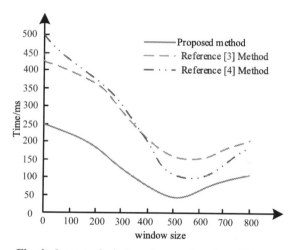

Fig. 4. Impact of window size on allocative efficiency

Figure 4 shows that the running time of allocating mobile online education digital resources using different methods decreases with the increase of window size. Compared with the other two methods, the allocative efficiency of this method is the highest in different window sizes, which indicates that the allocation efficiency of this method is higher than that of the other two methods.

To sum up, the mobile online education digital resource allocation method based on ant colony algorithm proposed in this paper first extracts features from the mobile online education digital resources to determine the key nodes, then uses the ant colony algorithm to initialize the pheromone of resource allocation, and determines the path selection rules. Finally, by dynamically updating the pheromone, the mobile online education digital resource allocation is realized. This method can extract information from digital resources quickly, and has a high degree of balance in the integration of educational digital resources. The integration effect of digital resources is better, and the distribution accuracy and allocative efficiency of educational digital resources are better.

On this basis, the time consumption of three methods for allocating digital resources in mobile online education was tested, and the experimental comparison results are shown in Fig. 5.

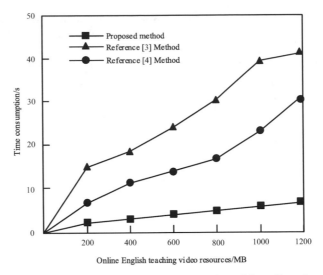

Fig. 5. Efficiency of digital resource allocation in mobile online education

Analyzing Fig. 5, it can be seen that the allocation time of mobile online education digital resources in the method of reference [3] is 30 s, the allocation time of mobile online education digital resources in the method of reference [4] is 42 s, and the recommended allocation time of mobile online education digital resources in the method of reference [3] is 7 s. The mobile online education digital resource allocation method proposed in this article has a short time consumption and good efficiency in the allocation algorithm.

5   Conclusion

This article proposes a mobile online education digital resource allocation method based on ant colony algorithm. The experimental results show that the method in this paper has the advantages of short time consumption, excellent output effect, high allocation accuracy and high allocative efficiency. The ant colony algorithm based digital resource allocation method for mobile online education has great application prospects and practical value. According to the needs of learners and the characteristics of resources, the algorithm simulates the behavior of ants, determines the priority of each resource through pheromone and distance, and provides the best resource allocation scheme for learners. This algorithm can handle multi-objective, complexity, and variability issues in educational digital resource allocation, and can better meet the needs and requirements of learners in the learning process.

In summary, this method will effectively improve the resource usage and user experience of mobile online education platforms, and have a positive promoting effect on research and development in the field of mobile online education. In future research, we can further optimize the application of Ant colony optimization algorithms in mobile online education digital resource allocation. Multiple objectives in the allocation of digital resources for mobile online education can be considered, such as learners' personalized needs, resource quality, and learning effectiveness. By designing suitable multi-objective optimization models and algorithms, more comprehensive and balanced resource allocation results can be achieved.

References

1. Xia, W., Shen, L.: Joint resource allocation at edge cloud based on ant colony optimization and genetic algorithm. Wireless Pers. Commun. **117**(2), 355–386 (2021)
2. Samriya, J.K., Patel, S.C., Khurana, M., et al.: Intelligent SLA-aware VM allocation and energy minimization approach with EPO algorithm for cloud computing environment. Math. Probl. Eng. **2021**(6), 1–13 (2021)
3. Liu, Y.: Research on online ideological and political education resources classification method based on decision tree algorithm. China Comput. Commun. **34**(22), 241–243 (2022)
4. Gao, Y.: Research on the classification of college English teaching resources based on density clustering algorithm. China Comput. Commun. **34**(22), 67–69 (2022)
5. Camel, V., Maillard, M.N., Descharles, N., et al.: Open digital educational resources for self-training chemistry lab safety rules. J. Chem. Educ. **98**(1), 208–217 (2021)
6. Lu, L., Zhou, J.: Research on mining of applied mathematics educational resources based on edge computing and data stream classification. Mob. Inf. Syst. **2021**(7), 1–8 (2021)
7. Tang, W., Djuric, P.M.: Bachelor of science in electrical engineering online: a journey of challenges and triumphs. IEEE Signal Process. Mag. **38**(3), 115–121 (2021)
8. Alismaiel, O.A.: Using structural equation modeling to assess online learning systems' educational sustainability for university students. Sustain. **13** (2021)
9. Lee, J.C., Xiong, L.N.: Investigation of the relationships among educational application (APP) quality, computer anxiety and student engagement. Online Inf. Rev. **1**, 46 (2022)
10. Du, W., Zhu, H., Saeheaw, T.: Application of the LDA model to semantic annotation of web-based English educational resources. J. Web Eng. **4**, 20 (2021)
11. Asghari, S., Navimipour, N.J.: The role of an ant colony optimisation algorithm in solving the major issues of the cloud computing. J. Exp. Theor. Artif. Intell. **3**, 1–36 (2021)
12. Alahyari, A., Mousavizadeh, S., Bolandi, T.G., et al.: A novel resource allocation model based on the modularity concept for resiliency enhancement in electric distribution networks. Int. J. Energy Res. (2) (2021)
13. Breakstone, J., Smith, M., Ziv, N., et al.: Civic preparation for the digital age: how college students evaluate online sources about social and political issues. J. Higher Educ. **93** (2022)
14. Liu, M., Jiang, W.: Empirical research on the influence mechanisms of digital resources input on service innovation in China's finance industry. Sustain. **14** (2022)
15. Smith, R.A.: Pandemic and post-pandemic digital pedagogy in hospitality education for generations Z, alpha, and beyond. J. Hosp. Tour. Res. **45**(5), 915–919 (2021)
16. Doddavarapu, V.N.S., Kande, G.B., Rao, B.P.: Rotational invariant fractional derivative filters for lung tissue classification. IET Image Process. **15**(10), 2202–2212 (2021)
17. Hong, S., Kim, J., Yang, E.: Automated text classification of maintenance data of higher education buildings using text mining and machine learning techniques. J. Arch. Eng. **28**(1) (2022)
18. Mendoza-Mendoza, A., Visbal-Cadavid, D., De La Hoz-Domínguez, E.: Classification of university departments of industrial engineering in Colombia: a DEA application without explicit inputs. Xinan Jiaotong Daxue Xuebao/J. Southwest Jiaotong Univ. **56**(2), 469–480 (2021)
19. Park, S.Y., Kim, J.H., Kim, J.C., et al.: Classification of softwoods using wood extract information and near infrared spectroscopy. BioResources **16**(3), 5301–5312 (2021)
20. Wolkiewicz, M.: On-line detection and classification of PMSM stator winding faults based on stator current symmetrical components analysis and the KNN algorithm. Electronics **10** (2021)

A Machine Learning Based Security Detection Method for Privacy Data in Social Networks

Zhiyu Huang[✉] and Chenyang Li

Shenyang Institute of Technology, Shenyang 113122, China
huangzhiyu2004@163.com

Abstract. In order to improve the social Internet privacy data security detection effect and improve the data security detection efficiency, this paper proposes a social Internet privacy data security detection method based on machine learning. First, collect social Internet privacy data and construct N-Gram language model to realize the standardization of social Internet privacy data; Secondly, a semantic vector based representation model is used to obtain topic semantic vectors, and the obtained topic semantic vectors are matched; Finally, social Internet privacy data security risk detection is carried out by using the skew coefficient method in machine learning. The results show that the method in this paper effectively compresses the time consumption of security detection through machine learning. The time consumption of detection is only 5.3 s, and the accuracy of data detection can reach 99.5%. The method in this paper can effectively improve the efficiency of social Internet privacy data security detection and improve the detection accuracy, but the detection cost needs to be reduced.

Keywords: N-Gram language model · Semantic vector · Deviation coefficient · machine learning

1 Introduction

In recent years, with the rapid development of social network, in view of the social network data publishing technology applications in some studies and has made great progress in the social network to provide a large number of social network user data, the data is collected, grind Gui and release [1] for a variety of purposes. Some large Internet companies can combine large amounts of data to construct a clear behavioral map of an individual, and then predict their preferences and behaviors [2]. This data is very valuable in the consumer market, and can be used to proactively promote certain products or services to specific groups of people. However, a lot of user data in social networks also contains many users' personal privacy information, such as personal address information, friends, interests and hobbies, etc. In addition, it is difficult for network service providers to restrict the use of data at present. Because the user's identity, address, contact information and other information are currently in the hands of service providers, but how the people who deal with these data every day should use the

© ICST Institute for Computer Sciences, Social Informatics and Telecommunications Engineering 2024
Published by Springer Nature Switzerland AG 2024. All Rights Reserved
L. Yun et al. (Eds.): ADHIP 2023, LNICST 547, pp. 350–365, 2024.
https://doi.org/10.1007/978-3-031-50543-0_24

data to avoid privacy disclosure is a very worthy question. Once these released data is maliciously used, it will cause unpredictable loss and impact on individuals and society. Therefore, how to obtain personal privacy protection while releasing social network user data is a hot issue in current research.

In recent years, due to the rapid popularity of social network services, the amount of social network data has exploded in various application scenarios. Since all data is generated from people's daily behavior, almost all data is related to personal data. If these data are correctly shared or aggregated, they can be used to extract valuable information and knowledge. However, while sharing or mining data, there may be privacy breaches due to the failure to control personal data from users. Therefore, data mining for social network privacy protection has received great attention in recent research. Reference [3] proposed a social Internet privacy privacy data security detection method based on cloud computing, which uses encryption algorithms to encrypt users' privacy data to ensure the security of data during transmission and storage. Use access control policies to restrict access to private data. Using methods such as role-based access control (RBAC) or attribute based access control (ABAC) to manage and control user access to data. Classify and label private data based on sensitivity and privacy level, and set different access permissions and security policies for different data. Record user access and operations to private data, and generate audit logs. By monitoring and analyzing audit logs, abnormal behavior and security vulnerabilities can be detected in a timely manner. Regularly backup private data and establish a comprehensive data recovery mechanism to prevent data loss or damage. Establish a real-time security monitoring system to monitor and analyze data traffic and user behavior on social network platforms. Once abnormal activities or potential security threats are detected, timely alerts are issued and corresponding response measures are taken. Using network security devices such as firewalls and intrusion detection systems (IDS) to strengthen network security protection for social network platforms and prevent unauthorized access and attacks. Regularly conduct security assessment and Penetration test, find potential security vulnerabilities and risks, and timely repair and strengthen security measures. This method can effectively improve the accuracy of social Internet privacy privacy data security detection, but the efficiency of privacy data security detection is low. Reference [4] proposes a social network user data security protection method based on the Big data background, classifies and marks social network user data, classifies them according to sensitivity and privacy level, and sets different protection policies for different data. Automated methods can be used to identify and label sensitive information in user data. Desensitize some non sensitive attributes to protect users' personal privacy. Data anonymization user data to remove directly identifiable personal identity information to protect user privacy. The Differential privacy technology is used to realize the de identification of data and protect the personal privacy of users. Establish a comprehensive access control mechanism to restrict access to user data. Authorize different users and roles through the role and permission management system, ensuring that only authorized users can access the corresponding data. Establish a real-time security monitoring system to monitor and analyze data traffic and user behavior on social network platforms. Discovering abnormal behavior and potential security threats through the use of machine learning and data mining techniques. Regularly backup user data, establish a comprehensive data recovery mechanism, adopt distributed storage and disaster recovery technology, and improve data reliability and availability. However, this method has poor security detection efficiency.

Currently, many social network service providers have realized the importance of the personal data they collect for making business decisions, information discovery, or other research needs. However, these data typically contain private and sensitive information about individuals, which can be discovered through privacy breach attacks on published data [5]. Therefore, these data should be published in a manner that prohibits the disclosure of any personal privacy information and the disclosure of personal identity. In addition, social network service providers should publish data to the public without violating personal information protection guidelines. This has raised concerns about protecting personal privacy when publishing data. Therefore, the publication of privacy protection data on social networks At present, the issue of privacy protection for data release on social networks is mainly how to release data in a way that protects personal privacy. With the deepening of the research, more valuable research results have been obtained at home and abroad. The research results can be divided into two categories: one is to achieve the purpose of privacy protection based on K-anonymity model. A common method is to use non-specific information to replace sensitive and specific information, that is, information generalization [6]. The other is to use probability or statistical methods to protect data privacy while keeping the statistical characteristics and classification attributes of the final data unchanged. Such as clustering, randomization, sampling, data exchange, and data perturbation for data publication.

To solve the problem of poor security detection efficiency and poor detection accuracy, this paper proposes a social Internet privacy data security detection method based on machine learning, constructs N-Gram language model to realize social Internet privacy data standardization, uses semantic vector based representation model to obtain topic semantic vector, and matches the obtained topic semantic vector, Social Internet privacy data security risk detection based on the method of partial coefficient in machine learning.

2 Machine Learning Based Privacy Data Security Detection Method for Social Networks

2.1 Access to Social Network Privacy Data

The data set used in this paper comes from public data sets and web crawler. There are more than 4 million data sets collected, of which about 3.8 million and 200000 are public and crawlers respectively. The log data collected on social networking sites generally includes user ID, number of followers, number of followers, number of followers, timestamp, and message text. The public experimental data used in this article comes from Baidu AIStudio, which is an artificial intelligence learning and training community based on the Baidu deep learning platform Feijiang, providing a variety of massive data for users to use [7]. All types of data are freely uploaded by professionals in various fields, and the platform will conduct review and supervision, with certain standardization of the data. The data we downloaded on this platform is taken from Weibo. Weibo is the largest social media platform in China, with over 550 million active users in 2020. Users can register Weibo accounts, create personal information, establish friendly contacts with other users, upload photos, share interesting things, etc. The platform has a certain degree of user stickiness and strong research value.

After a user registers an account on Weibo platform, the system will assign a unique ID number to the user as the user ID, and the user ID needs to be passed when the platform uses the open API to obtain the corresponding data. Therefore, we obtain user data by ID, but the user ID of the platform is not continuous. In order to reduce sampling deviation, We start by randomly generating 10,000 ids. The API was used to obtain the personal information of these ids. Among them, 1,547 ids returned correct information, indicating that these ids did not correspond to users. For the accounts that returned the correct information, we used the API to get the dynamic content published in the last week.

Next, it is necessary to determine whether obtaining account information is abnormal. For this purpose, two classmates were invited to do manual marking.

These two volunteers are both graduate students in our laboratory and familiar with the dynamic content of Weibo. Volunteers determine whether it is abnormal behavior based on the content posted by the user. If an account repeatedly forwards, posts useless information, or points to websites such as shopping, phishing, or pornography in a short period of time, then the main body of the account can be determined to be abnormal based on the abnormal behavior.

Volunteers need to manually identify all IDs, considering that manual identification may result in errors, and therefore abandon controversial accounts. In the end, 1045 accounts were identified as abnormal, and the 120780 message text associated with them was also marked as abnormal. The remaining 8067 IDs were identified as normal accounts, and the 16450 message content associated with them was identified as normal.

2.2 Standardized Processing of Social Network Privacy Data Based on N-Gram Language Model

The original data on the social network platform exists in the form of text, and the content of each message in the serialized data is text. In the previous section, the linguistic probability model of the (n-1) Markov chain in the collected messages is presented according to the probability judgment statement of the occurrence of n word. Its basic idea is to divide the text content into sliding Windows of length N [8] according to byte size. Each byte fragment is called a *gram*, and the occurrence frequency of all *gram* is counted, and the threshold is set in advance for filtering, so as to form a list of *gram*, that is, the feature space of text vectors. Each *gram* in the result is a feature vector dimension.

This model assumes that the n word is only related to the occurrence of the first $n-1$ words, and is independent of other words. The product of the probability of each word's occurrence is used as the probability of the entire sentence. The commonly used models include the binary model Bi Gram and the ternary model Tri Gram. The specific usage details will be introduced below.

If there is a sequence composed of m words, hoping to obtain the probability $p(w_1, w_2, ..., w_i)$, according to the chain rule, we can obtain

$$p(w_1, w_2, ..., w_m) = p(w_1) * p(w_2|w_1) * p(w_3|w_1, w_2)...p(w_m|w_1, w_2, ..., w_{m-1}) \quad (1)$$

Among them, w_i represents the probability of a certain word appearing. This paper uses the Markov hypothesis to calculate this probability, taking into account that the current word is only related to a limited number of words before it, so there is no need to go back to the first word in the calculation, which can greatly reduce the calculation of the above formula. After simplification, the above formula can be expressed as:

$$p(w_1, w_2, ..., w_m) = p(w_i | w_{i-n+1}, ..., w_{i-1}) \tag{2}$$

When n = 1, the unigram model can be represented as:

$$p(w_1, w_2, ..., w_m) = \prod_{i=1}^{m} p(w_i) \tag{3}$$

When n = 2, bigram model can be expressed as:

$$p(w_1, w_2, ..., w_m) = \prod_{i=1}^{m} p(w_i | w_{i-1}) \tag{4}$$

With the above model concept, the above conditional probability value can be calculated from the given training corpus by using Bayes theorem.

For the bigram model:

$$p(w_i | w_{i-1}) = \frac{C(w_{i-1} w_i)}{C(w_{i-1})} \tag{5}$$

For the N-Gram model:

$$p(w_i | w_{i-n-1}, ..., w_{i-1}) = \frac{C(w_{i-n-1} w_i)}{C(w_{i-n-1} w_{i-1})} \tag{6}$$

In this paper, 2-g model is used. The specific modeling method is designed to define a sliding window with length of 2 and move back step by step. After each move, 2 words in the window are taken as a sequence, and finally the sequence is converted into a numerical vector. Since the size of the sliding window is set to 2 and there are a maximum of 256 characters in the text, there will be 2**16 2-g combinations [9]. A common practice is to use a one-hot vector (set to 1 if present, 0 otherwise). To preserve more information, it is also possible to replace 1 and 0 with the number of times each 2-g appears in the text. Now, each message content can be transformed into a vector. Since the content of each text is different, the length of each text is naturally inconsistent. To ensure that the input data has the same length, you can fill it with specific characters to ensure that the input content length is standardized.

2.3   Semantic Vector Based Representation Model

The original text is composed of a series of symbols, which is a set of symbols with certain semantics. The text is recorded by human beings. Due to the diversity of human language, it is quite difficult for machines to understand the text, and many obstacles need to be overcome to accurately understand natural language. Since machines cannot understand natural language like human beings, they need to model text semantics and realize semantic analysis and understanding through solving mathematical models [10]. The model based on semantic vector maps the text vector processed above to the potential semantic space, and represents the semantic relationship through the new text vector.

For the construction of the original text vector, first extract the subject words from the preprocessed message content text vector. In this paper, we use Latent Dirichlet Allocation (LDA) topic model to extract topic words, and construct D * Z dimension text topic matrix and Z * W dimension topic word matrix according to the distribution of vocabulary content.

All parameters in PLSA model represent random variables, without considering prior distribution. Since the parameters in PLSA correspond to multinomial distribution, naturally a better choice is Drichlet distribution. We can add prior distribution before the parameters to transform PLSA model into LDA model, so that the model has strong generalization ability. At the same time, the prior distribution is added to make LDA model have a complete Bayesian framework. $\vec{\alpha}$ is a K-dimensional vector, used to represent the hyperparameter of Document-Topic distribution, and $\vec{\theta}_m$ is an K-dimensional vector composed of $p(z|d = i)$. $\vec{\beta}$ is a K-dimensional vector used to represent the hyperparameter of the Topic-Word distribution, and $\vec{\theta}_k$ is an K-dimensional vector composed of $p(w|z = i)$.

Similar to the Unigram Model, an V_i-sided die, with one word for each side, is used to represent the i document. Therefore, the probability model diagram can be decomposed into two processes, as shown in Table 1.

Table 1. LDA model generation process

LDA model generation process	
1	$\alpha \rightarrow \theta_m \rightarrow z_{m,n}$ This process means that the topic distribution θ_m of the document in Chapter m is generated by sampling from the Dirichlet distribution α, and then the topic $z_{m,n}$ of the nth word of the document in Chapter m is sampled from the topic polynomial distribution θ_m
2	$\beta \rightarrow \varphi_k \rightarrow w_{m,n}|k = z_{m,n}$. This process means that the word distribution φ_k with the subject $z_{m,n}$ is generated by sampling from the Dirichlet distribution, and then the word $w_{m,n}$ is generated by sampling from the polynomial distribution D

In the whole model, the joint distribution of all implicit variables and visible variables is:

$$p(w_m, z_m, \theta_m, \phi | \alpha, \beta) = \prod_{n=1}^{N_m} p(\theta_m | \alpha) p(\phi | \beta) p(w_{m,n} | \varphi z_{m,n}) \tag{7}$$

In the above equation, all documents of $\Phi = \{\Phi_k\}_{k=1}^{K}$ are shared, and the maximum likelihood estimation of the word distribution in the final document is summed by z_m and the above equations θ_m and Φ are integrated:

$$p(w_m|\alpha, \beta) = \int_{\theta_m} \int_{\phi} \sum_{z_m} p(w_m, z_m, \theta_m, \phi|\alpha, \beta) \tag{8}$$

With the joint probability distribution, Gibbs Sampling algorithm can be used for sampling. First, all the words in the document are traversed and assigned A topic, i.e., $z\_(m, n) = k \sim Mult(1/K)$, where K represents the total number of topics and $z_{m,n}$ represents the n topic in the document of article m. After that, corresponding n_m^k, n_m, n_k^k, n_k represents the number of occurrences of theme k in document m, the number of topics in document m, the number of words t corresponding to topic k, and the total number of topic k. The most important Gibson sampling formula is obtained according to the existing joint probability distribution $p(\vec{w}, \vec{z})$

$$p(z_i = k|z_{\neg i}, w) \propto \frac{(n_{k,\neg i}^{(t)} + \beta_t)(n_{m,\neg i}^{(k)} + \alpha_k)}{(\sum_{t=1}^{V} n_{k,\neg i}^{(t)} + \beta_t)} \tag{9}$$

After multiple iterations, the output Topic Word parameter matrix $\varphi_{k,t}$ and Document Topic matrix $\theta_{n,k}$ converge, as shown in Eq. (10). During the inference stage, keeping $\varphi_{k,t}$ unchanged and updating $\theta_{n,k}$ can obtain potential topic semantic vectors corresponding to the new text.

$$\varphi_{k,t} = \frac{(n_k^{(t)} + \beta_t)}{(n_k + \beta_t)} \tag{10}$$

$$\theta_{m,k} = \frac{(n_m^{(t)} + \alpha_k)}{(n_m + \alpha_k)} \tag{11}$$

The topic semantic vector is obtained, and the topic semantic vector is matched.

2.4 Semantic Model Based on Text Similarity

After the above operations, the theme words in each original text can be obtained. Furthermore, the theme words in the Document Topic matrix and Word Topic matrix can be mapped with the corresponding words in the text one by one to the user's behavioral concept, and the words and theme words can be corresponding to the user's actual behavioral concept, and they can be regularized. Based on the mapping relationship constructed between concepts, the analysis of the relationship between words and themes can be transformed into analysis between conceptual levels. At this point, what is represented in the matrix is no longer a simple distribution relationship between Word, Topic, and Document, but a semantic relationship at the conceptual level. This relationship matrix better represents the relationship between log text and user behavior concepts, which can reduce useless features and dimensions.

The closer the similarity between the words in the text and the extracted topic, the better the current word reflects the meaning of the document and is therefore worth keeping. Common methods for calculating text similarity include reference method and word shift distance method. The former first calculates the average embedding value of all words in the sentence, and then calculates the cosine similarity between the two sentences to judge. The latter uses word embedding between two texts, and then calculates the minimum distance needed to move a word from one text to another text in a semantic space vector. In this paper, text2vec package is used for judgment. This library uses Tencent AI Lab open source library to collect a variety of high-quality Chinese word data, which can achieve lower granularity of words and higher accuracy, and is very suitable for Chinese similarity calculation.

Assuming $z_{m,0}$ is the theme word of document d_m in article m, $n_{m,0}$ is the number of occurrences of theme $z_{m,0}$ in text d_m, $w_i \in (w_1, w_2, ..., w_n)$ is any word in text D, and the semantic increment between word w_i and the theme word $z_{m,0}$ in article m is $SI_{m,i} = si_{m,i} * n_{l,0}$. Where $SI_{m,i}$ is the semantic similarity between the word w_i and the d_m topic $z_{m,0}$.

3 Social Network Privacy Data Security Risk Detection

3.1 Analysis of Chaotic Sequence Encryption Process

Chaos phenomenon is a kind of random process in nonlinear deterministic system. Two particularly close original values are introduced into the same chaotic function to carry out iterative calculation. After the calculation in a specific stage, the numerical sequence has no similarity. This encryption method belongs to the deterministic system, but it is difficult to predict it, hidden in the jumbled system but can not be decomposed.

The characteristics of chaotic signal, such as non-periodicity, continuous broadband spectrum and similarity to noise, make it possess natural hidden characteristics, highly sensitive to original conditions, and long-term unpredictability, making it difficult for data to be subjected to malicious damage and attack.

Logistic mapping represents nonlinear chaotic equation, and its mapping process is as follows:

$$X_{n+1} = bX_n(1 - X_n) \quad X_n \in [0, 1] \tag{12}$$

In the equation, b represents the control parameter variable. After specifying the specific value of b, a clear time series $X_1, X_2, \cdots, X_n$ can be iteratively calculated using the random original value $X_0 \in [0, 1]$.

The security key of chaotic sequence encryption depends on chaotic key stream. In chaotic encryption system, random sequence $\{x_i\}$ generated by chaotic system is regarded as bitwise operation of key stream $\{k_i\}$ and plaintext data stream $\{m_i\}$, and then ciphertext data stream $\{c_i\}$ is obtained. The plaintext data stream is binary, while the key stream $\{k_i\}$ is obtained by data processing of the chaotic sequence $\{x_i\}$. The initial chaotic data $\{x_i\}$ is processed by computer technology, and the randomness of chaotic sequence is completed under the premise of limited computational accuracy.

Chaos encryption cipher as a sequence cipher. The encryption and decryption ends of the chaotic sequence cipher system are two completely independent and equal chaotic systems, and there is no coupling relationship between the two systems. Clear text data is encrypted at the encryption end and directly transmitted to the decryption end. The decryption end can perform decryption after receiving all data. The encryption method of chaotic sequence cipher is relatively flexible, which can effectively utilize the characteristics of chaotic signals to obtain complex encryption functions. The chaotic sequence cipher system is shown in Fig. 1.

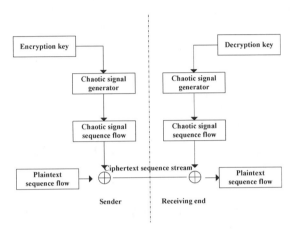

Fig. 1. Chaotic encryption process

3.2 Privacy Encryption of Social Network Privacy Data in Chaotic Sequence

Transforming the original data through chaotic sequences allows unauthorized users to obtain encrypted data, but due to the inability to decrypt it, the method for determining the information content is still unclear. Maximize data security and prevent data leakage. Simultaneously deprivarizing data before encryption can more effectively improve data security.

Record the sequence probability distribution function generated by the logistic mapping of Eq. (9) pattern as:

$$\rho_{(x)} = \begin{cases} \dfrac{1}{\pi} \dfrac{1}{x(1-x)} & 0 < x < 1 \\ 0 & else \end{cases} \tag{13}$$

Some important statistical features of the chaotic sequence generated by Logistic mapping can be easily obtained by using $\rho_{(x)}$. For example, the average time of x, which is also the mean value of the trajectory points of the chaotic sequence, is described as:

$$\bar{x} = \lim_{N \to \infty} \frac{1}{N} \sum_{i=0}^{N} x_i = \int_0^1 x\rho_{(x)}dx = 0.5 \tag{14}$$

For the cross correlation function, if two original values x_0 and y_0 are selected separately, the sequence cross correlation function is:

$$c(l) = \lim_{N \to \infty} \frac{1}{N} \sum_{i=0}^{N-1} (x_i - \bar{x})(y_{i+1} - \bar{y})$$

$$= \int_0^1 \int_0^1 \rho_{(x,y)}(x_i - \bar{x})(T'(y) - \bar{y})dxdy = 0 \tag{15}$$

The autocorrelation function ACF of the sequence is equal to the delta function.

Since the encryption object of privacy removal is A numeric quantity, the sequence $\{x_k\}$ composed of real numbers is a pseudo-random sequence formed by integers to achieve the purpose of privacy removal.

In order to make the sequence more random and improve its encryption rate, it is most appropriate to choose an interval M of 6 for random values. The value of Y_k is 3, 4, and 5 decimal places after the decimal point of X_k, which will enhance the resistance to selective plaintext attacks.

In order to improve the fitness of plaintext attacks, data processing is carried out on X_n to make the relationship between X_n and X_{n+1} more complex, and then prevent attackers from using a simple derivation process to solve the μ value. This article uses interval retrieval to place a numerical value between X_n and X_{n+1}, and the correlation between X_n and X_{n+1} is transformed into:

$$X_{n+1} = \mu \times (\mu \times x_n \times (1 - x_n)) \times (1 - \mu \times x_n \times (1 - x_n)) \tag{16}$$

Equation (13) is a quadratic equation with one variable. Add N values between X_n and X_{n+1}, then the correlation between X_n and X_{n+1} is a system of degree N with one variable. N μ solutions can be approximated by Newton stepwise approach method. At the same time, if N is large enough, a small change in X_n will cause X_{n+1} to make a change that affects x'_n. Set the value of x'_n to three digits behind the decimal point.

In order to avoid exposing the redundancy of plaintext information, encryption systems need to fully and uniformly use the ciphertext space, and chaotic sequences also need to clarify the statistical distribution state of the encrypted ciphertext data, that is, whether the plaintext data is completely covered up, which is also a key criterion for weighing the effectiveness of encryption methods. If the key sequence is completely random, then the encrypted ciphertext data follows a uniform distribution. Treat the plaintext and key sequence as a single byte data stream, and arbitrarily select a byte of plaintext m and key k. Assuming that the values of m are not equal, that is, the probability of a certain bit appearing as 0 or 1 is different, and the occurrence of a data bit appearing as 0 or 1 is an independent event. If the probability of 1 appearing in the i bit of the plaintext is p, the key sequence conforms to white noise under ideal conditions, that is, the probability of generating 0 or 1 is the same, both are 0.5. So the probability of 1 appearing in the i bit of ciphertext c after encryption is:

$$P_{(C_i=1)} = P_{(m_i=1,k_i=0)} + P_{(m_i=0,k_i=1)}$$
$$= P \times 0.5 + (1 - P) \times 0.5 = 0.5 \tag{17}$$

Therefore, it can be seen that the probability distribution of encrypted ciphertext is symmetrical.

3.3   Privacy Data Security Detection Methods

Social network privacy data security risk assessment is to take an all-round evaluation and measurement of data risk. Privacy risk refers to the possibility of data leakage and subsequent negative impact.

In the set pair analysis theorem, the correlation number describes the same, different and inverse relation analytic expression of two sets which are correlated with each other, restricted and highlighted. Assuming that Z is a non-empty set sum, then the relation number of $A=\{< z, a_A(z), b_A(z), c_A(z) > |z \in Z\}$ obtained is:

$$\mu_A(z) = a_A(z) + b_A(z)i + c_A(z)j \tag{18}$$

In the formula, $a_A(z)$, $b_A(z)$, and $c_A(z)$ are the support, uncertainty, and opposition of element A belonging to F within Z.

The above equation contains the same, different, and opposite factors, so it is also called a ternary connection number. By conducting in-depth calculations on the uncertainty value b_i within the ternary connection number, the general form obtained is:

$$\mu = a + b_1 i_1 + b_2 i_2 + \cdots + b_n i_n + cj \tag{19}$$

Partial correlation number is a function that represents the change level of deviation trend in machine learning, reflecting the development trend and change of the homologous and different anticorrelation form. In the process of social network privacy data security risk assessment, homogeneity means that privacy risk assessment and expected standard risk are close to the same change pattern, that is, in the low risk area of privacy assessment; Equilibrium means that the gap between privacy risk assessment and expected standard risk level is slightly higher, and it is in the intermediate risk area of privacy assessment. The backlash represents the opposite state of privacy risk assessment and expected standard risk. In the high risk area of privacy assessment.

The scoring function of the five-element correlation number μ is expressed as:

$$S(\mu) = (a + b) - (d + e) \tag{20}$$

The exact function of the five element connection number μ is:

$$H(\mu) = a + b + d + e \tag{21}$$

The score function $S(\mu)$ and the exact function $H(\mu)$ are similar to the mean and variance in data statistics. The higher the score function is, the greater the data implied risk is. The lower the scoring function, the smaller the risk index. When the scoring function is equal, the higher the precision function is, the greater the data risk is, and vice versa. Therefore, this paper uses score function and precision function to carry out hierarchical prediction of hidden risks of social network privacy data.

Weight refers to the numerical value of the impact of each indicator to be evaluated during data privacy risk assessment. This article combines subjective and objective weighting to design a minimum binary weighting method for the potential of five element partial connection numbers.

Regarding privacy risk attribute G_i, describe its r-order partial connection potential matrix as:

$$P_i^{(r)} = \begin{pmatrix} p_{11}^{(r)} & p_{11}^{(r)} & \cdots & p_{1n}^{(r)} \\ p_{21}^{(r)} & p_{22}^{(r)} & \cdots & p_{2n}^{(r)} \\ \vdots & \vdots & \vdots & \vdots \\ p_{m1}^{(r)} & p_{m2}^{(r)} & \cdots & p_{mn}^{(r)} \end{pmatrix} = (p_{ij})_{nm} \tag{22}$$

For two unequal r-order partial relation number potential matrices $P_j^{(r)}$ and $P_k^{(r)}$, the deviation is described by $D_i^{(r)}(\omega)$, which is specifically denoted as:

$$D_i^{(r)}(\omega) = \left(\sum_{i=1}^{n} \sqrt{\left(P_{ij}^{(r)} - P_{ik}^{(r)} \right)^2} \right)^{\frac{1}{2}} \tag{23}$$

Equation (6) only represents the state where the key levels of weights are equal. In general, the weight key levels of data privacy risk indicators are not equal to each other. To understand the applicability of the evaluation results within the evaluation indicators composed of different weight vectors W_i, it is necessary to establish a minimum binary function:

$$\min \ D^{(r)}(\omega) = \sum_{i=1}^{m} D_i^{(r)}(\omega) = \sum_{j=1}^{n} \left(\sum_{i=1}^{n} \omega_j^{(r)} \sqrt{\left(P_{ij}^{(r)} - P_{ik}^{(r)} \right)^2} \right)^{\frac{1}{2}} \tag{24}$$

$$s.t. \sum_{j=1}^{n} \omega_j^{(r)} = 1, \ \omega_j^{(r)} \geq 0, \ j = 1, 2, \cdots, n$$

The Lagrange function is constructed to calculate the above formula, and the analytical formula of minimum h-partial weighting is obtained as follows:

$$\omega_j^{(r)} = \frac{\left[\sum_{i=1}^{m} \left(\sqrt{\left(P_{ij}^{(r)} - P_{ik}^{(r)} \right)^2} \right)^h \right]^{\frac{1}{h}}}{\sum_{j=1}^{n} \left[\sum_{i=1}^{m} \left(\sqrt{\left(P_{ij}^{(r)} - P_{ik}^{(r)} \right)^2} \right)^h \right]^{\frac{1}{h}}} \tag{25}$$

A machine learning based method for detecting privacy data security in social networks has been implemented.

4 Experiment

4.1 Experimental Design

According to the calculation steps of the method in this paper, Logistic is taken as an example to explore the real utility of chaotic sequence encryption, and carry out the simulation of power marketing data privacy encryption and decryption.

Figure 2 is a schematic diagram of the ASC code value distribution of ciphertext characters after two groups of slightly different original keys are successively encrypted for the same plaintext. The dotted line represents the plaintext, and the square dotted line indicates the ciphertext character, $\mu = 3.74$, which is obtained after encryption and the value of x_0 is 0.60001. As can be seen from the figure, the plaintext of the same text will change obviously when the key changes slightly, reflecting the sensitivity of the ciphertext to the key.

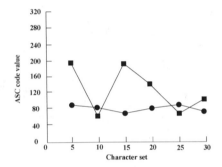

Fig. 2. ASC code value distribution diagram of ciphertext characters

Figure 2 shows the distribution of the number of iterations of the encryption method in this article. From the figure, it can be seen that the number of iterations is concentrated between 220 and 620, which reduces the number of iterations and improves the calculation speed and cost of the encryption method (Fig. 3).

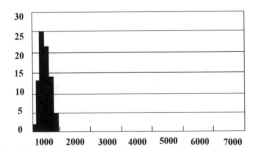

Fig. 3. Schematic diagram of the distribution of iteration times in this method

4.2 Experimental Results

In order to further verify the reliability of the security detection performance of social network private data, the proposed method is simulated and compared with literatures [3] and [4], and the security detection operation is conducted on the social network private data of an enterprise. The test environment is a Pentium 3.0GHzCPU with 2GB of memory. The three method performance pairs are obtained as shown in Table 3.

Table 2. Performance comparison of security detection methods for social network privacy data

Number of iterations	Social network privacy data security detection time/s		
	Method of reference [3]	Method of reference [4]	Textual method
1000	28.6	39.2	0.5
2000	35.1	48.9	0.9
3000	69.0	69.0	1.6
4000	92.3	78.1	2.9
5000	120.5	89.9	5.3
6000	138.9	125.3	8.1

From Table 2, it can be seen that when the number of iterations is 1000, the social network privacy data security detection time of the method in reference [3] is 28.6 s, the social network privacy data security detection time of the method in reference [4] is 39.2 s, and the social network privacy data security detection time of the method in this article is only 0.5 s; When the number of iterations is 5000, the social network privacy data security detection time of the method in reference [3] is 120.5 s, the social network privacy data security detection time of the method in reference [4] is 89.9 s, and the social network privacy data security detection time of the method in this article is only 5.3 s; The detection efficiency of the method proposed in this article is better than that of two references, which proves the superiority of the method in detecting privacy data security in social networks. This is because this method realizes the standardization of social Internet privacy data through N-Gram language model; The social Internet privacy data security risk detection based on the skew coefficient method in machine learning effectively reduces the time consumption of social Internet privacy data security detection, and improves the detection efficiency.

In order to further verify the security detection capabilities of different methods for social network privacy data, the accuracy of social network privacy data was tested, and the results are shown in Table 3.

According to Table 3, when the number of iterations is 1000, the accuracy of social network privacy data for the method in reference [3] is 66.2%, the accuracy of social network privacy data for the method in reference [4] is 72.1%, and the accuracy of social network privacy data for the method in this paper is 99.3%; When the number of iterations is 3000, the accuracy of social network privacy data for the method in reference [3] is 70.1%, the accuracy of social network privacy data for the method in reference [4]

Table 3. Precision of social network privacy data

Number of iterations	Social network privacy data accuracy/%		
	Method of reference [3]	Method of reference [4]	proposed method
1000	66.2	72.1	99.3
2000	68.9	76.5	98.6
3000	70.1	78.9	99.5
4000	73.8	68.5	96.0
5000	79.2	70.3	98.2
6000	66.0	69.2	99.1

is 78.9%, and the accuracy of social network privacy data for the method in this paper is 99.5%; The accuracy of the method presented in this paper for submitting network privacy data is much higher than other methods, indicating that the method presented in this paper has high data accuracy. This is because the method used in this article uses a semantic vector representation model to obtain topic semantic vectors, and matches the obtained topic semantic vectors to achieve accurate detection of privacy data.

5   Conclusion

This paper proposes a machine learning-based privacy data security detection method for social networks. Collecting privacy data of social network, constructing N-Gram language model to realize standardized processing of privacy data of social network; A representation model based on semantic vector is used to obtain the topic semantic vector, and the obtained topic semantic vector is matched. Finally, the partial contact number method based on machine learning is used to detect the security risk of social network privacy data. Experimental results show that the security detection time of social network privacy data under the proposed method is only 5.3 s, and the detection accuracy of social network privacy data is 99.5%, which proves the superiority of the proposed method for the security detection efficiency of social network privacy data.

References

1. Qiao, Y.: Research on social network security and privacy data fusion method based on cloud computing. Software **43**(06), 109–111 (2022)
2. Liu, C., Du, J., Zhou, N.: A cross media search method for social networks based on adversarial learning and semantic similarity. Chin. Sci. Ser. F **51**(05), 779–794 (2021)
3. Niu, N., Zhou, S., Lu, R., Yan, S., Zhang, M., Wang, C.: Attribute based signature encryption scheme based on cloud computing in medical social networks. J. Electron. Inf. Technol. **45**(03), 884–893 (2023)
4. Zhang, H.: Analysis of social network user data security protection based on big data background. Comput. Knowl. Technol. **18**(28), 69–71 (2022)

5. Zhu, P., Hu, J., Lv, S., et al.: Research on blockchain based privacy data protection methods for social networks. Inf. Sci. **39**(3), 94–100 (2021)
6. Jia, R., Wang, X., Fan, X.: Research on the influencing factors of personal information security and privacy protection behavior of social network users. Modern Intell. **41**(09): 105–114+143 (2021)
7. Wu, X., Liu, Q., Zhu, C.: Research on the application of collaborative public opinion fraud detection methods in social networks. J.Zhengzhou Univ. (Eng. Ed.) **43**(02), 7–14 (2022)
8. Zhu, D., Zhang, X., Gu, C.: Probability prediction of social network information leakage nodes based on EDLATrust algorithm. J. Tsinghua Univ. (Nat. Sci. Ed.) **62**(02), 355–366 (2022)
9. Zheng, Z., Wu, X., Wang, H., Liu, K., Shen, Z.: PTPM protection method for trajectory privacy in mobile social networks. Small Micro Comput. Syst. **42**(10), 2153–2160 (2021)
10. Lv, J., Zhang, Z., Xu, Y.: Blockchain based social network digital rights management protection method. Comput. Eng. Des. **42**(6), 1562–1570 (2021)
11. Li, Q., Hu, Y., Zhou, Q., Zhou, G.: K-anonymity based data privacy social network protection scheme. Modern Inf. Technol. **6**(09), 89–91 (2022)
12. Jianghui, F.: Data fusion method of social network security privacy based on cloud computing. J. Univ. Jinan: Nat. Sci. Ed. **35**(1), 5–16 (2021)
13. Zheng, J., Yang, L.: Large social network differential privacy algorithm based on Singular value decomposition. Comput. Technol. Dev. **32**(3), 126–131 (2022)
14. Zhang, Y., Zhang, J.: Research on the protection of social Internet privacy in MapReduce model based on K-means method. Wirel. Internet Technol. **19**(20), 162–165 (2022)
15. Zhu, Y.: A method to eliminate redundancies in browsing behavior data of social network users based on Random forest. J. Ningxia Normal Univ. **42**(1), 73–78 (2021)

An Online Integrated Classification Algorithm for Innovation and Entrepreneurship Teaching Data Based on Decision Tree

Juanjuan Zou[✉]

Chongqing Vocational Institute of Engineering, Jiangjin 402260, China
17726637816@163.com

Abstract. In order to improve the accuracy of data classification, reduce misclassification rates, and improve classification efficiency, a decision tree based online integrated classification algorithm for innovation and entrepreneurship teaching data is proposed. Establish a prototype system for online integration of innovation and entrepreneurship teaching data, and based on the results of data integration, preliminarily construct a decision tree model. Then, the fuzzy decision tree obtained by combining fuzzy theory with decision tree is used to solve the problems of data imbalance and missing data types. In order to further reduce the misclassification rate of data, the differential grey wolf optimization algorithm is used to optimize the decision tree, and the decision tree is improved through operations such as feature selection and decision tree pruning to obtain the optimal classification results of innovation and entrepreneurship teaching data. The experimental results show that the proposed method has high data classification accuracy, low misclassification rate, and high classification efficiency, which verifies the effectiveness of the method.

Keywords: Decision Tree · Innovation and Entrepreneurship · Data Classification · Differential Grey Wolf Optimization · Pruning Decision Tree

1 Introduction

With the continuous development of information technology and the Internet, people's demand for data is becoming increasingly urgent, especially in the field of innovation and entrepreneurship education. Data analysis and application have become important teaching content [1]. The online integration and classification of data are key technologies for achieving data sharing and utilization. Therefore, the background significance of studying the online integration and classification of innovation and entrepreneurship teaching data is to improve data utilization efficiency and accuracy, provide more comprehensive and accurate data support for students [2, 3]. However, due to the presence of a large number of redundant and irrelevant features in the dataset, this can affect the accuracy of classification algorithms and make the classification results more complex and inexplicable. Therefore, for each classification problem, it is necessary to correctly select features, eliminate redundant information, and improve data availability.

© ICST Institute for Computer Sciences, Social Informatics and Telecommunications Engineering 2024
Published by Springer Nature Switzerland AG 2024. All Rights Reserved
L. Yun et al. (Eds.): ADHIP 2023, LNICST 547, pp. 366–378, 2024.
https://doi.org/10.1007/978-3-031-50543-0_25

In the above context, relevant scholars have proposed some data classification methods, among which reference [4] proposes an imbalanced data classification algorithm based on undersampling and cost sensitivity. Firstly, before each iteration of the AdaBoost algorithm to train the base classifier, the majority class samples are sorted by weight from largest to smallest; Then normalize the weight of the sampled majority class samples and form a temporary training set with the minority class samples to train the base classifier; Secondly, in the weight update stage, a higher misclassification cost is assigned to minority classes, resulting in faster weight increase for minority class samples and slower weight increase for majority class samples, thus achieving data classification. The experimental results show that there is a problem of low data classification efficiency. Reference [5] proposed an unbalanced data classification method based on probability threshold Bagging algorithm, which combined threshold mobile technology technology with Bagging integrated algorithm, used the original distributed training set for training in the training phase, complete data classification. The experimental results show that this method has good classification advantages in handling imbalanced data, but there is a problem of inaccurate data classification when the data volume is large. Reference [6] proposes a data classification method based on the improved ID3 algorithm. This classification method determines the equilibrium coefficient through the modified information gain, optimizes the information gain obtained by the ID3 algorithm using the equilibrium coefficient, obtains the root node of the decision tree based on the optimized information gain, divides the nodes, classifies the attributes, constructs the decision tree, and achieves data classification through the decision tree. Through examples, it has been proven that this classification method can achieve control over multivalued bias and avoid selecting attributes with more values as branch nodes, but there is a problem of high misclassification rate. Reference [7] proposed an unbalanced data classification method based on the cost sensitive Activation function XGBoost, and introduced the cost sensitive Activation function to change the gradient change of the Loss function of samples under different prediction results, to solve the problem that the misclassified minority samples cannot be effectively classified in the XGBoost iteration process due to the small gradient change. The results show that the algorithm has a high detection rate for minority class samples, indicating that its classification effect for minority class samples is good. However, when the data volume is large, the accuracy of data classification is not high.

In order to solve the problems of low accuracy, low efficiency, and high misclassification rate in data classification of the existing methods mentioned above, a decision tree based online integrated classification algorithm for innovation and entrepreneurship teaching data is proposed. The main research content of this article's method is as follows:

(1) Establish a prototype system for online integration of innovation and entrepreneurship teaching data, improving data storage and access efficiency.
(2) Based on the online integration results of innovation and entrepreneurship teaching data, a decision tree model is constructed, and a decision tree based on differential grey wolf optimization is adopted. The data category weight modification strategy is used to reduce the weight of unimportant data and increase the weight of important data, in order to obtain the optimal result of data classification.

(3) Using data classification accuracy, efficiency, and misclassification rate as experimental indicators, compare the proposed method with traditional methods and draw relevant conclusions.

2 Online Integration of Innovation and Entrepreneurship Teaching Data

Before classifying innovation and entrepreneurship teaching data, the first step is to integrate the data, which aims to improve the coverage of the data. Integrating data from different sources can cover more comprehensive data and help improve the practical application value of data classification results. This article will use open-source software to study the online integration method of innovation and entrepreneurship teaching data, and design a prototype system to achieve online integration of innovation and entrepreneurship teaching data.

The prototype system adopts a browser/server (B/S) mode, with Windows as the operating system, and uses Ajax technology to develop web front-end interactive programs. The server side programs are developed using Apache network servers and PHP language. The framework structure of the system is shown in Fig. 1.

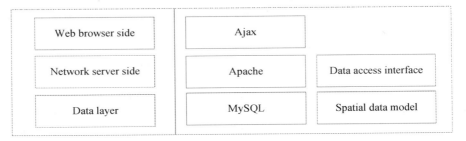

Fig. 1. Schematic diagram of the prototype system framework structure

As shown in Fig. 1, the system consists of three parts: a web front-end, a network server, and a database. The web front-end is responsible for collecting data, submitting the collected data to the server, and visualizing the data from the server. After receiving the data uploaded by the web front-end, the network server performs formatting on the data. The database is responsible for storing the collected data and providing data access services.

The prototype system is divided into four modules according to its functions: interaction, data collection, data storage, and data access, each with a detailed functional design. Among them, the interaction and data collection functions are mainly implemented on the web front-end. The functional structure design of the system is shown in Fig. 2.

As shown in Fig. 2, users can easily collect data through a browser, and the collected data content is completely customized by the user, enhancing the interactivity of the data and enabling ordinary users to transform from data users to data providers, enriching the sources of data. The use of free open source software greatly reduces the cost of

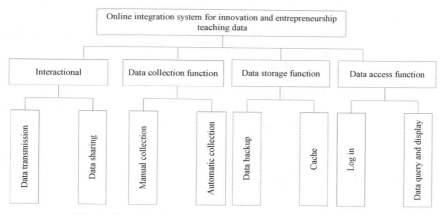

Fig. 2. Functional structure diagram of the prototype system

data collection and management. In addition, a MYSQL based database has been built to improve the efficiency of data storage and access. Provided a data access interface, enabling the collected data to be more widely used, thereby achieving online integration of innovation and entrepreneurship teaching data.

3 Classification of Innovation and Entrepreneurship Teaching Data

Based on the online integration results of innovation and entrepreneurship teaching data, conduct research on the classification of innovation and entrepreneurship teaching data.

3.1 Decision Tree Principle

Decision tree is a commonly used machine learning algorithm for classifying and predicting data. Its principle is to classify data through a series of branch conditions [8]. The decision tree can be seen as dividing some regions in the data sample space, with each region corresponding to a category. This division process is the process of constructing the decision tree.

The process of constructing a decision tree is mainly divided into three steps: selecting attributes, partitioning data, and constructing the tree.

(1) Selecting attributes: At each node of the decision tree, it is necessary to select an optimal attribute as the partitioning criterion to minimize the differences between different categories of data in the current node.
(2) Partition data: After determining the partition attributes, the data needs to be divided into different subsets based on the values of the attributes. This process generally involves determining multiple attributes, which are equivalent to the question of 'yes'. Continue to select attributes for partitioning in each subset until all data in the subset belongs to the same category.

(3) Building a tree: Based on the partitioning results, construct a tree structure to represent the classification model. Decision trees can be constructed and optimized using algorithms such as depth first, breadth first, and pruning to achieve better performance and generalization ability.

Decision trees are widely used in data classification, such as financial fraud detection, disease diagnosis, sentiment analysis, etc. Due to the simple and easy to understand process of constructing decision trees, clear classification rules can often be generated, making it easier to analyze and interpret data.

3.2 Construction of Decision Tree Model

According to the decision tree principle, when conducting online integrated classification of innovation and entrepreneurship teaching data, it is necessary to consider the problem from a global perspective, find the optimal data content, and improve the accuracy and efficiency of data classification. The decision tree algorithm is an important part of the current development process of artificial intelligence technology. It can comprehensively mine sample data without rules and orders, form the corresponding mathematical analysis model, obtain the most basic data classification rules, and then predict and classify various data sets. In the framework of blockchain, a diversified decision tree model can be constructed based on the distribution characteristics and actual situation of each node to synchronize and efficiently classify data [9]. To ensure the effectiveness of constructing the decision tree model, it is necessary to first generate candidate splitting points, specify the best data splitting time in each node, and then treat the best data classification points as the core part to achieve the purpose of online integrated classification of innovation and entrepreneurship teaching data. Finally, the model is iteratively updated to optimize its performance. In this process, node splitting is the basis for information gain rate analysis and information entropy processing in the online integration and classification of innovation and entrepreneurship teaching data, so it is necessary to focus on node splitting methods, as shown in Fig. 3.

In the overall practical operation process, the number of innovation and entrepreneurship teaching data and training samples waiting for classification should be set to N, consistent with the number of nodes in the blockchain. The overall number of training set types should be Z, labeled as $N_1, N_2, ..., N_z$. Assuming that any attribute of the training set is R, this attribute involves attribute data values, labeled as $N_1, N_2, ..., N_z$, and the number of samples that correspond to attribute data values R_1 is N_{R_i}, under the framework of blockchain, the information entropy of the training sample set in the innovation and entrepreneurship teaching data set can be calculated according to formula (1):

$$D(W) = \sum_{i=1}^{Z} t_i \log_2 t_i \tag{1}$$

Add attribute R to the feature data set and calculate according to formula (2):

$$D(R, W) = \sum_{i=1}^{Z} \frac{N_{R_i}}{N_i^R} \log_2 \frac{N_{R_i}}{N_i^R} \tag{2}$$

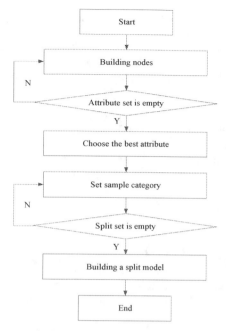

Fig. 3. Schematic diagram of data node splitting

After completing the above operation, use formula (3) to represent the information gain rate:

$$E(R, W) = \frac{\Delta(R, W)}{\sum\limits_{i=1}^{Z} \frac{N_{R_j}}{N_i^R} \log_2 \frac{N_{R_j}}{N_i^R}} \tag{3}$$

In essence, scientific and reasonable selection will directly affect the application efficiency of the decision tree model. If there is no information entropy attribute or information gain efficiency attribute in the process of building the model, it will lead to classification errors. Therefore, in practical work, it is necessary to ensure the integrity of information entropy and information classification attributes of the overall decision tree model, and scientifically optimize and process various decision tree models, Prevent misclassification from occurring at the fundamental level. Meanwhile, due to the possibility of local node splitting and obstruction during the application process of the model, in order to avoid affecting the normal data splitting of other nodes, it is necessary to focus on optimizing the decision tree algorithm technology. Within the overall framework of the blockchain, buffer empty nodes should be pre-set, and the number of node attributes used for data splitting should be set to m. During the optimization process, $m + 1$ sub node should be set, Once there is a problem of insufficient attributes in the model, this sub node can split the data into empty nodes based on the currently known attributes to avoid classification interruption.

3.3 Decision Tree Based on Differential Grey Wolf Optimization

In the data classification problem of innovation and entrepreneurship teaching, traditional classification algorithms cannot handle the uncertain and imprecise information in attributes well, and fuzzy theory can handle this to some extent. Therefore, the fuzzy decision tree obtained by combining fuzzy theory with decision trees has better applicability. When encountering two problems in innovation and entrepreneurship teaching data: data imbalance and missing data types, it can lead to model misjudgment and reduced classification accuracy. To solve this problem, a data category weight modification strategy is adopted, which reduces the weight of unimportant data and increases the weight of important data. The change in weight is based on the best choice of differential grey wolf optimization [10].

3.3.1 Feature Selection

On the basis of splitting the decision tree results, a decision tree is further constructed using the differential grey wolf optimization algorithm for the classification of innovation and entrepreneurship teaching data. The construction of a decision tree is usually divided into three processes: feature selection, decision tree generation, and decision tree pruning. Feature selection is an important step before constructing a decision tree. If features are randomly selected, the learning efficiency of the established decision tree is very low. The generation of a decision tree starts with an empty tree, where the samples to be classified enter from the root of the tree. At each node of the tree, different paths are selected to gradually descend to the bottom by judging a certain attribute of the samples, and their category is determined. To prevent overfitting of the decision tree, an attempt is made to eliminate redundant nodes after constructing the entire tree, which is called pruning [11]. The pruning process involves checking nodes with the same parent node to determine whether the information gain after merging them will be less than a specified value. If so, merge these nodes.

The usual criteria for feature selection are based on information gain theory, which provides criteria for selecting candidate attribute lists between each decision node. Select the feature with the highest information gain by calculating the information gain of each feature:

$$G(Z, R_t) = P(Z) - P_{R_t}(Z) \tag{4}$$

In the formula, R_t represents the sample features; $P(Z)$ represents the entropy of the feature; $P_{R_t}(Z)$ represents the information gain rate, and the expression for both is as follows:

$$P(Z) = \sum_{j=1}^{m} \frac{m(F_j, Z)}{|Z|} \log_2 \frac{m(F_j, Z)}{|Z|} \tag{5}$$

$$P_{R_t}(Z) = \sum_{h \in C(R_t)}^{m} \frac{Z_h^{R_t}}{|Z|} P\left(Z_h^{R_t}\right) \tag{6}$$

In the formula, $m(F_j, Z)$ represents the number of objects belonging to class F_j in the training set; $C(R_t)$ represents the Finite field of feature R_t; $Z_h^{R_t}$ represents the cardinality of the set of objects with a value of h for feature R_t. The information gain ratio refers to the ratio of information gain to the entropy of the value of feature R_t in the training dataset Z, which can be expressed as:

$$V(Z, R_t) = \frac{G(Z, R_t)}{Q(Z, R_t)} \tag{7}$$

In the formula, $Q(Z, R_t)$ represents the potential information generated by dividing P into k subsets, and the mathematical definition equation is:

$$Q(Z, R_t) = \sum_{h \in C(R_t)}^{k} \frac{\left|Z_h^{R_t}\right|}{|Z|} \log_2 \frac{\left|Z_h^{R_t}\right|}{|Z|} \tag{8}$$

3.3.2 Decision Tree Generation

The generation of a decision tree starts from the root node, calculates the information gain of all features at the node, selects the feature with the highest information gain as the node feature, and establishes sub nodes based on different values of the feature. For sub nodes, the above method is used for recursion. When the information gain of all features is small or there are no features to choose from, the decision tree construction is completed.

3.3.3 Decision Tree Pruning

The purpose of pruning decision trees is to prevent overfitting. Some unnecessary classification features are removed from the generated decision trees by optimizing the loss function to reduce the overall complexity of the model. The method of pruning is to start from the leaf nodes of the tree, shrink upward, and gradually judge [12, 13]. If the loss function corresponding to the whole decision tree is smaller after cutting off a certain feature, then cut off the branches of the feature set.

Decision tree pruning is generally realized by minimizing the overall loss function of the decision tree. The loss function can be expressed as:

$$U_a(T) = U(T) + a|T| \tag{9}$$

In the formula, T represents any subtree; $|T|$ represents the number of nodes in the subtree; $U(T)$ represents the prediction error of the training data; a is used to measure the complexity of the model and the degree of fit of the training data.

3.3.4 Decision Tree Optimization

The main purpose of this article is to optimize the decision tree classifier during the process of constructing a decision tree. Therefore, a decision tree optimization strategy based on Differential Grey Wolf Optimization (DGWO) is proposed, with the accuracy of the decision tree classifier as the objective function. By optimizing and modifying the weight of data categories, the optimal decision tree is constructed.

The differential grey wolf optimization algorithm basic idea is to introduce the idea of differential evolution on the basis of GWO, and by introducing crossover and mutation operations, it increases the algorithm's local search ability and the ability to jump out of local optima [14, 15]. In addition, DGWO also introduces the mechanism of Tabu search to make the algorithm more easily jump out of the local optimal solution, thus obtaining better global search ability.

The specific implementation process of the DGWO algorithm is as follows:

(1) Initialize the population: randomly generate a certain number of gray wolf individuals and calculate their fitness.
(2) Updating of gray wolf individuals: According to the location and fitness of gray wolf individuals, calculate the direction and distance of updating of gray wolf individuals, and update the location of gray wolf individuals.
(3) Differential evolution operation: Randomly select a certain number of gray wolf individuals as the mutation population according to a certain probability, and generate a certain number of offspring individuals by selecting the parent individual and mutation operation. By comparing the fitness of the offspring and the parent, the individuals with better fitness were selected to update the gray wolf population.
(4) Cross operation: According to a certain probability, randomly select two gray wolf individuals for cross operation. The cross operation can be performed by linear interpolation.
(5) Tabu search: Tabu search mechanism is introduced to avoid the algorithm falling into local optimal solution. This is generally achieved by setting the length of taboos, maintaining taboo tables, and other methods.
(6) Judgment of termination conditions: after multiple iterations, terminate the algorithm by judging whether the fitness in the population converges or reaches the preset number of iterations.

In a word, DGWO algorithm introduces differential evolution and Tabu search functions on the basis of grey wolf optimization algorithm, which can search the solution space more accurately and avoid local optimal solution. In addition, due to its simple principle and easy implementation, the algorithm has good application prospects in various optimization problems.

Based on the differential grey wolf optimization algorithm, the objective function for optimizing the classification of innovation and entrepreneurship teaching data is:

$$Y = \max(accuracy) \tag{10}$$

In the formula, *accuracy* represents the accuracy of the decision tree classifier.

At present, the GWO algorithm has solved many engineering problems, but it also has certain limitations and lacks the ability to find global optimal solutions that affect the convergence speed of the algorithm. The model has high complexity and is essentially nonlinear. Due to the linear decrease in convergence factor, it cannot truly reflect the actual search process. In order to overcome the limitations of conventional GWO algorithms in terms of efficiency, exploration, and development characteristics, this paper proposes a differential grey wolf optimization algorithm. On the basis of the conventional GWO algorithm, update the optimal solution position as follows:

$$X(\theta + 1) = \left(\frac{X_1 + X_2 + X_3}{3}\right) - \left(\frac{X_1' + X_2' + X_3'}{3}\right) \tag{11}$$

In the equation, $X_1' X_2'$ and X_3' represent the first, second, and third optimal solutions, respectively. Based on the optimization results, obtain the optimal classification results of innovation and entrepreneurship teaching data, and provide data reference for student entrepreneurship.

4 Experiments and Result Analysis

To verify the effectiveness of the online integrated classification algorithm for innovation and entrepreneurship teaching data based on decision trees, experimental research was conducted.

4.1 Experimental Environment Configuration

The hardware environment used in this experiment is Intel i5–6500 CPU, 12G system memory, and 500G available hard disk space. The software environment used in this experiment is Windows 10. Matlab is mainly used to normalize experimental results for easy plotting.4.2Experimental results.

4.2 Analysis of Experimental Results

In order to enhance the credibility of the experimental results, the methods of reference [6] and reference [7] were used as comparative methods for comparative analysis with the proposed methods. Firstly, the data classification accuracy of the three methods was compared, as shown in Fig. 4.

By analyzing Fig. 4, it can be seen that as the number of data samples increases, the data classification accuracy of the three methods shows a gradually decreasing trend. Among them, the highest data classification accuracy of the proposed method is 94.5%, while the highest data classification accuracy of the methods in reference [6] and [7] are 84.1% and 79.8%, respectively, which are significantly lower than the proposed method. This indicates that the proposed method has a high accuracy in data classification and can accurately classify different types of innovation and entrepreneurship teaching data.

Secondly, the data misclassification rate was used as an experimental indicator to further validate the data classification effect of the proposed method, as shown in Fig. 5.

Analyzing Fig. 5, it can be seen that as the number of data samples increases, the data misclassification rates of all three methods show a gradual increasing trend. Among them, the highest data misclassification rate of the proposed method is only 3.8%, while the highest data misclassification rates of the methods in reference [6] and [7] are 11.9% and 13.3%, respectively, which are significantly higher than the proposed method. From the above comparison results, it can be seen that the proposed method can effectively reduce

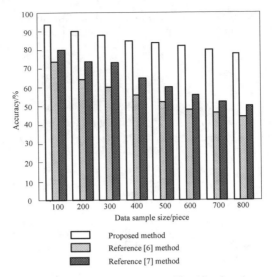

Fig. 4. Comparison Results of Data Classification Accuracy

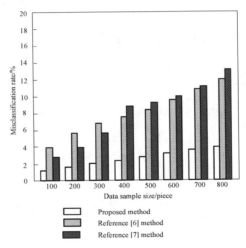

Fig. 5. Comparison Results of Data Misclassification Rate

the data misclassification rate, further verifying the reliability of its data classification results.

Finally, the data classification efficiency was used as an experimental indicator to compare the application effects of the proposed method, reference [6] method and reference [7] method. The results are shown in Fig. 6.

From Fig. 6, it can be seen that the data classification time of the proposed method is significantly lower than that of the methods in reference [6] and [7], with a classification time consistently below 15 s. However, the data classification time of the methods in

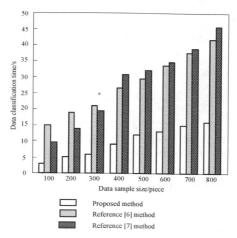

Fig. 6. Comparison Results of Data Classification Efficiency

reference [6] and [7] has reached over 40 s. This indicates that the data classification efficiency of the proposed method is higher, and it can achieve rapid division of innovation and entrepreneurship teaching data in a shorter time.

5 Conclusion

Aiming to improve the accuracy of data classification, reduce misclassification rates, and improve classification efficiency, a decision tree based online integrated classification algorithm for innovation and entrepreneurship teaching data is proposed. Establish a prototype system to achieve online integration of innovation and entrepreneurship teaching data through this system; Based on the results of data integration, a decision Tree model is constructed, and a fuzzy decision tree is obtained by combining the fuzzy theory with the decision tree to solve the problem of data imbalance and missing data types. The difference gray wolf optimization algorithm is used to optimize the decision tree, and the decision tree is improved through feature selection, decision tree pruning and other operations to obtain the best innovation and entrepreneurship teaching data classification results. The experimental results show that the proposed method has a high data classification accuracy, with a maximum value of 94.5%, a low misclassification rate, and a high classification efficiency, indicating that the method can improve the data classification effect. In the future, research will be conducted on how to effectively integrate and integrate innovation and entrepreneurship teaching data from different sources and formats, in order to better support teaching decision-making and evaluation.

References

1. Zhang, X., Zhou, X., Zhao, C., Shao, L.: Unbalanced data classification based on hesitant fuzzy decision tree. Comput. Eng. **45**(08), 75–79+91 (2019)
2. Chen, L., Fei, H., Ding, H., Cheng, L., Zhai, J.: A data sampling method based on double decision tree. Comput. Eng. Sci. **41**(01), 130–135 (2019)

3. Lu, X., Chen, Y., Xiong, Z., Liao, B.: A fast parallel decision tree algorithm for big data analysis. J. Yunnan Univ. (Nat. Sci. Edn.) **42**(02), 244–251 (2020)
4. Wang, J., Yan, J.: Classification algorithm based on undersampling and cost-sensitiveness for unbalanced data. J. Comput. Appl. **41**(01), 48–52 (2021)
5. Zhang, Z., Wu, D.: An imbalanced data classification method based on probability threshold Bagging. Comput. Eng. Sci. **41**(06), 1086–1094 (2019)
6. Meng, Y., Zhou, Q., Shi, H., Ma, N.: Data classification method based on improved ID3 algorithm. Comput. Simul. **39**(05), 329–332+417 (2022)
7. Li, J., Wang, X.: XGBoost for imbalanced data based on cost-sensitive activation function. Comput. Sci. **49**(05), 135–143 (2022)
8. Zheng, J., Li, X., Liu, S., Li, D.: Improved random forest imbalance data classification algorithm combining cascaded up-sampling and down-sampling. Comput. Sci. **48**(07), 145–154 (2021)
9. Bao, H., Fan, X.: Simulation of dynamic classification for unbalanced big data in cloud computing environment. Comput. Simul. **37**(08), 311–314+461 (2020)
10. Liu, P., et al.: Remotely sensed data classification by collaborative processing of Landsat, radarsat-2 and topography information. Remote Sens. Technol. Appl. **34**(06), 1269–1275 (2019)
11. Liu, X., et al.: Imbalanced data classification algorithm based on ball cluster partitioning and undersampling with density peak optimization. J. Comput. Appl. **42**(05), 1455–1463 (2022)
12. Chen, L., Tang, X.: Improved sine cosine algorithm for optimizing feature selection and data classification. J. Comput. Appl. **42**(06), 1852–1861 (2022)
13. Liang, Y., Liu, X., Li, Q., Bai, Y., Ma, Y.: Classification method for unbalanced and small sample data in judicial documents. J. Comput. Appl. **42**(2), 118–122 (2022)
14. Li, A., Han, M., Mu, D., Gao, Z., Liu, S.: Survey of multi-class imbalanced data classification methods. Appl. Res. Comput. **39**(12), 3534–3545 (2022)
15. Zhou, E., Gao, S., Shen, Z.: Classification algorithm of imbalanced data based on rotation balanced forest. Comput. Eng. Des. **43**(02), 458–464 (2022)

High Quality Resources Sharing of College Students' Career Guidance Course Teaching Based on Decision Tree Classification Algorithm

Meiling Ou(✉)

Chongqing Vocational Institute of Engineering, Jiangjin 402260, China
m19912412058@163.com

Abstract. In order to improve the quality and efficiency of sharing high-quality teaching resources and achieve ideal results, a decision tree classification algorithm based method for sharing high-quality teaching resources in college student employment guidance courses is proposed. Firstly, before resource sharing and transmission, design a security key for public information and encrypt the public information of high-quality teaching resources. Secondly, the decision tree classification algorithm is used to construct a classification standard for electronic archives, and regional partitioning is performed to extract resource sharing classification codes. At the same time, establish a database of teaching resources for employment guidance courses to store information related to high-quality teaching resources for college students' employment guidance courses. Finally, establish a blockchain based teaching resource security sharing model to achieve information sharing of high-quality teaching resources. Experimental analysis shows that the proposed method can complete the high-quality resource sharing task of college student employment guidance course teaching within 5–8.5 ms after application, with a significant advantage in resource sharing efficiency.

Keywords: Decision Tree Classification Algorithm · College Student · Career Guidance Courses · Teaching. High Quality · Resource Sharing

1 Introduction

The employment of college students has always been the focus of social attention. Classroom teaching of employment guidance course is an important way to promote the employment guidance work in colleges and universities [1]. For colleges and universities, it is particularly urgent to focus on the research on the current teaching situation of career guidance courses. Employment guidance is a kind of social service work [2] produced with the rapid development of economy and the continuous differentiation of occupations. Employment guidance focuses on the whole process of career selection, career preparation and career acquisition, and runs through the whole process of college education. The career guidance course refers to a quality improvement course for college students to choose, prepare for and obtain a career according to social needs

© ICST Institute for Computer Sciences, Social Informatics and Telecommunications Engineering 2024
Published by Springer Nature Switzerland AG 2024. All Rights Reserved
L. Yun et al. (Eds.): ADHIP 2023, LNICST 547, pp. 379–393, 2024.
https://doi.org/10.1007/978-3-031-50543-0_26

and their own characteristics on the premise that it is included in the teaching plan and university curriculum system. In this study, courses involving "career development planning", "career guidance", "career development and employment", "career planning and employment" and other contents are uniformly defined as "career guidance courses" [3].

College students' employment guidance and career planning are two different concepts that are related to and obviously different from the needs of the object at different stages and goals in college. Career planning refers to the object's cognition, formulation, determination and development of its lifelong career development, while employment guidance directly refers to the object's employment [4]. Employment guidance is derived from vocational guidance." Career guidance" is "the process of providing consultation, guidance and help for job seekers to obtain employment, maintain employment stability, develop their careers and employers to reasonably employ people". With the reform and development of the graduate employment system in China, the employment guidance in the new era has changed a lot compared with the traditional graduate employment education. The employment guidance in this article is different from the vocational guidance activities such as career counseling, career introduction and so on that are carried out in the society for all kinds of people with needs. It also does not include a series of employment services and management related to college graduates in colleges and universities. The employment guidance in this paper refers to an educational activity carried out for college students throughout the whole process of college education. Therefore, college students' employment guidance is to help college students to achieve smooth employment, with colleges and universities as the main body, college students as the object, college students' career guidance personnel as the actual operator, colleges and universities' advantageous educational resources as the carrier, college students' career planning as the core, and improving college students' employability and employment quality as the goalSelf realization is an educational activity that runs through the whole process of the university.

There are a large number of high-quality teaching resources in college students' career guidance courses, so we need to adopt reasonable sharing methods to achieve the goal of sharing massive high-quality teaching resources. Decision tree classification algorithm is a method to approximate the value of discrete function [5]. It is a typical classification method. First, it processes data, generates readable rules and decision trees using inductive algorithms, and then analyzes new data using decisions. In essence, a decision tree is a process of classifying data through a series of rules. In order to realize the ideal goal of high-quality resource sharing of college students' career guidance course teaching, this paper introduces the decision tree classification algorithm to carry out the research on high-quality resource sharing of college students' career guidance course teaching.

2 Design of High-Quality Resource Sharing Method for College Students' Career Guidance Course Teaching

2.1 Design Public Information Security Key of Teaching Resources

In order to ensure the security and privacy of high-quality teaching resources of college students' career guidance courses in the sharing process, it is necessary to encrypt the resources before sharing and transmission, and establish a security key for public information. First set a safe information parameter q_a, compare this parameter with a large prime number k_a And jointly establish the common attribute set of the two.

$$U_t = \{xr_{a1}, xr_{a2}, xr_{a3}\} \tag{1}$$

Among them, U_t Represent the general structure of attribute set circular mapping generated by combining security parameters with large prime numbers; xr_{a1} Indicates the previous cell of the selected attribute cell; xr_{a2} Indicates the selected attribute unit in the attribute set; xr_{a3} Indicates the next cell of the selected attribute cell. At this time, the attribute group in the collection is formed, and the global security public key should also be set in the attribute group, so that the cloud center of data transfer can encrypt any public key. At this time, the output structure of the security public key is:

$$G_{xaq} = \{G_{a1g}, G_{a2g}, G_{ang}\} \tag{2}$$

Among them, G_{xaq} It can represent the attribute unit in any security public key; G_{a1g} Represents the first attribute unit in the security public key, similarly G_{ang} Represents the last attribute unit in the security public key. Each security public key can generate its own security private key. The algorithm determines the user's identity and provides a random number as the generation attribute of the private key. Its structure is:

$$G_{xaq} = \{\delta_i, \forall xaq_i \in G : k_i^\delta\} \tag{3}$$

Among them, G_{xaq} Represents any output unit in the security public key; δ_i Indicates the random number of the verification code that needs to be provided by the user during identity determination; k_i^δ Represents the obtained private key structure. The obtained private key can be used as a ciphertext for covert data transmission G_{xaq} The random vectors are generated in plaintext and converted into ciphertext [6]. Convert the key into the public key of the data information in the computer, and transfer it from the transmitter to the cloud of the central processing unit.

2.2 Extracting Resource Sharing Classification Code Based on Decision Tree Classification Algorithm

After the public information security key conversion of teaching resources is completed, next, use the decision tree classification algorithm to build a classification standard for electronic files, divide them into regions, extract the classification code of resource sharing, and lay the foundation for sharing high-quality resources in the follow-up college students' career guidance courses [7].

First, build a decision tree centered on electronic archives. This decision tree needs to give full examples of all collected samples, then calculate the overlapping part, and judge the error of decision classification through mathematical methods. Assume that the number of sample sets is x_i, training samples are $x_i = \{x_1, x_2, ..., x_n\}$, where x_i express n Any one of the training samples. The characteristic value in the sample is ζ_i Each sample has a characteristic value, then the set of characteristic values can be expressed as:

$$\zeta_i = \{\xi_1, \xi_2, ..., \xi_n\} \tag{4}$$

Among them, ζ_i It represents any sample feature in the characteristic value. In the feature classification of archive resources, there are usually three categories. The decision tree classifier shown in Fig. 1 can be established through the decision tree.

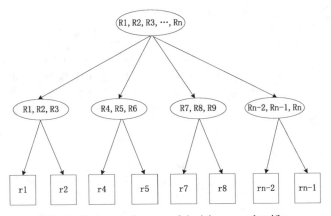

Fig. 1. Structure diagram of decision tree classifier

The decision tree classifier established through Fig. 1 will have a feature to summarize each information gain node. At this time, the classification expectation of high-quality teaching resource samples is:

$$E_s = \sum_{i=1}^{n} \frac{P_i}{\log_2 (P_i)} \tag{5}$$

Among them, E_s The automatic classification expectation of decision tree representing the digitalization of electronic archives resources; P_i Indicates that any sample is classified as a category i Probability of. In the feature classification of high-quality teaching resources at this time, the information gain of resource A can be expressed as:

$$T_A = (E_s, E_f) \tag{6}$$

Among them, T_A Indicates the classification of resource A, E_s Represents the first attribute code of the decision tree classification; E_f Represents the second attribute code of the decision tree classification. Combining the above two attribute codes, we can get the specific classification position of the high-quality teaching resources in the decision tree.

2.3   Establish a Database of Teaching Resources for Career Guidance Courses

After the above resource sharing classification code is extracted based on the decision tree classification algorithm, the specific classification position of high-quality resources for college students' career guidance course teaching in the decision tree is obtained. Next, establish a database of teaching resources for career guidance courses, and store the data with the blockchain through the call of the contract layer. Store interactive data on the blockchain and teaching resources for big data on IPFS.

And then store the resource address of IPFS on the blockchain [8]. First of all, design the entity map of college students' career guidance course attributes, as shown in Fig. 2.

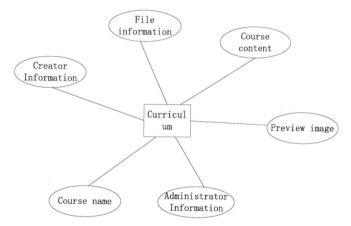

Fig. 2. Entity Chart of Undergraduate Employment Guidance Course Attributes

Through the course attribute in Fig. 2, the database entity of college students' employment guidance course attribute is obtained. On this basis, a database is established to store the basic information and operation information related to high-quality resources of course teaching. The database of teaching resources for career guidance courses is shown in Table 1.

According to the field classification in Table 1, store the relevant information of high-quality teaching resources for college students' career guidance courses, set the courses as the primary key, and ensure the security of resource storage. The sharing of high-quality teaching resources is divided into five layers: user layer, application layer, service layer, contract layer and data layer. The data layer is used to store high-quality resource data. The contract layer is responsible for the realization of resource storage and sharing functions. The service layer is the interface between contract and application, and the application layer and user layer are the realization of user operations [9]. This paper introduces the design and implementation of smart contracts in detail, which can solve the problem of safe storage of high-quality teaching resources and provide basic guarantee for resource sharing.

Table 1. Database of Teaching Resources for Employment Guidance Courses

Field Name	explain	type
KC-MC	Course name	VARCHAR
KC-JS	Course Introduction	VARCHAR
KC-HYDZ	Course contract address	VARCHAR
KC-CJZZHDZ	Account address of course creator	VARCHAR
KC-JGTDZ	Course Structure Map Address	VARCHAR
KC-ZYDZ	Course resource address	VARCHAR
KC-XZRS	Number of course downloads	INT
KC-PLNR	Comments	VARCHAR
KC-GXCS	Course sharing times	INT

2.4 Sharing of High-Quality Teaching Resources

After the establishment of the database of teaching resources for career guidance courses, the problem of safe storage of high-quality teaching resources has been solved. On this basis, this paper establishes a blockchain based security sharing model of educational resources (EduMSFB). The framework uses the consensus mechanism, encryption algorithm, smart contract, cloud storage, public and private key encryption scheme, authority authentication management and other technologies of the blockchain to complete the information sharing, secure storage and authority management of high-quality teaching resources [10].

First of all, through the analysis of the high-quality resource structure of college students' career guidance course teaching and the demand for safe sharing of resources, the basic structure of the sharing model framework is designed, which is divided into three parts: the user layer, the service layer, and the management layer. The user layer can use the functions of the service layer to carry out corresponding operations, while the management layer mainly manages the sharing process to achieve safe sharing of resources. The framework structure of high-quality teaching resource sharing designed in this paper is shown in Fig. 3.

As shown in Fig. 3, in the sharing of high-quality teaching resources, the user layer includes not only the service objects (resource users) of resource sharing, but also resource providers, such as large institutional databases and individual resource sharing. Therefore, users can be both resource consumers and resource providers.

The service layer mainly includes the main functions of resource sharing, including uploading, browsing, downloading and other functions of resources. Users conduct corresponding operations through the corresponding function modules of the service layer to meet their own needs.

In the established sharing framework of high-quality teaching resources, the management layer, through the relevant technologies of the blockchain and through multiple institutional members and users to jointly form the nodes of the chain, jointly manage

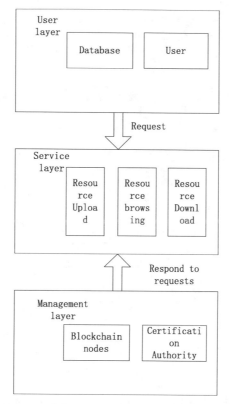

Fig. 3. Framework for sharing high-quality teaching resources

and maintain teaching resources, so as to carry out free circulation and exchange of educational resources within a certain range, realize their safe sharing and break information barriers. The certification authority in the blockchain conducts systematic management on behalf of all nodes, but all operations of the certification authority will be broadcast in all nodes through the consensus mechanism to achieve consistency of data and operations; The identity authentication service is carried out through the identity management mode of the blockchain network itself. All nodes entering the resource sharing must be authenticated. Only through authentication can the corresponding user operations be carried out.

The users of teaching resource sharing framework include resource providers and resource users. According to the identity of resource sharers, their sharing methods are also different. First, the resources in the institutional database are shared through the centralized sharing mode, but the shared information will be backed up and managed through the blockchain, and the safety and reliability of transactions can be effectively guaranteed through the supervision of the certification authority; For ordinary users, the decentralized sharing mode is used to share resources, effectively taking advantage of the decentralized characteristics of the blockchain network.

The high-quality resource sharing process of college students' career guidance course teaching designed in this paper is shown in Fig. 4.

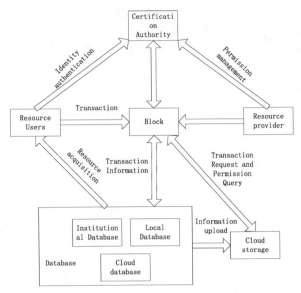

Fig. 4. High quality resource sharing process of college students' career guidance course teaching

As shown in Fig. 4, in the process of resource sharing, institutional data resource sharing is to upload the resource index of its database to the blockchain, where only upload records, transaction information, etc. are saved, and the index table is placed in the database function cloud database.

For individual resource sharing, the resource provider uploads the index of shared resources to the blockchain, and stores the source data in the local database or ECS of the database module. Users of resources browse the resources through the functions in the service layer. After purchasing the resources, they can obtain the address information and decryption key of the real data through the certification authority according to the obtained permissions, and decrypt and use the obtained resources; When downloading the resources in the institutional database and accessing the address information of the corresponding resources obtained through the blockchain, the institutional resource provider judges whether the resource requester has access to the resources by querying the access permission list of the blockchain, and then responds to the operation.

Writing appropriate access control protocols through smart contracts can make the behavior of the certification authority more fair and effective, and the sharing process of resources can be more secure through user identity authentication, identity management, and permission management, ensuring the privacy of user information and the transparency of transactions. After obtaining the resource address, the user must have access to the resource at the same time before downloading the resource. At the same time, all data on the chain are encrypted, and the corresponding information can be seen through the certificate issued by the certification authority.

Set the database teaching quality resource sharer as A1, the individual teaching quality resource sharer as A2, and the teaching quality resource user as U. Use the transaction process between A1, A2, and U to specifically design the process of the whole employment guidance course teaching quality resource sharing. The steps are as follows.

1. Institutional database A1, personal resource sharer A2 and resource user U are authenticated through the blockchain and become members of the educational resource sharing framework.
2. The resource upload process is divided into two parts: (1) The institutional database A1 uploads resources, and the shared resources uploaded by A1 are divided into two parts. The index information of the shared resources is stored in the blockchain, and the resource address information corresponding to each index is put into cloud storage and encrypted. (2) The resources uploaded by personal resource sharer A2 are also divided into two parts. The information about shared resources stored on the blockchain includes resource processing permissions, content summaries, time information, and real data addresses. The other part is the source data of shared resources, which is encrypted and stored in the database. All resource sharers can set their personal transaction records as encrypted or public according to their own needs. If encrypted, they can use their uploaded public key for encryption.
3. The transaction of the certification authority and the transaction in the blockchain are confirmed through the consensus mechanism. The nodes confirm the transaction information through broadcasting, update the local ledger information, and check the effectiveness of the local ledger and whether it has been tampered with at regular intervals.
4. The resource user U purchases resources and sends a request to the blockchain, including the information of the resource requester, the summary of the purchased resources, etc. The request is broadcast to all nodes through the blockchain consensus mechanism and confirmed, and the transaction is handed over to the certification authority for the resource user U's permission modification.
5. According to the transaction information, the certification authority unlocks the permission of the corresponding resources, decrypts the resource request information, obtains the resource address information through the blockchain, and sends it to the resource user U. The resource user U obtains the shared resources according to the resource address to complete the sharing process.
6. When resource sharer U obtains the address information in institutional database A1 and requests resources from institutional database A1 through the resource address, resource database A1 first queries the transaction information and corresponding permission information of the corresponding resource user U in the blockchain network, and then opens the interface after confirmation so that resource users can download shared resources.

3 Experimental Analysis

3.1 Experiment Preparation

In order to test the feasibility of the above method of sharing high-quality resources for college students' career guidance courses based on the decision tree classification algorithm, the experimental analysis is carried out as shown below. First of all, based on the demand of high-quality teaching resource sharing, build the operating environment required for resource sharing. The building steps are as follows: Download JDK 6.0 or above from the official website Oracle, install and configure it; Download Eclipse development tools; Download Struts -2.5.10.1, spring framework -4.3.5. RELEASE, hibemate core -5.2.9. Final framework; Download and install the Tomcat server and integrate it into the Eclipse development environment; Download Maven; To download and install Hadoop, it is necessary to set up environments such as HADOOP-HOME and HADOOP-CONF-DIR; Download and install HBase and ZooKeeper; Install and deploy SolrCloud. The teaching high-quality resource sharing testing platform consists of four servers. The specific allocation plan is: one application server, with Tomcat deployed on it, one NameNode serving as the Master, and two DataNodes serving as the Slave.

The teaching quality resource sharing test platform consists of four servers. The specific allocation scheme is: one application server, on which Tomcat is deployed, one is the NameNode acting as the master, and two are the DataNode acting as the slave. Following the above process,, build the high-quality teaching resource sharing operating environment required for this experiment. In addition, the underlying environment is based on HDFS, specifically storing massive data of various types on HDFS, and the teaching resource sharing system under big data manages the database based on MySQL. On top of that, other tools and products related to big data technology are introduced, such as HBase, SolrColoud, and Zookeeper. In order to realize the rapid retrieval of high-quality teaching resources of college students' career guidance courses, this experiment optimized the retrieval of HBase, combined with the index mechanism of Solr, and achieved the perfect integration with HBase index. The index structure of high-quality resource sharing for college students' career guidance course teaching is shown in Fig. 5.

According to the shared index architecture shown in Fig. 5, some columns (or all columns) of the HBase table are indexed into the Solr index in near real time, which has high flexibility and scalability, and optimizes and customizes the retrieval scoring rules of Solr, so as to meet the needs of massive teaching resource retrieval in this article. After completing the above experimental preparation, the next step is to test the proposed high-quality resource sharing method for career guidance course teaching.

3.2 Result Analysis

In order to make the experimental results more intuitive and convincing, this paper introduces the principle of comparative experimental method. Set the teaching quality resource sharing method of college students' career guidance courses based on the decision tree classification algorithm proposed in this paper as the experimental group, and set the traditional teaching quality resource sharing method as the control group, and make a comparative analysis of the application effects of the two sharing methods.

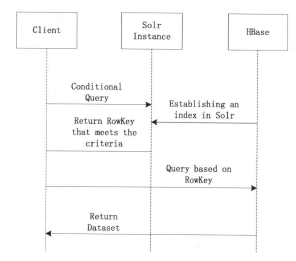

Fig. 5. Index structure for sharing high-quality teaching resources

The traditional method of sharing high-quality teaching resources is based on cloud computing technology, combined with cloud computing based shared learning mode and high-quality resource sharing platform for information-based teaching.

The completion time of high-quality teaching resource sharing task of college students' employment guidance course is selected as the evaluation index of this experiment. The shorter the completion time, the higher the efficiency of high-quality teaching resource sharing. The sharing function of high-quality teaching resources includes browsing, uploading, downloading and retrieving high-quality teaching resources of college students' employment guidance courses. Table 2 shows the testing process of the sharing function of high-quality teaching resources.

According to Table 3, complete the function test of sharing high-quality resources of college students' career guidance course teaching. On this basis, the MATLAB simulation analysis software is used to measure the time required for each resource sharing function to complete the operation after the application of the two sharing methods, and compare them to determine the application effect and feasibility of the proposed methods. The comparison results are shown in Fig. 6.

From the comparison results in Fig. 6, it can be seen that two methods for sharing high-quality resources in the teaching of college student employment guidance courses show performance differences. Among them, after the application of the teaching high-quality resource sharing method based on the decision tree classification algorithm proposed in this article, the completion time of various resource sharing functions is significantly shorter than traditional methods. It can complete the task of sharing high-quality resources in college student employment guidance course teaching within 5-8 ms, and display the resource list by category Uploading and downloading of resources. Users determine that they are looking for a certain resource and can search through resource sharing. Through parallel retrieval, shared search results are returned to users in a short amount of time.

Table 2. Test process of sharing function of high-quality teaching resources

function	input	Expected output	Whether it is normal
Browse Resources	On the main page, select resources under different categories and view them by discipline	Display resource list by category	normal
Download Resources	Click a resource to enter the details page of the resource, and click "Download" to download the resource locally	Successful download and prompt that the download is complete	normal
Upload Resources	Select the upload button, click the "Browse" button, find the local file to upload, click "Upload", and upload the local resources to the sharing platform	Upload successfully and prompt that the upload is complete	normal
Retrieve Resources	Select the Search button to retrieve the corresponding high-quality teaching resources according to the needs	Retrieved successfully	normal

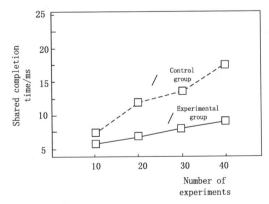

Fig. 6. Comparison Results of Evaluation Indicators of Teaching Quality Resource Sharing Methods

Establish a scoring system that allows users to rate high-quality teaching resources for college student employment guidance courses based on decision tree classification algorithms. Evaluate the superiority of the decision tree classification algorithm based method in terms of resource sharing quality and popularity by collecting user rating data and comparing the average resource ratings of the experimental group and the control group. At the same time, collect user feedback to improve and optimize resource sharing methods and provide a better user experience.

The design of the scoring system is as follows:

Step 1. Set scoring standards: Determine a set of scoring standards from both richness and applicability. Use a 1–10 point scoring system to represent different levels of ratings.

Step 2. Scoring Interface: Provide users with a scoring interface that allows them to easily rate shared resources. The interface is intuitive, concise, and clearly instructs users on how to rate.

Step 3. Resource Display: On the rating interface, display relevant information about high-quality teaching resources for college student employment guidance courses, including resource names, descriptions, authors, etc. Ensure that users have a full understanding of the content and purpose of resources in order to make accurate evaluations.

Step 4. Scoring process: On the scoring interface, provide users with the option to choose a rating. Users can rate each aspect based on the set scoring criteria.

The partial scoring results are shown in the table below:

Table 3. Scoring Results

user	experimental group		control group	
	richness	Applicability	richness	Applicability
1	7 points	8 points	4 points	5 points
2	6 points	7 points	5 points	6 points
3	7 points	9 points	5 points	7 points
4	8 points	9 points	3 points	5 points
5	5 points	7 points	4 points	6 points
6	8 points	8 points	6 points	5 points
7	7 points	9 points	5 points	6 points
8	6 points	7 points	4 points	3 points
9	7 points	8 points	5 points	6 points
10	7 points	8 points	3 points	3 points
Average score	7 points	8 points	4.4 points	5.2 points

From Table 3, it can be seen that the average resource score of the experimental group is 7 points in terms of richness and 8 points in terms of applicability. The average resource score of the control group was 4 points in terms of richness and 5.2 points in terms of applicability. It shows that the method based on decision tree classification algorithm can provide a better experience of sharing teaching resources of college students' career guidance courses.

4 Conclusion

With the development of educational informatization, online teaching resources in schools are constantly enriched, providing students with diverse learning methods, but also posing challenges to traditional resource management. In order to expand the scope of resource sharing, improve the efficiency of resource utilization, and solve the problems of uneven distribution of teaching resources among schools and massive data resource management and retrieval, this paper designs a method of high-quality teaching resource sharing in the Big data environment. Firstly, design a security key for public information and encrypt the public information of high-quality teaching resources. Secondly, the decision tree classification algorithm is used to construct a classification standard for electronic archives and extract resource sharing classification codes. At the same time, establish a database of teaching resources for employment guidance courses to store information related to high-quality teaching resources for college students' employment guidance courses. Finally, establish a blockchain based teaching resource security sharing model to achieve information sharing of high-quality teaching resources. The experimental results show that the proposed method can complete the task of sharing high-quality resources in the teaching of college student employment guidance courses within 5–8.5 ms after application. Through the research in this article, the efficiency of sharing high-quality resources in the teaching of college student employment guidance courses has been effectively improved, enabling users to quickly and accurately share and retrieve the resources they want, with good scalability and flexibility. In the context of the continuous in-depth development of education informatization, the sharing method of high-quality teaching resources in the Big data environment will continue to be promoted. The sharing of high-quality teaching resources in the future will be more intelligent, cross school linkage, focus on data security protection, and expand the scope of sharing, providing students with more personalized and diverse learning resources, and promoting the sustainable development of educational informatization.

References

1. Ma, H.: Design and application of teaching resources sharing platform for physical education major based on internet. J. Phys. Conf. Ser. **1992**(2), 022197 (2021). https://doi.org/10.1088/1742-6596/1992/2/022197
2. Feng, J., Zhang, W., Tsai, S.B.: Construction of a multimedia-based university ideological and political big data cloud service teaching resource sharing model. Math. Probl. Eng. **2021**, 1–12 (2021). https://doi.org/10.1155/2021/9907630
3. Liu, H.: Application and strengthening strategies of network resources in the construction of teaching platform. J. Phys. Conf. Ser. **1915**(4), 042056 (2021). https://doi.org/10.1088/1742-6596/1915/4/042056
4. Tucker, B.V., Kelley, M.C., Redmon, C.: A place to share teaching resources: Speech and language resource bank. J. Acoust. Soc. Am. **149**(4), A147–A147 (2021)
5. Yao, S., Li, D., Yohannes, A., et al.: Exploration for network distance teaching and resource sharing system for higher education in epidemic situation of COVID-19 - ScienceDirect. Procedia Comput. Sci. **183**, 807–813 (2021)

6. Rodrigues, T.C., Leitão, F.O., Thomé, K.M., Cappellesso, G.: Sharing economy practices in agri-food settlements: integration of resources, interdependence and interdefinition. J. Cleaner Prod. **294**, 126357 (2021). https://doi.org/10.1016/j.jclepro.2021.126357
7. Syarifah, L.S.: Career guidance in preparing students with disabilities for get in to the worke in preparing students with disabilities for get in to the work. Jurnal Pendidikan Dasar **9**(1), 135–145 (2021)
8. Ma, J., Sun, L., Zhang, T.: Thoughts on online teaching of employment guidance course for civil aviation majors under the COVID-19 epidemic--based on the teaching practice of college students enrolled in 2018 in the Civil Aviation Management Institute of China. J. Phys. Conf. Ser. **1931**(1), 012001 (2021). https://doi.org/10.1088/1742-6596/1931/1/012001
9. Osborn, D.S., Brown, C.A., Morgan, M.J.: Expectations, experiences, and career-related outcomes of computer-assisted career guidance systems. J. Employ. Couns. **58**(2), 74–90 (2021). https://doi.org/10.1002/joec.12158
10. Cheng, S., Liu, G., Ren, C., et al.: Remote resource sharing simulation of interactive experiment platform based on blockchain. Comput. Simul. **006**, 039 (2022)

Evaluation of Word-of-Mouth Influence of Cross-Border E-commerce Products Based on Social Network Data Analysis

Weiwei Zhang[1], Yuanting Lu[1], Lingming Cao[2(✉)], and Hui Li[3]

[1] Computer Information and Engineering College, Guizhou University of Commerce, Guiyang 550014, China
[2] Internet of Things Engineering, Hehai University, Changzhou 211100, China
caolingming@hhu.edu.cn
[3] Hunan Industry Polytechnic, Changsha 410000, China

Abstract. To improve the performance of word-of-mouth impact assessment for cross-border e-commerce products, a method based on social network data analysis is proposed for evaluating the word-of-mouth impact of cross-border e-commerce products. Eliminate abnormal users on social networks and establish an evaluation index system for the reputation and influence of cross-border e-commerce products. Based on this system, the basic indicators for evaluating the word-of-mouth impact of cross-border e-commerce products are determined, and an indicator weight judgment matrix is constructed using Analytic Hierarchy Process to calculate the weight of the word-of-mouth impact evaluation indicators for cross-border e-commerce products. Based on the weight calculation results, a word-of-mouth impact evaluation algorithm for cross-border e-commerce products was designed to achieve the evaluation of word-of-mouth impact of cross-border e-commerce products. Case analysis demonstrates that this approach effectively evaluates the impact of word-of-mouth for cross-border e-commerce products with a 95% improvement in evaluation efficiency and reliability, boosting overall performance.

Keywords: Social Network Data Analysis · Cross-Border E-Commerce · Product Reputation · Effect · Assessment · Weight Calculation

1 Introduction

With the upgrading of China's consumption structure, the domestic market's demand for overseas goods continues to release. In addition to the favorable policy background, import cross-border e-commerce has developed rapidly. At present, import and retail cross-border e-commerce is mainly based on the B2B model, but in recent years, the transaction scale and proportion of import and retail cross-border e-commerce have continued to expand. According to the statistics of Intelligent Research Consulting, the transaction scale of import and retail cross-border e-commerce has rapidly increased

L. Yun et al. (Eds.): ADHIP 2023, LNICST 547, pp. 394–408, 2024.
https://doi.org/10.1007/978-3-031-50543-0_27

from 47.8 billion yuan in 2012 to 408.2 billion yuan in 2016, from 19.9% to 34% of the import cross-border e-commerce transaction scale, and is expected to increase to 40% in 2018. With the expansion of consumer demand and the tightening of regulatory policies, import and retail cross-border e-commerce has gradually developed from C2C mode to B2C mode [1]. In 2016, the transaction scale of B2C mode accounted for 56.4% of import and retail cross-border e-commerce.

In domestic research, Chen Yufen [2], conducted risk analysis on four links involving commodity quality, and built a risk identification, evaluation index system, risk measurement. The quality risk assessment system of imported B2C cross-border e-commerce goods is divided into five parts, namely, risk rating and risk cause tracing, and takes Hangzhou imported B2C cross-border e-commerce enterprises as an example for empirical research. The problem of how to maintain the reputation of cross-border e-commerce platform reputation defenders when facing one or more losers was attempted to be solved by Zhan Haoling and others [3], by establishing a multi-objective Stackelberg game model with incomplete information of defenders and losers. In this model, we analyzed the prior probability, damage effect, and the impact of input costs of both sides on the maintainer's strategy in the reputation network game model of cross-border e-commerce platforms, and put forward relevant suggestions. In foreign research, Zhang X et al. [4] mainly studied the mechanism and model of cross-border e-commerce green supply chain based on customer behavior. The green supply chain partners select 24 secondary indicators of the evaluation system as input vectors. Wang F et al. [5] examined the differences in the impact of OL and WOML on consumer decision-making in three stages of online shopping through the theoretical approach of motivation reinforcement. The results indicate that when consumers purchase products with high participation, word-of-mouth has a greater impact on the consumer decision-making process than OL, while when consumers purchase products with low participation, OL has a greater impact on the consumer decision-making process than word-of-mouth. Miremadi A et al. [6] published a topic titled "Evaluation of the Role of Electronic Word of Mouth (EWOM) in Customer Perception and Behavior in Electronic Stores". Then, research hypotheses and objectives were proposed, and samples required for methodology, statistical population, and analytical techniques were introduced to achieve the assumed objectives and results. Miranda S et al. [7] used a mixed method of symmetry and asymmetry. Through SEM, it was found that consumers with high suspicion of behavior, perfectionism oriented towards others, and a lack of recognition of imperfections have a lower tendency to provide positive word-of-mouth due to their higher perception of social risks.

This paper evaluates the word-of-mouth influence by analyzing social network data. Firstly, by removing abnormal users from social networks, the accuracy and reliability of evaluations can be improved. Secondly, a cross-border e-commerce product word-of-mouth influence evaluation index system has been constructed, which includes a series of indicators that can reflect the product's word-of-mouth influence, such as the number of comments and likes. Then, by calculating the weights of evaluation indicators, the importance of each indicator in the evaluation of word-of-mouth influence can be determined. Finally, an algorithm for evaluating the word-of-mouth influence of cross-border e-commerce products was designed, and the product's word-of-mouth influence score was obtained by integrating the weights and values of various indicators. Through this study, the word-of-mouth influence of products can be objectively and accurately evaluated. This helps cross-border e-commerce platforms and merchants understand the reputation and performance of products in the market, thereby improving product improvement and market promotion, enhancing product competitiveness and sales performance.

2 Design of Evaluation Method for Word-of-Mouth Influence of Cross-Border E-commerce Products

2.1 Eliminate Abnormal Users of Social Networks

The degree of co-occurrence of two commands in social network user login is called tight density. Here, we use the sliding window to measure the tight density of cross-border e-commerce product word-of-mouth evaluation commands. Social networking [8, 9] user history operations are recorded as a sequence of shell commands. The command sequence is collected through the sliding window, and the word-of-mouth evaluation commands often used together will appear in a window. Assuming that the sliding windows are numbered and the windows that appear in each command are counted, each word-of-mouth evaluation command will correspond to a feature vector and represent the window that has already appeared, then the similarity between vectors is the tightness between social network users' word-of-mouth evaluation commands.

Set two parameters related to sliding window: sliding window size s, which is the number of commands that the window can hold. Window change step l, which is the number of sliding commands per window. When the window is sliding, count the number of times each command appears in the window. Set the training data length to L, all commands in training data are Σ, the number of different commands is $m = |\Sigma|$, the number of windows is $n = \frac{L-s}{l} + 1$. The login characteristic data of social network users can be expressed as $m \times n$ Matrix of W, the expression is:

$$W = \begin{bmatrix} w_{11} & w_{12} & \cdots & w_{1m} \\ w_{21} & w_{22} & \cdots & w_{2m} \\ \cdots & \cdots & \cdots & \cdots \\ w_{n1} & w_{n2} & \cdots & w_{nm} \end{bmatrix} \tag{1}$$

Among them, W is the vector values in the columns in indicate the frequency of the corresponding commands in different windows, W is the vector value of each line in indicates the frequency of different commands in the corresponding window.

Setting and W pass the civil examinations i. The corresponding command listed is s_i, to measure two random commands s_i and s_j. For the purpose of compactness W. Two matrices are generated, namely:

$$B = \begin{bmatrix} b_{ij} \end{bmatrix} \tag{2}$$

$$B^* = \begin{bmatrix} b^*_{i'j'} \end{bmatrix} \tag{3}$$

Among them, B and B^* represents a binary matrix, b_{ij} indicates a command s_j in the window ϖ_i whether it appears or not. The same command appears more frequently in the window, which can be seen as the result of the alternate appearance of the command and its twin commands, $b^*_{i'j'}$ representative order s_j. Twin command s_j in window ϖ_i whether it appears or not.

According to the above analysis and calculation, we can get the formula for calculating the compactness between the two commands of social network users' evaluation of cross-border e-commerce product reputation:

$$Clo(s_i, s_j) = \begin{cases} \dfrac{b_i \cdot b^*_{i'}}{\|b_i\| \cdot \|b^*_{i'}\|}, & s_i, s_j \in \Sigma \cap i = j \\ 0, & s_i \notin \Sigma \text{ or } s_j \notin \Sigma \\ \dfrac{b_i \cdot b_j}{\|b_i\| \cdot \|b_j\|}, & s_i, s_j \in \Sigma \cap i \neq j \end{cases} \qquad (4)$$

The average closeness of word-of-mouth evaluation commands in the window can be called the window convergence degree. Assuming that the word-of-mouth evaluation commands in the window are frequently used by legitimate users, the higher the window convergence degree is; Otherwise, the window aggregation degree is lower.

According to formula (4), social network users evaluate the compactness between the two orders of cross-border e-commerce product reputation $Clo(s_i, s_j)$ the calculation formula can effectively detect the word-of-mouth evaluation order by social network users with low degree of aggregation, and eliminate such users as camouflage users.

2.2 Build an Index System for Evaluating the Word-of-Mouth Influence of Cross-Border E-commerce Products

On the basis of fully understanding the concept and characteristics of word-of-mouth influence of cross-border e-commerce products, this paper uses literature search, process analysis and field research methods to identify the word-of-mouth influence in the four links of overseas procurement, warehousing logistics, platform sales and product after-sales under the bonded warehouse and overseas warehouse models. This constitutes a list of word-of-mouth influence of cross-border e-commerce products. At the same time, utilizing the distinct implications of diverse effects and the accessibility of diverse impact indicator information, while adhering to the principles of inclusiveness, feasibility, comparability, and relevance in constructing the indicator framework, a word-of-mouth impact assessment indicator framework for cross-border e-commerce products has been devised, demonstrated in Table 1. The specific construction process of the evaluation indicator system of word-of-mouth influence is shown in Fig. 1.

Based on the principles of comprehensiveness, operability, comparability and pertinence in the construction of the indicator system, the evaluation indicator system of the reputation influence is constructed.

2.3 Calculate the Weight of Cross-Border E-commerce Products' Reputation Impact Evaluation Indicators

Based on the evaluation index system of word-of-mouth influence of cross-border e-commerce products, the analytic hierarchy process [10–13] is used to calculate the

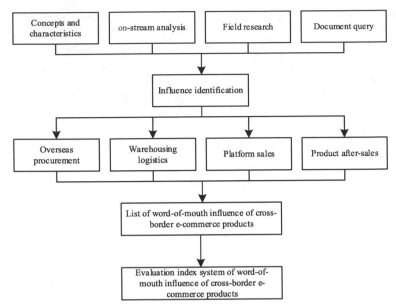

Fig. 1. Construction process of evaluation indicator system for word-of-mouth influence of cross-border e-commerce products

weight of the evaluation index of word-of-mouth influence of cross-border e-commerce products. The specific calculation process is as follows:

Step 1: Determine the basic indicators for evaluating the reputation influence of cross-border e-commerce products.

Users can obtain relevant information about cross-border e-commerce products on various social networks. According to the analysis of social network data, the basic indicators for evaluating the word-of-mouth influence are determined, namely:

$$N = \begin{cases} \sqrt{N_w} + \zeta \\ \sqrt{(X_a B_b)^2} + \zeta \end{cases} \tag{5}$$

where, N_w indicates the quality indicators of cross-border e-commerce products, X_a represents the security base in transportation, B_b represents the safety amplitude in transportation, ζ represents the weight coefficient.

Step 2: Build indicator weight judgment matrix.

After determining the basic indicators for evaluating the word-of-mouth influence of cross-border e-commerce products, the redundant information of the evaluation indicators at different levels is normalized by comparing the evaluation indicators at the same level, so as to improve the accuracy of the weight calculation of the word-of-mouth influence evaluation indicators of cross-border e-commerce products. If the evaluation indicators of different levels are redundant i and j the comparison results of v_{ij}, then the indicator weight judgment matrix can be expressed as:

$$v_{ij} = \frac{1}{v_{ji}} \tag{6}$$

Table 1. Evaluation index system of word-of-mouth influence of cross-border e-commerce products

Target layer	Primary indicators	Secondary indicators
Evaluation index system of word-of-mouth influence of cross-border e-commerce products	Overseas procurement	Cognition of product quality standards
		Supplier qualification and credit level
		Supplier supply defect rate
		Product acceptance mode
	Warehousing logistics	Warehouse sanitation and safety compliance rate
		Sampling inspection frequency of product quality in stock
		Rationality of product logistics, transportation and packaging
		Operation error rate of logistics personnel
	Platform sales	Number of platform entry assessment items
		Monitoring frequency of platform public opinion
		Lack of product traceability information
		Lack of information displayed online
	After sales	Number of after-sales guarantee items
		One time solution rate of after-sales quality problems
		Collection frequency of product after-sales quality information

Step 3: Process coincidence indicators.

When evaluating the word-of-mouth influence of cross-border e-commerce products, we need to make up for the shortcomings of influence evaluation indicators from different perspectives, which will lead to overlap of evaluation indicators. In order to ensure the rationality of evaluation indicators in the evaluation of reputation influence of cross-border e-commerce products, according to $n_1 + n_2 \leq 2\sqrt{n_1 n_2}$ based on the inequality principle, redundant information of coincidence index is removed.

Step 4: Calculate the weight of the evaluation index of word-of-mouth influence at all levels.

First level evaluation indicators for the reputation influence of cross-border e-commerce products A_i to build a judgment matrix U_0 and secondary evaluation indicators B_{ij} judgment matrix of U_{i1}, U_{i2}, U_{i3} and U_{i4}, i.e.:

$$U_0 = \begin{bmatrix} 1 & 4 & 7 \\ 1/4 & 1 & 4 \\ 1/7 & 1/4 & 1 \end{bmatrix} \tag{7}$$

The square root method is used to calculate the weight vector of the evaluation index of word-of-mouth influence of cross-border e-commerce products. The formula is:

$$\omega_i = \frac{x_i}{\sum\limits_{j=1}^{n} x_j} + y_i \tag{8}$$

Among them, x_i represents the square root result of the product of each element in the judgment matrix, y_i represents the weight of matrix elements.

According to the weight vector of cross-border e-commerce product reputation influence evaluation indicators [14], the weight set of cross-border e-commerce product reputation influence evaluation indicators is obtained, namely:

$$\Omega_z = v_{ij}\omega_i \tag{9}$$

According to the above calculation steps, the weight value of the cross-border e-commerce product reputation influence evaluation index is calculated.

3   Design the Algorithm for Evaluating the Word-of-Mouth Influence of Cross-Border E-commerce Products

Assume that the nodes for the evaluation of the reputation influence are B_i, the importance of reputation influence is Z, the impact of cross-border e-commerce product reputation is O, the momentum of cross-border e-commerce products' reputation influence nodes is N_iR. The word-of-mouth influence is analyzed hierarchically [15], which is described as follows:

$$B_iR = \frac{AS \times OS \times Z}{\Omega_z} \tag{10}$$

In the formula provided, the level of AS information represents the reputation effect of cross-border trading items. On the other hand, OS displays the influence of social media data concerning the cross-border trading products' word-of-mouth effect [16–18]. Following the hierarchical analysis of word-of-mouth impact on cross-border e-commerce products, significant variations are apparent. The particle swarm optimization algorithm is leveraged to describe this connection. Thus, the ensuing formula determines the criteria for evaluating the word-of-mouth influence of cross-border e-commerce products.

$$AS = \frac{Q_i}{S^l} \tag{11}$$

Among them, Q_i indicates the degree of correlation between the word-of-mouth influence level and the word-of-mouth influence, S^I refers to the evaluation criteria.

Hypothesis N_i express the reputation influence target node of O_i, ΔO represents the evaluation time domain of the reputation influence, and its satisfaction $\Delta O = |t_c - t_d|$ the current time period for evaluating the reputation influence is t_c, the past period is t_d the time window for evaluating the reputation influence is $[t_d, t_c]$ [19]. Then the reputation influence O_i. Target node for N_i stay ΔO the reputation influence in the period can be divided into m influence links At_j, then the influence index of cross-border e-commerce product reputation is:

$$OS_{(t_i,N_i)} = \frac{\chi}{N_i} \cdot 10^{v_i} \cdot G^{n_i} \tag{12}$$

In the above formula, χ is the influence coefficient representing the reputation, G^{n_i} is Bias information indicating the reputation influence, set K_i for the related information of cross-border e-commerce products' word-of-mouth influence [20], it is necessary to consider the main influencing factors of cross-border e-commerce products' word-of-mouth influence and select the direct influence of cross-border e-commerce products' word-of-mouth v_i as the most influential factor, namely:

$$v_i = \max\left\{\phi K_{i1} \cdot v, \cdots, K_{in_i} \cdot v\right\} \tag{13}$$

Among them, $\phi K_{i1} \cdot v$ represents the smallest factor affecting the reputation of cross-border e-commerce products, $K_{in_i} \cdot v$ represents the biggest factor affecting the reputation of cross-border e-commerce products. The word-of-mouth promotion has a complete strategy, and the specific membership values [21] are S_C, S_I, S_A according to different levels of influence, we can get the degree of influence subordination of cross-border e-commerce product word-of-mouth:

$$S = \sqrt{\frac{S_C^2 + S_I^2 + S_A^2}{3}} \tag{14}$$

According to the factors affecting the reputation and the proportion of the reputation influence of cross-border e-commerce products, the evaluation weight of the reputation influence is obtained:

$$\omega = \sum_{j=1}^{m} v_j \tag{15}$$

Among them, m indicates the amount of word-of-mouth influence of cross-border e-commerce products, and the weight of word-of-mouth influence of each cross-border e-commerce product is v_j [22–24], any influence can be used j express. In period Δt Internal, define the threshold of word-of-mouth influence as $SR_{(\Delta t)}$, the value is calculated by the weight of the reputation influence and the reputation influence weight of each cross-border e-commerce product, namely:

$$SR_{(\Delta t)} = \sum_{j=1}^{n} N_i R_{(N_i,\Delta t)} \times \omega_i \tag{16}$$

Among them, n refers to the number of word-of-mouth influence of cross-border e-commerce products, $NR_{(N_i, \Delta t)}$ means that the reputation is in the period possible influence value in Δt.

Assuming that there are multiple influence coefficients within the reputation influence level $NR_{(N_i, \Delta t)}$. Indicates the risks in the process of evaluating the reputation influence of cross-border e-commerce products, ω_i express the reputation influence Evaluation coefficient of N_i [25].

According to the weight vector determined above V_i And judgment matrix S_j, calculate the normal vector on the evaluation set, and the formula is:

$$\psi_i = V_i \oplus S_j = (a_1, a_2 \cdots, a_n) \tag{17}$$

In the above formula, $\oplus$ It is a composite operator of cross-border e-commerce product word-of-mouth influence evaluation. On the basis of determining the calculation equation, it designs a cross-border e-commerce product word-of-mouth influence evaluation algorithm, and realizes the evaluation of cross-border e-commerce product word-of-mouth influence.

4 Example Analysis

4.1 Experimental Data

In order to verify the performance of the method in this paper in the evaluation of word-of-mouth influence of cross-border e-commerce products, eight products of a cross-border e-commerce enterprise were selected as the example analysis samples to evaluate the word-of-mouth influence of cross-border e-commerce products. The weight data of the word-of-mouth influence of eight cross-border e-commerce products are shown in Table 2.

Table 2 is from a cross-border e-commerce enterprise that has settled on the Amazon platform. The company's products cover various fields such as electronic products, fashion clothing, household products, beauty and skincare products, mother and baby products, and food.

4.2 Measure the Cumulative Contribution Rate of Cross-Border E-commerce Product Reputation Impact Assessment

Based on the weight data of word-of-mouth influence in Table 2, the evaluation methods in the article, the evaluation method based on the entire process in reference [2], and the evaluation method based on incomplete information in reference [3] are used for comparison, and the principle of cumulative contribution rate greater than 96% is adopted to screen out the main components of the evaluation indicators of word-of-mouth influence of cross-border e-commerce products. The cumulative contribution rates of the three methods are shown in Fig. 2.

According to the results in Fig. 2, the cumulative contribution rate of the first five cross-border e-commerce product word-of-mouth influence evaluation indicators is 96.12% when the evaluation method based on the whole process is adopted, while the

Table 2. Weight data of word-of-mouth influence of cross-border e-commerce products

Product category	A	B	C	D
Cognition of product quality standards	0.12	0.52	0.16	0.34
Supplier qualification and credit level	0.56	0.14	0.75	0.26
Supplier supply defect rate	0.74	0.35	0.97	0.18
Product acceptance mode	0.15	0.86	0.43	0.54
Warehouse sanitation and safety compliance rate	0.53	0.24	0.65	0.53
Sampling inspection frequency of product quality in stock	0.25	0.42	0.68	0.94
Rationality of product logistics, transportation and packaging	0.74	0.98	0.25	0.61
Operation error rate of logistics personnel	0.69	0.76	0.27	0.98
Number of platform entry assessment items	0.38	0.63	0.61	0.16
Monitoring frequency of platform public opinion	0.42	0.16	0.86	0.37
Lack of product traceability information	0.71	0.73	0.81	0.25
Lack of information displayed online	0.34	0.54	0.72	0.70
Number of after-sales guarantee items	0.29	0.29	0.37	0.68
One time solution rate of after-sales quality problems	0.64	0.65	0.93	0.24
Collection frequency of product after-sales quality information	0.23	0.14	0.18	0.06
Product category	E	F	G	H
Cognition of product quality standards	0.16	0.86	0.69	0.76
Supplier qualification and credit level	0.73	0.81	0.38	0.63
Supplier supply defect rate	0.54	0.72	0.42	0.16
Product acceptance mode	0.29	0.37	0.26	0.08
Warehouse sanitation and safety compliance rate	0.14	0.75	0.26	0.08
Sampling inspection frequency of product quality in stock	0.35	0.97	0.18	0.74
Rationality of product logistics, transportation and packaging	0.86	0.43	0.54	0.35
Operation error rate of logistics personnel	0.24	0.65	0.53	0.24
Number of platform entry assessment items	0.65	0.53	0.14	0.39
Monitoring frequency of platform public opinion	0.68	0.94	0.87	0.97
Lack of product traceability information	0.25	0.61	0.69	0.25
Lack of information displayed online	0.27	0.98	0.35	0.45
Number of after-sales guarantee items	0.61	0.18	0.47	0.67
One time solution rate of after-sales quality problems	0.05	0.29	0.18	0.84
Collection frequency of product after-sales quality information	0.47	0.73	0.37	0.91

cumulative contribution rate of the first four cross-border e-commerce product word-of-mouth influence evaluation indicators is 96.13% and 96.87% when the evaluation

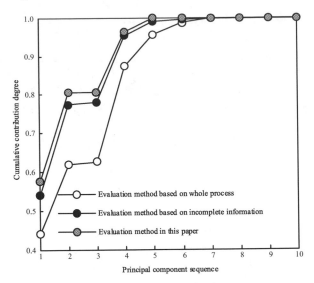

Fig. 2. Cumulative contribution rate

method based on incomplete information and the evaluation method in the text are adopted. Therefore, it can be seen that the evaluation method based on incomplete information and the evaluation method in the text have better dimensionality reduction effect. Compared with the evaluation method in the text, the evaluation method in the text based on incomplete information has fast and efficient performance, and is suitable for the impact evaluation of cross-border e-commerce product word-of-mouth.

4.3 Evaluation Results

According to the test results of cumulative contribution rate, the word-of-mouth influence evaluation scores and ranking of eight cross-border e-commerce products are obtained by using the evaluation method in the paper. The results are shown in Table 3.

From the results in Table 3, we can see that the cross-border e-commerce products ranking higher are B, H and F. Using the method in the article, we can determine the ranking of eight cross-border e-commerce products of a cross-border e-commerce enterprise in terms of word-of-mouth influence. Therefore, we can get the evaluation method in the article, which can rank the word-of-mouth influence based on the comprehensive scores of cross-border e-commerce enterprises. It has certain application value.

4.4 Comparative Analysis

In order to avoid the oneness of the experimental results, an evaluation method based on the whole process and an evaluation method based on incomplete information are introduced to compare and test the efficiency and reliability of the evaluation of word-of-mouth influence of cross-border e-commerce products. The results are as follows.

According to the results in Fig. 3, the evaluation efficiency of the word-of-mouth influence using the evaluation method in the article is higher than that based on the whole

Table 3. Score and ranking of reputation influence evaluation of cross-border e-commerce products

Product category	Comprehensive score	ranking
A	−0.25843	5
B	0.17854	1
C	−0.368	6
D	−0.7642	8
E	−0.179	4
F	−0.0762	3
G	−0.7102	7
H	0.0238	2

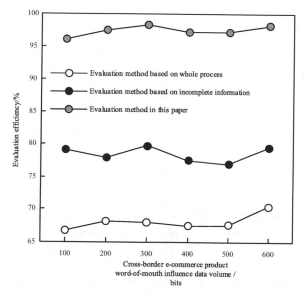

Fig. 3. Efficiency of reputation impact assessment of cross-border e-commerce products

process and incomplete information, because the method in the article can eliminate abnormal users of social networks according to the analysis of social network data, Avoid abnormal users of social networks affecting the evaluation results of word-of-mouth influence of cross-border e-commerce products, so as to improve the evaluation efficiency of word-of-mouth influence of cross-border e-commerce products.

The results in Fig. 4 show that compared with the evaluation method based on the whole process and the evaluation method based on incomplete information, the reliability of the method in this paper for the evaluation of the word-of-mouth influence can reach more than 95%, because the method in this paper designs the evaluation algorithm for the

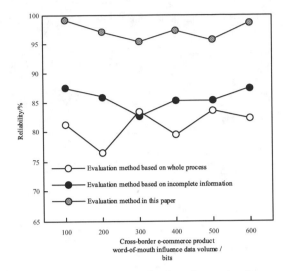

Fig. 4. Reliability of reputation influence evaluation of cross-border e-commerce products

word-of-mouth influence by calculating the weight of the evaluation index of the word-of-mouth influence of cross-border e-commerce products, This makes the evaluation of reputation influence more reliable.

5 Conclusion

This paper proposes a method for evaluating the word-of-mouth influence based on social network data analysis. Through case analysis, it is found that this method can evaluate the word-of-mouth influence and improve the evaluation performance. Although this research has achieved some results, there are still many shortcomings. In the future research, we hope to introduce the gray correlation analysis to calculate the correlation coefficient between different evaluation indicators, so as to improve the economic benefits of cross-border e-commerce products. The method of evaluating the reputation and influence of cross-border e-commerce products based on social network data analysis has important research value and application prospects under existing technological conditions, but there are also some limitations, such as the reliability of data. The data on social networks may have fraudulent behavior, such as likes and purchases, which can mislead the accuracy of the analysis results. The next step will focus on solving this problem.

References

1. Fang, C., Liu, H.: The influence of shopping website characteristics on brand loyalty under B2C mode-taking taobao as an example website quality. IOSR J. Comput. Eng. **23**(1), 31–35 (2021)
2. Chen, Y.F.: Research on risk assessment system for imported commodity quality in the whole process of B2C cross-border e-commerce. J. Bus. Econ. **12**, 5–16 (2019)

3. Zhan, H.L., Yan, W., Song, B.: Reputation network game of cross-border e-commerce platform based on incomplete information. Math. Pract. Theory **51**(19), 57–69 (2021)
4. Zhang, X., Liu, S.: Action mechanism and model of cross-border e-commerce green supply chain based on customer behavior. Math. Probl. Eng. **2021**(3), 1–11 (2021)
5. Wang, F., Wang, M., Wan, Y., et al.: The power of social learning: how do observational and word-of-mouth learning influence online consumer decision processes? Inf. Process. Manag. **58**(5), 102632 (2021)
6. Miremadi, A., Kenarroudi, J., Ghanadiof, O.: Evaluation on role of electronic word of mouth (EWOM) ads in customers' emotions and choices in e-shops. Int. J. Ind. Mark. **6**(1), 56 (2021). https://doi.org/10.5296/ijim.v6i1.18561
7. Miranda, S., Duarte, M.: How perfectionism reduces positive word-of-mouth: the mediating role of perceived social risk. Psychol. Mark. **39**(2), 255–270 (2021)
8. Zhang, J., Guo, Y.: Social network user identity recognition method based on preference logic. Comput. Simul. 39(4), 450–453,505 (2022)
9. Song, B., Yan, W., Zhang, T.: Cross-border e-commerce commodity risk assessment using text mining and fuzzy rule-based reasoning. Adv. Eng. Inf. **40**(APR), 69–80 (2019)
10. Zhao, H.: A cross-border e-commerce approach based on blockchain technology. Mob. Inf. Syst. **2021**(4), 1–10 (2021)
11. Shi, W., Lai, J.H.K., Chau, C.K., Wong, P., Edwards, D.: Analytic evaluation of facilities performance from the user perspective: case study on a badminton hall. Facilities **39**(13/14), 888–910 (2021). https://doi.org/10.1108/F-10-2020-0119
12. Chao, R.F., Yi, F., Liang, C.H.: Influence of servicescape stimuli on word-of-mouth intentions: an integrated model to indigenous restaurants. Int. J. Hospitality Manag. **96**, 102978 (2021). https://doi.org/10.1016/j.ijhm.2021.102978
13. Bai, J., Xu, X., Duan, Y., et al.: Evaluation of resource and environmental carrying capacity in rare earth mining areas in China. Sci. Rep. **12**(1), 6105 (2022)
14. Abdou, E.H.E., Ebada, H.A., Salem, M.A., et al.: Clinical and imaging evaluation of COVID-19-related olfactory dysfunction. Am. J. Rhinol. Allergy **37**(4), 456–463 (2023)
15. Huang, X., Wang, X., Liu, F.: Evaluation of learners' online learning behaviour based on the analytic hierarchy process. Int J. Continuing Eng. Educ. Life-Long Learn. **31**(3), 311–324 (2021)
16. Moisescu, O.I., Dan, I., Gic, O.A.: An examination of personality traits as predictors of electronic word-of-mouth diffusion in social networking sites. J. Consum. Behav. **21**(3), 450–467 (2021)
17. Shahmirani, M., Rashti, A., Ramezani, M., et al.: Application of fuzzy modelling to predict the earthquake damage degree of buildings based on field data. J. Intell. Fuzzy Syst. Appl. Eng. Technol. **41**(2), 2717–2730 (2021)
18. Xie, M., Zhao, C., Hu, D.: Performance evaluation index of radar signal sorting based on pulse train. J. Signal Process. **38**(11), 2350–2358 (2022)
19. Tang, M.C., Wu, P.M.: Reconciling the effects of positive and negative electronic word of mouth: roles of confirmation bias and involvement. Online Inf. Rev. **46**(1), 114–133 (2021)
20. Furrer, O., Kerguignas, J.Y., Landry, M.: Customer captivity, negative word of mouth and well-being: a mixed-methods study. J. Serv. Mark. **35**(6), 755–773 (2021). https://doi.org/10.1108/JSM-07-2020-0311
21. Alsheikh, D.H., Aziz, N.A., Alsheikh, L.H.: Influencing of e-word-of-mouth mediation in relationships between social influence, price value and habit and intention to visit in Saudi Arabia. Humanit. Soc. Sci. Lett. **10**(2), 186–197 (2022). https://doi.org/10.18488/73.v10i2.3010
22. Kesten, K.S., Moran, K., Beebe, S.L., et al.: Drivers for seeking the doctor of nursing practice degree and competencies acquired as reported by nurses in practice. J. Am. Assoc. Nurse Pract. **34**(1), 70–78 (2021)

23. Nguyen, O.D.Y., Lee, J.J., Ngo, L.V., et al.: Impacts of crisis emotions on negative word-of-mouth and behavioural intention: evidence from a milk crisis. J. Prod. Brand. Manag. **31**(4), 536–550 (2022)

24. Mandal, S., Sahay, A., Terron, A., Mahto, K.: How implicit self-theories and dual-brand personalities enhance word-of-mouth. Eur. J. Mark. **55**(5), 1489–1515 (2021). https://doi.org/10.1108/EJM-07-2019-0591

A Remote Access Control Method for Electronic Financial Management Data Based on Object Attribute Matching

Xue Yuan$^{(\boxtimes)}$ and Zimin Bao

Department of Management, Xi'an Jiaotong University City College, Xi'an 710018, China
xycy272@163.com

Abstract. In response to the shortcomings of existing data remote access control methods such as long private key generation time and high packet loss rate in remote control, this paper proposes an electronic financial management data remote access control method based on object attribute matching. Firstly, analyze the connotation of role-based access control and determine the object attributes for accessing electronic financial management data; Secondly, set the access permission category for electronic financial management data files and match the object attributes of external access control; Once again, the weight assignment method of double coefficient of variation is used to assign weights to electronic financial management data, and a secure access control model is designed based on attribute encryption; Finally, a recursive algorithm is used to decrypt and control remote access, completing the design of a remote access control method for electronic financial management data based on object attribute matching. Through experiments, it has been proven that the average private key generation time of the proposed method for accessing electronic financial management data is about 205ms, and the packet loss rate of remote control is always below 1.0%, indicating good application performance.

Keywords: Object Attributes · Access Control · Attribute Matching · Weight Assignment · Decryption Control

1 Introduction

With the rapid development of information technology and the popularization of electronic financial management, more and more enterprises and organizations tend to store financial data in electronic systems. These data include sensitive data such as financial statements, transaction records, and customer information. However, remote access to these financial data also brings a series of security risks and challenges, making the research and implementation of remote access control methods for electronic financial management data crucial [1].

Firstly, the security of financial data is crucial for the operation and reputation of enterprises and organizations. Unauthorized access may lead to data leakage, tampering,

L. Yun et al. (Eds.): ADHIP 2023, LNICST 547, pp. 409–421, 2024.
https://doi.org/10.1007/978-3-031-50543-0_28

or loss, causing significant losses to the enterprise. Therefore, effective access control methods are needed to ensure that only authorized users can remotely access financial data, and encryption and other security measures are taken to protect the confidentiality and integrity of the data. Secondly, the demand for remote access is an important factor driving the development of remote access control methods for electronic financial management data [2]. Finally, the development of access control technology provides technical support for remote access control methods for electronic financial management data. Technologies such as authentication, permission management, and encrypted communication can be used to ensure that only authorized users can remotely access financial data and protect the security of data during transmission and storage.

Access control not only ensures the security of information and data, but also serves as an important means of authentication and authorization for visitors. Therefore, relevant scholars have designed a series of access control technologies. Currently, role-based access control (RBAC) [3] and attribute based access control (ABAC) [4] are widely used. In RBAC, the system sets different roles in advance and assigns corresponding permissions to them. Users and roles adopt a many to many relationship. Due to the relatively closed early system, there were not many types of roles, which effectively reduced management costs. At the same time, the existence of roles also facilitates the development of security policies by enterprises. In relatively small network environments, RBAC performs well, but with the expansion of network scale and the rise of cloud computing IoT, RBAC is becoming increasingly difficult to meet demand. A large number of network nodes make network management extremely complex. RBAC is difficult to achieve fine-grained access. Administrators cannot predict in advance how many kinds of users will apply for access, nor can they fully consider the role permission correspondence. The attribute based access control (ABAC) scheme provides a new idea. It binds the control policy and the user attribute together. Only the attribute set satisfies the control policy can the resource be accessed. Some researchers have proposed an access control model of role matching in distributed workflow environment. Based on different tasks of workflow, this model can search for one or more groups of role sets with relevant task execution permissions from system roles, and then carry out matching optimization by referring to the environment, time constraints and inheritance relationship between roles, and finally select the optimal role set for users [5]. However, this method may be affected by the surrounding environment, which prolongs the private key generation time and affects the effect of remote access control. Some researchers have proposed a smart grid electricity data access control method based on the ABAC access control model. A dynamic context authorization strategy based on the ABAC access control model was implemented in the WSO2 system using XACML policy language, which can meet the context based dynamic access characteristics of smart grid environments [6]. However, in the practical application of remote control, this access control method may lead to an increase in the system's packet loss rate and reduce the effectiveness of remote access control.

Therefore, in order to further improve the security of access control mechanism, this paper combines the access control method based on attributes and roles, and proposes a remote access control method for electronic financial management data based on object attribute matching.

2   Role-Based Access Control and Attribute-Based Access Control

2.1   Role-Based Access Control

RBAC, or Role Based Access Control, is an access control mechanism widely used in computer security. It controls users' access to resources by assigning roles to users. In RBAC, a role is a set of permissions that can be divided based on a user's job responsibilities or positions, and each role can be authorized to one or more users. Users can only access the permissions of their roles when using resources, without specifying whether they have specific access permissions. This approach greatly simplifies administrator tasks and improves system security by encapsulating the permissions of each user [7].

RBAC can help organizations effectively manage access to resources, dictate who can do what and when, ensure that the system is only used by authorized personnel, and facilitate administrators' management of users and their roles. RBAC also reduces human error and administrative costs, improving system reliability and security. This mechanism has been widely used not only in computer operating system and network system, but also in many other fields, such as enterprise, medical, education, finance and so on. In RBAC, roles are mainly divided into subject and object, which refer to users (or processes), and object refers to permissions and resources (or objects). Subject obtains authorization through roles, so that it can operate objects [8]. In RBAC, a principal can have multiple roles, and a role can contain multiple permissions; An object can also be accessed by multiple roles, and a permission can be granted to multiple roles. This mechanism makes the management of roles, permissions, and resources more flexible and efficient, while also improving the security of the system. In a database system, administrators can define multiple roles, such as super administrators, system administrators, regular users, etc. Authorize these roles to different user groups or individuals, and different users can use their respective roles for authorization operations.

2.2   Attribute-Based Access Control

ABAC (Attribute-Based Access Control) is an access control mechanism widely used in computer systems. Unlike RBAC, ABAC controls users' access to resources through attributes rather than assigning roles to users. In ABAC, users, resources, and environments can all be assigned attributes that can be used in access control decisions. The access request contains user attributes and resource attributes. The system determines whether the user is authorized to access resources based on these attributes. This method can not only achieve more accurate and dynamic access control, but also adapt to different scenarios and application requirements [9].

ABAC is a flexible access control mechanism that can adapt to complex and ever-changing application scenarios, and can more finely control user access to resources, thereby improving system security. As a result, ABAC has been widely applied in many large organizations and enterprises, cloud computing environments, the Internet of Things, and industrial automation fields.

2.3 Access Control Based on Role Attributes

RABAC is a role and attribute based access control model, which refers to the extension of role based access control and attributes. It is an extended access control model that combines RBAC and ABAC. In RBAC, access control decisions are only based on the user's role, without considering the user's attributes; ABAC focuses on user attributes and corresponding user permissions. In RABAC, access control decisions are not only based on user roles, but also on user attributes. In the RABAC model, roles are no longer the only authorization criteria, and users also need to meet certain attribute conditions to obtain access to a resource [10].

At the heart of the RABAC model is the integration of user attributes and role assignment. Users can have multiple attributes, and each attribute can be associated with one or more roles. Each role contains not only a set of operation permissions, but also a set of attribute restrictions. When a user initiates a resource access request, the system makes access control decisions based on the attributes and roles of the user. RABAC is more flexible in defining access control policies by taking into account user attributes than the RBAC model. By introducing attribute conditions, the RABAC model can simplify the complexity of user access control management by using the abstraction mechanism of roles, and at the same time, give full play to the important role of user attributes in access control, improve the degree of refinement of access control, and better meet the actual demand for access control.

Therefore, this paper combines the characteristics of RBAC and ABAC and applies the RABAC model in remote access control of electronic financial management data based on object attribute matching, so as to better meet the current complex network environment and diversified authorization needs.

3 Design of Remote Access Control Method for Electronic Financial Management Data

3.1 Object Attribute

The object owner has supervisory authority and is able to proactively select data visitors, which to some extent ensures users' control over the privacy of electronic financial management data.

The data contributor sets the access permission level for the electronic financial management data file, and calculates its weighted value based on this access permission level, which allows visitors to obtain the lowest weighted value of the electronic financial management data file and store it in the file. When there is an access request, the weighted calculation results of the application access permissions of the data visitor are compared with the weighted permission values calculated from many electronic financial management data files in the database that have already been set with access permissions, and the electronic financial management data files that meet the access permissions of the data visitor are selected. The object matching information flow diagram is shown in Fig. 1.

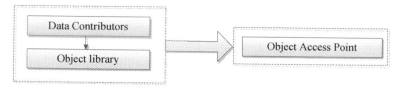

Fig. 1. Object matching information flow diagram

The specific steps are as follows:

(1) The data contributor sets the minimum access permission value of the data to be uploaded, calculates the weighted permission value, writes it into the file, and then uploats the data to the Object Library (OL);
(2) Receive the weighted permission value of the data visitor from the second part and also pass it into the object database OL;
(3) According to the permission threshold set by the contributor, the Object database queries and screens the database, and sends the relevant object data meeting the permission of the visitor to the Object Access Point (OAP) in the second part.

3.2   External Access Control Object Attribute Matching

If you want to access electronic financial management data, when designing External Access Control (EAC), you need to use attribute centered RABAC access control, which takes the role of the object as an attribute to express a user's permission. Due to the fact that various external institutions and types of data visitors belong to cross domain data access, a unified external interface is provided for external visitors. In the institutional system in which it is located, if there is already an access control mechanism (whether it is Discretionary Access Control (DAC), Mandatory Aces Control (MAC), RBAC, ABAC, or other access control mechanisms), the permissions it possesses are represented through the EAC section at the unified external interface, and then transferred to the Level Management Table (LMT), Perform permission ID comparison and matching on the attribute values of the object to access electronic financial management data.

The electronic financial management data file access permission is set as "top secret", "confidential", "secret", "sensitive" and "open" five categories. The access control permission of Top Secret is the highest. The access control permission of Secret, Secret, and Sensitive is decreasing. The access control permission of Public is not required. Write the preceding five categories to the LMT. The level authorization table Settings are shown in Table 1.

The access permission of electronic financial management data files is inclusive. If a user has access permission of "top secret" files, that is, permission ID = 1, he has the right to access four categories of files: "confidential", "secret", "sensitive" and "open". Similarly, the user has "confidential" file access permission, that is, if the permission ID = 2, the user has the right to access the three categories of "secret", "sensitive" and "public" files.

Table 1. Level authorization table

User class	Privilege level	Permission ID
Government department	Top secret	1
Research institutes	Confidential	2
Medical institution	Secret	3
Business organization	Sensitive	4
Individual users	Open	5

If a user wants to access electronic financial management data across permissions, they need to use access control mechanisms within their own domain to obtain their personal corresponding permissions. According to their object attributes, the highest level of access permissions that can be obtained by their organization cannot exceed. The specific access control information flow diagram is shown in Fig. 2.

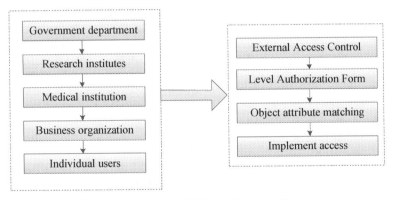

Fig. 2. Access control information flow diagram

The specific steps are as follows:
① The user obtains permissions through the access control system in the domain and accesses the accessible system through the external access control module.
② The EAC formats user permissions, generates unified files, and sends them to the level authorization table.
③ The LMT compares the attribute permissions in the file, matches the object attribute permissions, informs the visitor of the permission result, and generates an ID corresponding to the permission for the visitor, and writes the ID to the permission file.

3.3   Double Coefficient of Variation Weight Assignment Method

In RABAC, as the amount of electronic financial management data increases, changes in access permissions for various types of data will increase the complexity of matching access permissions. To solve this problem, this article uses the double coefficient of variation weight assignment method to assign weights to electronic financial management data, simplifying the difficulty of matching object attributes and improving matching speed.

At the present stage, there are differences in the representation and dimensionality of each object attribute value, but in the access control attribute matching, these attribute values need to be evaluated and compared horizontally and vertically. In order to improve the practical application effect, it is necessary to normalize the data with dimensionless quantization. Permission values that are already in the range of $[0,1]$ are no longer processed. Instead, the permission value not within the range of $[0,1]$ is normalized using formula (1). The calculation formula is as follows:

$$
e_{ij} = \begin{cases} \dfrac{a_{ij} - (a_{ij})_{\min}}{(a_{ij})_{\max} - (a_{ij})_{\min}}, & a_{ij}\text{Forward incremental} \\[3mm] \dfrac{(a_{ij})_{\max} - a_{ij}}{(a_{ij})_{\max} - (a_{ij})_{\min}}, & a_{ij}\text{Forward decreasing} \end{cases} \tag{1}
$$

Among them, a_{ij} is the original value; e_{ij} is the quantified value; $(a_{ij})_{\max}$ is the maximum original value; $(a_{ij})_{\min}$ is the minimum value of the original value.

The weight of the coefficient of variation weight assignment method is directly obtained by calculating the information of each indicator. To eliminate the differences between different dimensions of attribute values before calculation, the electronic financial management data is first quantified and processed. Assuming that the quantified electronic financial management data matrix is $Y = (y_{ij})_{m \times n}$, m represents the number of evaluation plans; n represents the number of indicators, so the calculation of the average value $\bar{y}_j$ of each column vector and the standard deviation S_j of each column vector is shown in Formulas (2) and (3):

$$
\bar{y}_j = \frac{1}{m} \sum_{i=1}^{m} y_{ij} \tag{2}
$$

$$
S_j = \sqrt{\frac{1}{m} \sum_{i=1}^{m} (y_{ij} - \bar{y}_j)^2} \tag{3}
$$

The calculation formulas of variation coefficient V_j and weight W_j of each index are shown in Formulas (4) and (5):

$$
V_j = \frac{S_j}{\bar{y}_j} \tag{4}
$$

$$
W_j = \frac{V_j}{\sum_{j=1}^{m} V_j} \tag{5}
$$

By applying the double variation coefficient weight assignment method, the corresponding calculation of variation coefficient weights was carried out, and the weights of each attribute were obtained; By combining the attribute values of the visitor's object role, the weighted permission values for visitors to access electronic financial management data can be obtained.

3.4 Attribute Encryption Secure Access Control Model

After assigning the weights for visitors' access to electronic financial management data, in order to ensure the security of access to electronic financial management data, this paper designs an Attribute Encryption secure access control model Based on Attribute-based Encryption (ABE).

Suppose $A = \{A_1, A_2, ..., A_n\}$ is the set of all attributes, then the object attribute UA is the non-empty subset of A. The attribute set with the number of attributes n can define 2^n attribute subsets in total, so it can identify 2^n users at most.

Access structure AS is a non empty subset of property set $A = \{A_1, A_2, ..., A_n\}$, $AS \subseteq 2^{\{A_1, A_2, ..., A_n\}}$. When the object attribute user appears in the access structure AS, it matches as an authorized user, otherwise it is an unauthorized user.

Assume that G and GT are multiplicative cyclic groups of prime order p, g can be used as the generator of G to bilinear map G. Randomly define M group elements $\{s_1, s_2, ..., s_M\} \in G$ and associate them with the attribute set $A = \{A_1, A_2, ..., A_n\}$, and randomly select two indices $a, b \in Z_p$ to obtain the public key set by the owner of electronic financial management data as follows:

$$PK = \left\{g, e(g, g)^a, g^b, h_1, h_2, ...h_M\right\} \tag{6}$$

The master key can be calculated as follows:

$$MK = g^a \tag{7}$$

The owner of the electronic financial management data sets a unique ID for the file that needs to be encrypted and stored in the cloud storage center, randomly selects a symmetric key SYK, uses the symmetric key SYK to encrypt the electronic financial management data file, and sets the access structure of the electronic financial management data file. The encrypted ciphertext can be expressed as:

$$CT(t) = Ency \times p(t) \times (PK, \{SYK, K_u, K_w\}, t) \tag{8}$$

where, t represents the timestamp of the encryption process; K_u indicates the write permission corresponding to the signature key, which applies to writable users. Users sign data after performing write operations. K_w indicates the read permission corresponding to the authentication key. It is used to verify the signature result for read-only users.

The owner of electronic financial management data can specify attribute set A_w for each user in the user permission list, and then calculate the private key of the corresponding object attribute user:

$$K_\sigma = keyGen(MK, A_w) \tag{9}$$

After obtaining the object attribute user private key K_σ, encrypt it based on the accessing user's public key PK and send it to the user who needs access.

The owner of electronic financial management data uploads the data file to the cloud for storage, and stores the user permission list of the file share in the cloud, which includes the user ID that the file can access, the valid status of the file, the user list UL of the file share, and the user list of deleted permissions.

3.5 Remote Access Decryption Control

Select any index $r \in Z_p$ for each object attribute user of access control and calculate:

$$D_y = g^a \times H(j)^{rj} \tag{10}$$

$$D_j' = g^{rj} \tag{11}$$

Compose the algorithm key through $D = g^{(\alpha+\beta)/\beta}, \forall j \in S : D_y = \times H(j)^{rj}$ and $D_j' = g^{rj}$.

During the process of remote access decryption control, perform decryption calculations on the ABE algorithm. Convert to recursive algorithm $Decrypt - Node(CT, SK, x)$ as:

$$Decrypt - Node(CT, SK, x) = \frac{e(D_i, C_x)}{e(D_i', C_x')} = e(g, g)^{r(0)} \tag{12}$$

where, x represents the non-leaf node of the decision tree. If S_x represents the set of decision root node children z of any size k_x, it can be stored as F_z through recursive algorithm $Decrypt - Node(CT, SK, z)$, and $F_z \neq \perp$, then it can be calculated:

$$F_x = \prod_{z \in S_x} F_z^{\Delta i, S_x(0)} = e(g, g)^{r(0)} \tag{13}$$

The recursive algorithm is used to determine the key solving process of the ABE algorithm, and the recursive function is used from the root node of the decision tree.

If the attribute set S satisfies the access decision tree T, then:

$$A = Decrypt - Node(CT, SK, r) = e(g, g)^{rs} \tag{14}$$

Thus, the process of remote decryption access control based on ABE algorithm is completed, and remote access control of electronic financial management data based on object attribute matching is realized.

4 Experiment

In order to verify the application performance of the remote access control method of electronic financial management data based on object attribute matching proposed in this paper, a comparative experiment was set up for testing.

4.1 Experimental Environment Settings

Select the financial data of a certain enterprise as the research object, extract the electronic financial management data separately, and import it into a new database. The total amount of existing electronic financial management data is 56.4G. After data cleaning and eliminating duplicate and redundant electronic resource data, there is a remaining 50.0G. Using the Python random sampling algorithm, 80% of the data is used as the training dataset and 20% as the testing dataset. The training dataset is 40.0G, and the testing dataset is 10G. This article implements attribute access control based on dynamic user trust using Matlab tools in the Win10 system environment. The specific parameters of the experimental environment are shown in Table 2.

Table 2. Experimental environment

Project	Parameter
Mobile terminal	ZUK Z2125
CPU	Inter Core i7 2.35Hz
ROM	64GB
RAM	6GB
Android version	6.0.1
Operating system	Ubuntu desktop 11.10 × 64
Open Stack	Essex
Simulation software	MATLAB 2019a
Virtual machine	VMware Workstation 6.5.2

In the experiment, setting the initial credibility of all users to the same value can ensure that the same initial conditions and evaluation criteria are adopted for all users during the application performance verification process to maintain fairness. Meanwhile, for new users, the experimental system may not have enough information to evaluate their level of trust. Therefore, setting their initial credibility to a lower value can reflect their lower level of trust in the system. For certain sensitive operations or critical data, the system may need to exercise more cautious access control. By setting the initial credibility of all users to a lower value, stricter control and review of potential risks can be added. Therefore, in order to simulate a lower initial trust level when verifying application performance and provide a fair and cautious evaluation standard, the initial trust level of all users is set to 0.3, where the proportion of direct credibility is 0.6.

In order to ensure the fairness of experimental testing, the access control method based on RBAC proposed in literature [5] and the access control method based on ABAC proposed in literature [6] are compared in this paper, and the remote access control method based on object attribute matching proposed in this paper is tested together.

The generation time of the private key and the packet loss rate of the remote access control are used as test indicators. The shorter the generation time of the private key, the lower the packet loss rate of the remote access control, and the better the application performance of the access control method.

4.2   Comparison of Private Key Generation Time

Three different methods were used to calculate the generation time of private keys during access control of electronic financial management data. The comparison results of private key generation time for different methods are shown in Fig. 3.

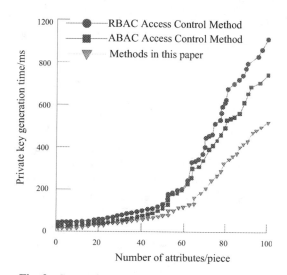

Fig. 3. Comparison of private key generation times

Observing Fig. 3, it can be seen that different methods increase the time for data owners to generate user private keys as the number of attributes changes, and the time for private key generation shows an increasing trend as the number of attributes increases. The average private key generation time of RBAC-based access control method and Abac-based access control method is about 298ms and 262ms, while the average private key generation time of the proposed method is about 205ms. When the number of attributes is less than 50, the performance of the comparison method has little difference, but when the number of attributes is more than 50, the private key generation time starts to show a significant gap. When the number of attributes reaches 60, the access control methods of RBAC and ABAC show a very obvious upward trend. Although the private key generation time of this method is gradually increasing, it is always lower than the two methods compared, and the private key generation time is shorter, indicating that this method has good application performance.

4.3   Packet Loss Rate for Remote Control

Three different methods were used to calculate the packet loss rate of remote control when accessing electronic financial management data. The comparison results of remote control packet loss rates using different methods are shown in Fig. 4.

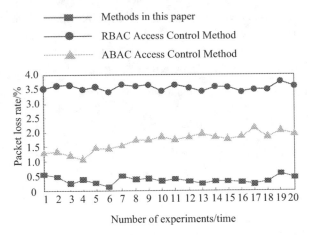

Fig. 4. Comparison of packet loss rates for remote control

Observing Fig. 4, it can be seen that after 20 experimental tests, the packet loss rate of remote control when accessing electronic financial management data based on RBAC is between 3.2% and 3.7%, and that of remote control when accessing electronic financial management data based on ABAC is between 1.0% and 2.0%. However, the packet loss rate of remote control is between 0.2% and 0.7% when accessing electronic financial management data, which is always lower than 1.0%, indicating that the access control security of the proposed method is high and it has good application performance.

5   Conclusion

Access control of data can help data owners implement coarse-grained and fine-grained access control policies, ensuring that sensitive data is only accessed and processed by authorized personnel, and avoiding the risk of malicious attacks or data leakage. In response to the shortcomings of existing remote access control methods for data, this article proposes a remote access control method for electronic financial management data based on object attribute matching.

(1) Determine the object attributes for accessing electronic financial management data, set the access permission category for electronic financial management data files, and match the object attributes of external access control; Assign weights to electronic financial management data, design a secure access control model based on attribute encryption, and use recursive algorithms to decrypt and control remote access, achieving remote access control of electronic financial management data.

(2) Through experiments, it has been proven that the average private key generation time of the proposed method for accessing electronic financial management data is about 205ms, and the packet loss rate of remote control is always below 1.0%, indicating good application performance.

Acknowledgement. 1. 2022 Annual Project of the "14th Five-Year Plan" Education and Science Planning of Shaanxi Province: Research on the ideological situation of university students and their guidance path in the era of convergent media (SGH22Y1721).

2. Research Projects of Xi'an Jiaotong University City College in 2021: Research on accounting information quality improvement methods based on blockchain technology (2021X29).

References

1. Wang, J., Zhang, W.: Attribute and RBAC based access control model and algorithm research. J. Chin. Comput. Syst. **43**(07), 1523–1528 (2022)
2. Liu, W., Sheng, C., She, W., et al.: Classified and hierarchical attribute access control method based on smart contract. Appl. Res. Comput. **39**(05), 1313–1318 (2022)
3. Pan, R., Wang, G., Huang, H.: Attribute access control based on dynamic user trust in cloud computing. Comput. Sci. **48**(05), 313–319 (2021)
4. Wei, D., Sheng, B., Xiang, W., et al.: Access control model in PDM system based on role and attribute. Mach. Des. Manuf. **12**, 259–263 (2019)
5. He, S., Ou, B., Liao, X.: Role matching access control model for distributed workflow. Comput. Sci. **45**(07), 129–134 (2018)
6. Shao, R., Tian, X.: ABAC access control scheme based on MQTT protocol in smart grid. Appl. Res. Comput. **39**(11), 3436–3443 (2022)
7. Zhou, C., Ren, Z.: Research of access control model combined attribute with role. J. Chin. Comput. Syst. **39**(04), 782–786 (2018)
8. Ge, L., Hu, Y., Zhang, G., et al.: Reverse hybrid access control scheme based on object attribute matching in cloud computing environment. J. Comput. Appl. **41**(06), 1604–1610 (2021)
9. Guo, X., Wang, Y., Feng, T., et al.: Blockchain-based role-delegation access control for industrial control system. Comput. Sci. **48**(09), 306–316 (2021)
10. Yu, B., Tai, X., Ma, Z.: Study on attribute and trust-based RBAC model in cloud computing. Comput. Eng. Appl. **56**(09), 84–92 (2020)

Research on Adaptive Tracking of University Funding Objects from the Perspective of Big Data

Yuliang Zhang[✉] and Xiaoyan Xu

Data and Information Center, Wuxi Vocational Institute of Commerce, Wuxi 214000, China
zhangyuliang2@wxic.edu.cn

Abstract. It is necessary to track the funding objects of colleges and universities adaptively to understand the effect of funding for poor students. For this reason, an adaptive tracking method for university funding objects is proposed from the perspective of big data. The tracking indicators are selected through the Delphi method, and the "background", "input", "response", and "output" form a pyramid tracking indicator system based on the CIRO evaluation model. The entropy weight method is used to calculate the weight of each tracking index of the system. Combining with the index membership degree, an adaptive tracking model of university funding objects is constructed to realize the analysis of funding effectiveness. The results show that the performance score of the three funded students has been increasing year by year, which indicates that the status of the three students has been significantly improved after receiving funding, but from the growth trend, the score of student 1 has entered a flat development period after getting better, which indicates that the status is stable and the funding can be gradually reduced. The score of student 2 shows a rising trend, but the rising trend is slow, so it is necessary to increase funding. Student 3 has been growing, and the score has not stabilized, so we can continue to maintain the current funding.

Keyword: University Funding Objects · CIRO · Tracking Index System · Weight · Adaptive Tracking Method

1 Introduction

In recent years, with the expansion of enrollment in universities, the number of financially disadvantaged students has been increasing. To provide greater financial support to economically disadvantaged university students and ensure their smooth access to education, efforts have been made to eliminate poverty through education, promote educational equity, and foster social harmony. By offering financial assistance to eligible students, universities can improve their learning conditions and outcomes, thus promoting social fairness. However, university aid recipients are diverse and complex, varying in terms of personal backgrounds, family economic situations, and learning needs. Aid recipients may experience changes over time, necessitating the establishment of an adaptive tracking approach to timely and accurately understand their evolving needs.

L. Yun et al. (Eds.): ADHIP 2023, LNICST 547, pp. 422–435, 2024.
https://doi.org/10.1007/978-3-031-50543-0_29

With the development and widespread application of information technology, universities have begun utilizing relevant technical means to enhance aid recipient management. Adaptive tracking techniques adjust tracking algorithms and models dynamically based on different circumstances and requirements, providing more accurate and real-time information about aid recipients. Currently, there are several main methods for adaptive tracking of aid recipients, including the Coyle model, Kaufman's five-level evaluation model, the CIRO model, the CIPP model, and Phillips' return on investment model. Compared to other models, the CIPP and CIRO models place greater emphasis on evaluating all elements of aid recipients and the implementation process of aid measures. Particularly, they prioritize the analysis of background factors related to aid needs, showcasing more advanced systematic thinking and ensuring the scientific rigor and integrity of tracking research. While the background evaluation, input evaluation, and outcome evaluation in both models are generally similar, the key difference lies in the third stage. In this stage, the CIPP model emphasizes comprehensive tracking and evaluation of aid measures throughout their implementation, providing timely feedback to organizers to facilitate adjustments and improvements and offer insights for subsequent aid organization. On the other hand, the CIRO model emphasizes relying on the subjective information of aid recipients, considering that the improvement of aid relies on collecting and utilizing the recipients' feedback. Based on the aforementioned analysis, this study believes that the CIRO model is more suitable for the research topic. Therefore, leveraging the CIRO model, a university aid recipient adaptive tracking model in the context of big data is constructed. The Delphi method is employed to select tracking indicators, and the CIRO evaluation model serves as the framework, forming a pyramid-shaped tracking indicator system consisting of "Background," "Input," "Reaction," and "Output." The weights of each tracking indicator in the system are calculated using the entropy weight method. Combining with indicator membership degrees, the university aid recipient adaptive tracking model is established to analyze the effectiveness of aid. The research on university aid recipient adaptive tracking in the context of big data provides new ideas and methods for aid recipient management. By leveraging data-driven decision support, real-time monitoring and prediction capabilities, personalized services and resource optimization, as well as intelligent decision-making and risk management, the efficiency and quality of university aid recipient management can be improved, better meeting the needs of aid recipients.

2   CIRO Based Adaptive Tracking Model for Funded Objects

The state has successively introduced a series of financial aid policies for students with financial difficulties from families, increased the amount of financial aid to college students, basically met the material needs of students with financial difficulties from families, and ensured that students with financial difficulties from families can enjoy the right and opportunity to receive higher education fairly. It reflects the concern and care of the Party and the state for students from poor families. However, we must also be soberly aware that there are also some unsatisfactory problems [3] in the funding work for poor college students. In order to ensure the effective implementation of funding measures, it is necessary to carry out adaptive tracking of funding objects. The CIRO model was proposed by Warr, Bird and Rackham in 1970. CIRO is the first letter of the four evaluation activities of Context Evaluation, Input Evaluation, Reaction Evaluation, and Output Evaluation in the model, which are translated as background evaluation, input evaluation, response evaluation, and result evaluation. See Table 1 for CIRO model.

Table 1. CIRO Model

Assessment activities	Content
Context evaluation	Determine funding needs and goals based on the current environmental background
Input evaluation	Collect and summarize relevant information and decide on funding inputs
Reaction evaluation	Collect and analyze feedback from funding recipients to improve funding activities
Outcome evaluation	Obtain the results after funding and compare them with previous goals as a reference for the next funding activity

2.1   Selection of Adaptive Tracking Indicators for University Funding Objects

The CIRO evaluation model is used as the framework to build an adaptive tracking indicator system for university funding objects, and four first level tracking indicators, namely "background", "input", "response" and "output", are determined. Under each indicator, several second level indicators are designed, and under each second level indicator, several third level indicators are designed, forming a pyramid shaped tracking indicator system. The following methods are mainly used to select and determine specific tracking indicators:

First, literature analysis. The author of this article has consulted a large number of research articles and documents on civil servant training, online training, performance evaluation, education and teaching, summarized and sorted out the content of the indicator design on training performance, referred to the research results of experts and scholars, and pre-selected the representative indicators that are strongly related to civil servant online training. On the basis of in-depth analysis, the hierarchy is determined, the classification is carried out, and the importance of different indicators is considered, the preliminary ranking is made, and the pre-selected indicator set [4] is constructed.

Second, expert consultation method. The index system preliminarily determined after pre selection has consulted the opinions of relevant experts through letters, interviews and other forms. Among these experts, there are teachers in charge of relevant work in colleges and universities, party schools and other institutions, as well as staff specially responsible for this work, which is quite representative. In the consultation process, experts do not have horizontal contact with each other, but only communicate with the author of this article in a two-way way, ensuring the independence and objectivity of expert opinions.

After repeated solicitation, induction, revision and summary, a preliminary tracking index system has been formed. The list of preliminary indicators is shown in Table 2.

Table 2. List of preliminary tracking indicators

Category	Index
Background	The importance attached by the school to funding
	Poor students' understanding of subsidy policies
	Attitudes of impoverished students towards participating in funding
	Pre funding needs survey
	Matching funding objectives with needs
Input	Recognition system for family economic difficulties
	Selection work system
	The distribution system of funding
	System for the use of funding
	The funding plan is reasonable and appropriate
	Supporting institutional construction
	Proportion of special funds invested in funding
	Diversity of funding methods
	Promotion of funding policies
Reaction	Legitimacy of funding organizations
	Funding policy construction and implementation status
	The completeness of funding policies
	Rationality of the use of funding resources
	Repayment status of student loans
	Employment situation of impoverished students
Output	Student learning ability
	Family status of impoverished students
	The Psychological Status of Poor Students
	Student's sense of responsibility
	Social public satisfaction

On the basis of the selected preliminary tracking indicators, further selection is needed to reduce the indicator dimension and build the final adaptive tracking indicator system for university funding objects. Delphi method is an expert investigation method with prescribed procedures, which absorbs the advantages of expert meeting method and overcomes the disadvantages of expert meeting method. Its basic principle is to put forward questions to the selected experts in the form of investigation and consultation, summarize and sort out the experts' opinions, and then anonymously feedback the obtained opinions to each expert, ask for opinions again, sort out again, and feed back again until the opinions are consistent [5]. The process of selecting tracking indicators by Delphi method is as follows:

Step 1: form an expert group and issue questionnaires. The questionnaire lists six basic evaluation indicators and supplementary evaluation indicators that need expert empowerment. In addition, all background information about the PPP project should be attached for experts' reference.

Step 2: The experts will score each indicator according to the information received, and explain how to use these information to propose scoring values.

Step 3: collect the questionnaires and conduct the first induction statistics. The coordination degree of experts' opinions is calculated according to the scores of experts on various indicators. The degree of coordination of expert opinions refers to whether the experts participating in the correspondence have differences on the indicators, which is usually expressed by the coefficient of variation and Kendall's coefficient of harmony W. The coefficient of variation indicates the coordination degree of experts on the relative importance of an indicator, the rationality of the calculation formula, and the operability of the collection method. The more the coefficient, the higher the coordination degree among experts [6]. It is generally believed that the coefficient of variation should be less than 0.25. The calculation formula is as follows:

$$\lambda_j = \frac{\left(\beta_j - \bar{x}_j\right)^2}{\bar{x}_j} \tag{1}$$

Among them,

$$\bar{x}_j = \frac{\sum\limits_{i=1}^{m} \eta_{ij}}{m} \tag{2}$$

$$\beta_j = \sqrt{\frac{\sum\limits_{i=1}^{m} \left(\eta_{ij} - \bar{x}_j\right)^2}{m - 1}} \tag{3}$$

where, λ_j On behalf of the j Coefficient of variation of item tracking index; η_{ij} On behalf of the i Experts j Scoring of item tracking indicators; $\bar{x}_j$ On behalf of the j Average value of item tracking indicators; β_j On behalf of the j Standard deviation of item tracking indicators; m Number of representative experts.

If the coefficient of variation is greater than 0.25, it means that the scoring opinions of various experts are quite different, and the second round of scoring and research will be carried out.

Step 4: distribute the scores of other experts and the results of summary statistics to all experts anonymously so that they can refer to and modify their own opinions. This process is repeated until experts no longer revise their opinions.

2.2 Weight Calculation of Tracking Indicators

Since there are many tracking indicators, each indicator is very important to the rationality of the weight value of the final evaluation result. In theory and practice, common weight calculation methods can be divided into subjective and objective methods.

(1) Subjective weighting method

The subjective weighting method is a method to collect the views of relevant professionals on various indicators through specific means, and determine the weight by integrating their views. The commonly used subjective weighting methods are: binomial coefficient method, classification scoring method, expert survey method, analytic hierarchy process, etc. [7]. The subjective weighting method is easy to operate, and can better reflect the ideas of experts or management. However, this method is very subjective and arbitrary, and may not completely objectively reflect the actual weights between indicators.

(2) Objective weighting method

The objective weighting method is a method that uses a specific calculation method to calculate and analyze the collected data to determine the weight of each indicator. The commonly used objective weighting methods include principal component analysis, factor analysis, correlation analysis, multi day scale planning, etc. The advantage of this method is that the weight does not depend on one observation and one score, and it is based on mathematical theory. However, its disadvantage is that the amount of sample data collected is large, the calculation method is relatively complex, and generally does not reflect the experts' views on the weight of different indicators [8]. Therefore, there may be a large discrepancy between the weight and the actual weight of the indicator.

In this paper, because there are many tracking indicators for the funded objects, including both qualitative indicators and quantitative indicators, considering the characteristics of subjective and objective weighting methods, it is difficult to ensure the accuracy of the indicator weights by using only one method, and it is also difficult to apply to the actual situation. Therefore, the research adopts the entropy weight method to determine the weight of each index, and make appropriate corrections according to the actual situation to maintain the scientificity, objectivity and rationality of the index. Entropy weight method is a method to determine the weight of indicators according to the amount of information contained in each indicator in the evaluation indicator system. Entropy weight method is an objective method to assign weight. The entropy weight method uses the entropy value in the information theory to reflect the degree of information disorder in different indicators, so as to measure the amount of information contained in an indicator, so as to determine the role of the indicator in target decision-making [9]. At present, entropy weight method has been applied in most fields, and the application fields are relatively wide. The specific steps of entropy weight method are as follows:

Step 1: Construct a judgment matrix according to the established indicator system. For the whole indicator system, if an indicator system has n funding objects and m tracking indicators, then n consists of n funding objects and m tracking indicators × The m-order judgment matrix is:

$$D = (d_{ij})_{nm} = \begin{bmatrix} d_{11} & d_{12} & \ldots & d_{1m} \\ d_{21} & d_{22} & \ldots & d_{2m} \\ \ldots & \ldots & \ldots & \ldots \\ d_{n1} & d_{n2} & \ldots & d_{nm} \end{bmatrix} \qquad (4)$$

where, d_{ij} No i Objects evaluated in the j Evaluation value under tracking indicators.

Step 2: preprocess the constructed judgment matrix.

The indicator system of this paper involves some very large indicators and very small indicators. According to the principle of preferential membership, different types of indicators have different standardized processing methods. In this paper, the extremely large indicators in the matrix are processed as follows:

$$g_{ij} = \frac{d_{ij} - \min d_{ij}}{\max d_{ij} - \min d_{ij}} \qquad (5)$$

The processing method for extremely small indicators in the matrix is:

$$g_{ij} = \frac{\max d_{ij} - d_{ij}}{\max d_{ij} - \min d_{ij}} \qquad (6)$$

The decision matrix is finally obtained by processing the constructed judgment matrix:

$$G = (g_{ij})_{nm} = \begin{bmatrix} g_{11} & g_{12} & \cdots & g_{1m} \\ g_{21} & g_{22} & \cdots & g_{2m} \\ \ldots & \ldots & \ldots & \ldots \\ g_{n1} & g_{n2} & \cdots & g_{nm} \end{bmatrix} \qquad (7)$$

Step 3: Calculate E_{ij}, i.e. j Index i The proportion of financial aid recipients. E_{ij} The specific calculation method of is as follows.

$$E_{ij} = \frac{g_{ij}}{\sum\limits_{i=1}^{n} g_{ij}} \qquad (8)$$

Step 4: Pass the calculated No j Index i Proportion of target indicators of financial aid E_{ij}, calculate the entropy weight of each index e_j. e_j The specific calculation method of is as follows:

$$e_j = \frac{\sum\limits_{i=1}^{n} E_{ij} \ln E_{ij}}{\ln n} \qquad (9)$$

According to the basic principle of logarithmic function, the value of logarithm cannot be 0. In this case, it is generally specified that when $E_{ij} = 0$ When, $E_{ij} \ln E_{ij} = 0$.

Step 5: According to the calculated entropy value e_j, find the j Weight of indicators s_j. Section j Weight of indicators s_j The calculation method of is as follows:

$$s_j = \frac{1 - e_j}{\sum_{j=1}^{m} (1 - e_j)} \tag{10}$$

It can be seen from the above formula that the formula meets $0 \leq s_j \leq 1, \sum_{j=1}^{m} s_j = 1$.

At the same time, it can be concluded from the formula that the entropy value e_j When increasing, the weight of the indicator s_j reduce; Entropy e_j Weight of indicators when decreasing s_j enlarge. Therefore, it can be summarized as follows: in the established indicator system, the smaller the entropy value of an indicator, the greater the change degree of the indicator value, the more information can be provided for the system, and the greater the role in the comprehensive evaluation, that is, the greater the weight of the indicator; When the entropy value of an indicator in the indicator system is larger, it indicates that the smaller the change degree of the indicator value is, the less information can be provided for the system, and the smaller the role it plays in the comprehensive evaluation, that is, the smaller the weight of the indicator. Weight s_j It reflects the information amount of each indicator, and can intuitively reflect the difference between each indicator. The larger the entropy weight value is, the greater the impact of this indicator on decision-making.

To make $\ln E_{ij}$ Meaningful, general assumption $E_{ij} = 0$ When, $E_{ij} \ln E_{ij} = 0$, but when $E_{ij} = 1$ When, $E_{ij} \ln E_{ij} = 0$, obviously not tangential, and contrary to the meaning of entropy, we can E_{ij} Revised and defined as:

$$\tilde{E}_{ij} = \frac{1 + g_{ij}}{\sum_{i=1}^{n} (1 + g_{ij})} \tag{11}$$

Then calculate the entropy weight of each index $\tilde{e}_j$,
Then we can calculate the j The weight of each indicator is removed.

$$\tilde{s}_j = \frac{1 - \tilde{e}_j}{m - \sum_{j=1}^{m} \tilde{e}_j} \tag{12}$$

In order to facilitate the subsequent unified operation, the weight obtained in the two cases s_j, $\tilde{s}_j$, uniformly called w_j.

2.3 Tracking Model Construction

Based on the weight of tracking indicators calculated above, a tracking model is constructed to analyze the effectiveness of funding.

(1) Tracking index membership

The determination of membership degree of tracking index is a key link in the application of fuzzy comprehensive evaluation. Because the evaluation of something is always complicated and fuzzy, it is inevitable that there will be personal subjective color when establishing the membership degree of the evaluation index, which virtually increases the complexity, fuzziness and diversity of the membership function. In the tracking of the effect of financial aid for poor college students, the effect grade is divided into "excellent", "good", "average" and "poor". Through empirical research, the evaluation value of a college can be determined, and the corresponding index value of the evaluation set can be determined. Its membership can be expressed as $V = \{v_1, v_2, v_3, v_4\}$. In particular, it is worth noting that the method of collecting data through questionnaires is used to determine the subordination matrix of the financial aid effectiveness of poor college students. There must be enough evaluators and the evaluators must have a good understanding of the financial aid work of poor college students to ensure that the degree of subordination is in line with the objective reality to the maximum extent [10]. The constructed tracking index membership matrix is as follows:

$$V = \left(v_{ij}\right)_{mn} = \begin{bmatrix} v_{11} & v_{12} & \ldots & v_{1n} \\ v_{21} & v_{22} & \ldots & v_{2n} \\ \ldots & \ldots & \ldots & \ldots \\ v_{m1} & v_{m2} & \ldots & v_{mn} \end{bmatrix} \tag{13}$$

Among them, v_{ij} Indicator j For comments i Membership, v_{ij} The value of is determined by questionnaire.

(2) Tracking model

combination w_j, v_{ij} Establish tracking model.

$$Y = W \cdot V = [w_1, w_2, \ldots, w_m] \begin{bmatrix} v_{11} & v_{12} & \ldots & v_{1n} \\ v_{21} & v_{22} & \ldots & v_{2n} \\ \ldots & \ldots & \ldots & \ldots \\ v_{m1} & v_{m2} & \ldots & v_{mn} \end{bmatrix} \tag{14}$$

Among them, Y Represent the corresponding results of the funded objects on evaluation set V; W Represents a set of weights.

The tracking set of funding effectiveness for poor college students can be expressed as V = [excellent, good, average, poor]. Because of the fuzziness and uncertainty of language comments, they can be expressed in the form of a percentage system, thus transforming qualitative analysis into quantitative analysis. The quantitative table is shown in Table 3 below.

Table 3. Classification of funding effectiveness

Grade	Fraction
Excellent	[90, 100)
Good	[75,90)
Commonly	[60,75)
Poor	[0,60)

Finally, the results are normalized to get the final tracked results.

3 Tracking Model Application Test

Taking three poor students in a university as an example, the tracking model studied is used to track the effectiveness of the funding to test the effectiveness of the tracking model studied.

3.1 CIRO Tracking Index System

After four rounds of expert scoring, the experts reached an agreement on the final tracking index system as shown in Table 4.

Table 4. CIRO tracking index system

Primary	Secondary	Third level	c_j
Effectiveness of university funding	Background	The importance attached by the school to funding	0.1263
		Poor students' understanding of subsidy policies	0.2153
		Pre funding needs survey	0.0263
		Matching funding objectives with needs	0.1262
	Input	Recognition system for family economic difficulties	0.0362
		The distribution system of funding	0.2023
		System for the use of funding	0.2156
		Proportion of special funds invested in funding	0.2325

(*continued*)

Table 4. (*continued*)

Primary	Secondary	Third level	c_j
	Reaction	Funding policy construction and implementation status	0.1548
		The completeness of funding policies	0.1632
		Rationality of the use of funding resources	0.1478
		Repayment status of student loans	0.1852
		Employment situation of impoverished students	0.2020
	Output	Student learning ability	0.1782
		Family status of impoverished students	0.069
		The psychological status of poor students	0.1585
		Student's sense of responsibility	0.1263
		Social public satisfaction	0.2214

3.2 Tracking Index Weight

Use the improved entropy weight method to calculate the weight of each indicator in Table 4, and the calculation results are shown in Fig. 1 and Table 5 below.

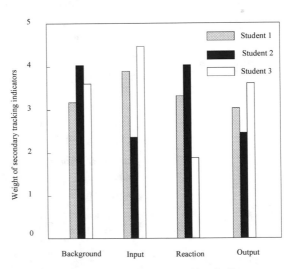

Fig. 1. Weight of secondary tracking indicators

Table 5. Weights of three-level tracking indicators

Third level	Student 1	Student 2	Student 3
The importance attached by the school to funding	0.5482	2.4862	1.0122
Poor students' understanding of subsidy policies	0.1485	3.4527	2.2254
Pre funding needs survey	0.2485	5.4865	4.5662
Matching funding objectives with needs	1.5672	1.4852	3.0214
Recognition system for family economic difficulties	2.3214	2.4752	5.0255
The distribution system of funding	1.2355	3.3312	2.0363
System for the use of funding	1.14682	3.2014	2.4874
Proportion of special funds invested in funding	2.4542	2.0141	0.3622
Funding policy construction and implementation status	1.6782	0.3969	3.6215
The completeness of funding policies	0.7822	0.5285	4.1256
Rationality of the use of funding resources	0.3755	2.3631	1.5245
Repayment status of student loans	2.6725	2.3244	3.1142
Employment situation of impoverished students	1.8452	2.0202	2.0235
Student learning ability	3.8545	3.3692	2.2687
Family status of impoverished students	2.8752	0.4587	1.3625
The psychological status of poor students	0.5458	1.2045	1.4782
Student's sense of responsibility	0.3255	2.5287	0.3034
Social public satisfaction	2.7845	3.1011	0.3969

3.3 Tracking Results

The tracking model is used to track the three funded students for four years, and the tracking effect is shown in Fig. 2 below.

As can be seen from Fig. 2, the performance score of the three funded students has been increasing year by year, which indicates that the three students have significantly improved their status in all aspects after receiving funding, but from the growth trend, the score of student 1 has entered a flat development period after improving, which indicates that the status is stable and the funding can be gradually reduced. The score of student 2 shows a rising trend, but the rising trend is slow, so it is necessary to increase funding. Student 3 has been growing, and the score has not stabilized, so we can continue to maintain the current funding.

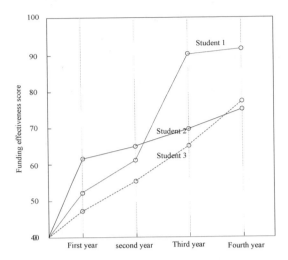

Fig. 2. Tracking Results of Funding Objects

4 Conclusion

The school provides financial aid to students. First, it investigates the family situation of students, and uses home visits and other forms to investigate the economic situation of students' families on the spot. Then it inspects the students themselves in various aspects at the school. In addition to the basic family situation, there are also various aspects of students' academic achievements, students' enthusiasm for learning, and students' popularity. After the investigation of these aspects, the school can report the basic information of applicants to the funding unit, and then apply for funding. School funding often ends at this stage, or it is usually after obtaining funding, the funding project will stagnate, and students can not get timely funding, which leads to poor results of funding, and funding is invalid. This requires colleges and universities to participate in the whole process of funding, so as to ensure the funding results of colleges and universities. Under the guidance of the concept of targeted poverty alleviation in the new era, university funding should not only provide autonomy for students in funding, but also provide funding and observation for other aspects of students' life and learning. It should also accurately target difficult target groups, and should not use the funding funds in places where they should not be used. In the new era, colleges and universities should not only set up a dynamic tracking system for students' growth and expand the scope of the school's funding, but also set up a hierarchical funding system to track the whole process of student funding, so as to ensure that the funded students are qualified groups. Through these measures, we can better ensure the quality of university funding by tracking university funding and its effect. For this reason, it is very necessary to carry out adaptive tracking of university funding objects from the perspective of Big data. According to the ideas and principles of CIRO model, Delphi method is applied, and relevant experts and scholars are consulted to determine the tracking index system of university funding objects for poor students. Calculate the weight of each tracking indicator in the system using the entropy weight method. Construct an adaptive tracking model for university funding objects based on indicator membership, and obtain relevant tracking results. Taking three impoverished students from a certain university as an example, the effectiveness of the study was demonstrated through empirical analysis. In the future, adaptive tracking research on university funding objects from the perspective of Big data can focus on data quality and privacy protection, prediction and early warning capabilities, personalized services and resource optimization, effect evaluation and decision support, as well as cross field cooperation and innovation. These research directions will further promote the development of the management of university funding recipients, improve the effectiveness of funding and student satisfaction, and promote the realization of educational equity and social harmony.

Acknowledgement. 2022 University Philosophy and Social Sciences Research Project: Research on the Realistic Dilemma and Technical Demands of Precise Funding in Universities from the Perspective of Big Data (2022SJYB1050).

References

1. Smith, C.M.: Does state allocation of university funding moderate effectively maintained inequality? Soc. Currents **9**(3), 245–264 (2022)
2. Charles, L., Shaen, C.: Generating stable university funding mechanisms: income contingent loan structure choice within the Irish education system. J. Educ. Finance **47**(1), 92–110 (2021)
3. El Gibari, S., Perez-Esparrells, C., Gomez, T., Ruiz, F.: Analyzing the impact of Spanish university funding policies on the evolution of their performance: a multi-criteria approach. Mathematics **9**(14), 1626 (2021). https://doi.org/10.3390/math9141626
4. Behnam, P.: University research funding: why does industry funding continue to be a small portion of university research, and how can we change the paradigm? Ind. High. Educ. **35**(3), 150–158 (2021)
5. Stenbacka, R., Tombak, M.: University-firm competition in basic research and university funding policy. J. Public Econ. Theory **22**(4), 1017–1040 (2020)
6. Han, S.F., Gao, J.P., Cao, Y., Zhu, R.F., Wang, Y.P.: Dermatology nurse prescribing in China: a delphi method. Front. Nurs. **10**(1), 95–114 (2023). https://doi.org/10.2478/fon-2023-0011
7. Yi, Z., Wenwen, X., Yingnan, W., Xianjia, W.: Sustainability assessment of water resources use in 31 provinces in China: a combination method of entropy weight and cloud model. Int. J. Environ. Res. Public Health **19**(19), 12870 (2022)
8. Yi, K., Lixia, Y., Lihong, L.: Research on the weight calculation of social benefit evaluation of Chinese film and TV enterprises based on fuzzy comprehensive evaluation method. J. Intell. Fuzzy Syst. **38**(6), 1–12 (2020)
9. Lee, C.L., Locke, M.: The effectiveness of passive land value capture mechanisms in funding infrastructure. J. Property Investment Finance **39**(3), 283–293 (2020)
10. Osidipe, A.: Funding effectiveness of TVET for decent employment and inclusive growth in Nigeria with perspectives from China. J. Educ. Pract. **10**(36), 46–61 (2019)

Research on Rapid Selection of University Funding Objects Based on Social Big Data Analysis

Xiaoyan Xu[✉] and Yuliang Zhang

Data and Information Center, Wuxi Vocational Institute of Commerce, Wuxi 214000, China
15861562560@163.com

Abstract. With the sharp increase in the number of college students, the number of students who need financial aid also increases.How to quickly and accurately select university funding objects has become the key to achieve the goal of funding education. Therefore, this paper proposes a research on rapid selection methods of university funding objects based on social big data analysis. Based on the principles of systematicness, objectivity, scientificity and feasibility, we will build an index system for the selection of university funding objects, deeply mine the index data for the selection of university funding objects in the big data of social communications, build a pre-processing framework for the selection of index data, re sample the index data for the selection of university funding objects based on the SMOTE algorithm, and eliminate the adverse effects of unbalanced data. Set up a model for selecting university funding objects, formulate rules for selecting university funding objects, and realize rapid selection of university funding objects. The experimental results show that after the application of the proposed method, the corresponding maximum accuracy rate of the selection results of university funding objects is 98%, the maximum recall rate is 91%, and the maximum F value is 0.96, which fully confirms that the proposed method has better application performance.

Keywords: Big Data Analysis Means · Colleges and Universities · Quick Selection · Social Big Data · Funding Target.

1 Introduction

Since the expansion of college enrollment, more and more students have entered the college gate. On the basis of the steady growth of the number of students, the corresponding number of students in need of financial assistance is also increasing.Some new characteristics of poverty are shown in students from economically disadvantaged families. According to the data of the National Student Aid Management Center, in 2022, the government, universities and the society set up various university student aid policies to subsidize 4387 ordinary colleges and universities nationwide890000 person times, with a subsidy of 115.03 billion yuan. On the one hand, the university funding work has

© ICST Institute for Computer Sciences, Social Informatics and Telecommunications Engineering 2024
Published by Springer Nature Switzerland AG 2024. All Rights Reserved
L. Yun et al. (Eds.): ADHIP 2023, LNICST 547, pp. 436–451, 2024.
https://doi.org/10.1007/978-3-031-50543-0_30

been carried out like a raging fire, obtaining abundant resource support, helping students with financial difficulties from families to have a solid material foundation and a variety of policy help. All difficulties caused by family financial difficulties have been well curbed. Large scale, diversified and deep difficulties are gradually reducing, and university funding work has made some progress. The uneven distribution of students from economically disadvantaged families and the diversification of causes of poverty require that student funding needs to be more precise [1]. On the other hand, the economic and social development has promoted the continuous upgrading of the industrial structure, the substantial improvement of people's living standards, and the change of production and lifestyle, making the traditional material poverty alleviation need to combine the characteristics and requirements of the new era, on the material basis, strengthen the improvement of the spiritual world. University funding is a specific way of poverty alleviation for students. Since the 18th National Congress of the Communist Party of China, General Secretary, on the basis of socialism with Chinese characteristics, summarized the experience and lessons of anti-poverty, put forward new ideas for poverty alleviation in theory and practice, and created a new program for poverty alleviation with Chinese characteristics.

The main task of education reform and development in the "13th Five Year Plan" is to help more students with difficult families, complete the task of poverty alleviation, and achieve the goal of full coverage of college education, so as to achieve education fairness. Especially for some areas with relatively backward development, we should increase poverty alleviation capacity, improve poverty alleviation effect, and achieve targeted poverty alleviation. In the plan, it is emphasized that in order to ensure the full realization of education fairness, a university management information platform based on students' basic information should be established, and the information management system of civil affairs, schools, poverty alleviation and other departments should be organically combined, so as to open a "green channel" for poor students in rural remote areas, increase funding, and enable more students to enjoy the right to receive education and enjoy educational achievements. We will lift ourselves out of poverty. Under the background of big data, it is more and more convenient to obtain data that can reflect students' consumption characteristics and learning status. This research makes full use of the behavior data of students in school provided by various platforms under the background of big data, and introduces data mining methods into the funding work. On the one hand, it is conducive to the connotation, quality and method of financial aid, Improve the accuracy of the identification of funding objects. On the other hand, it uses the means of informatization to accurately identify funding objects, and fairly and reasonably allocate funding resources, so as to better solve the problem of malicious tampering with information in the current university funding work [2]. Accurate identification of funding objects is conducive to improving the effectiveness of funding work and better realizing the purpose of funding education.

Based on existing research results, the commonly used methods for rapid selection of university funding targets are the campus one card based method [3] and the impoverished student consumption information based method [4]. The former uses the campus all-in-one card data to study the laws of students' Consumer behaviour using the methods of association analysis and clustering in data mining, so as to provide relevant suggestions for the distribution of school grants. The latter uses support vector machines and association rule methods in data mining to establish normal and abnormal consumption patterns for impoverished students based on data from the consumption information system of impoverished students in universities. Both of the above two methods have certain defects. The selection efficiency of university funding objects obtained is low, and the accuracy of selection results is poor, which affects the effect of funding education. Therefore, the research on rapid selection methods of university funding objects based on social big data analysis is proposed.

2 Research on Quick Selection Method of University Funding Objects

2.1 Construction of Index System for Selecting University Funding Objects

Under the background of big data, the construction of the index system for the selection of university funding objects should follow the following principles: systematic, objective, scientific and feasible, as shown in Table 1.

The indicator system for the selection of university funding objects is constructed according to the principles shown in Table 1, as shown in Table 2.

As shown in Table 2, the number of first level indicators, second level indicators and third level indicators is 3, 6 and 20, respectively, in the indicator system for the selection of university funding objects, which can comprehensively measure the actual situation of university funding objects and provide support for rapid and accurate selection of funding objects.

2.2 Preprocessing of Target Data for Funding Object Selection

Data mining is a technology that extracts useful information and patterns from a large amount of data, which can help understand data, discover hidden associations and trends, and provide support for subsequent analysis and decision-making. Data mining plays an important role in the preprocessing of funding target selection indicator data. It can also help clean data, select features, transform and integrate data, and discover association rules and patterns in the data. These steps help improve data quality, simplify analysis and decision-making processes, and provide more accurate and effective support for subsequent funding target selection.

Based on the above constructed index system for the selection of university funding objects, mine the index data for the selection of university funding objects in the big data of Social Communications, and preprocess them to provide assistance for subsequent research.

Data mining technology mainly includes five steps, as shown below:

(1) Data integration and cleaning: In the era of big data, data sources are diversified, and the data format is diverse and disorderly. Therefore, cleaning the acquired knowledge is of great significance for improving the accuracy of experiments.

Table 1. Principles for Selection of University Subsidies

Principle	Content description
systematicness	A complete index system for the selection of university funding objects should have sufficient coverage, including all aspects of students' information, from students' basic information, all-in-one card consumption, personal daily performance to academic achievements, so as to establish scientific management guidelines.The selected variables should be able to reflect the students' consumption level, consumption potential, poverty index, learning willingness, learning achievements, etc., without too many variables, leading to the prediction model being too complex.Therefore, no important indicator should be omitted in the selection of variables, and the situation should be reflected as completely, comprehensively and systematically as possible to avoid generalizing
objectivity	The indicators selected according to the research content can truly and objectively reflect the essential attributes of things.In the precise selection model of university funding objects, the important thing of the indicator system is to reflect the quality rather than the quantity.The selected variables should truly reflect the students' consumption level and learning attitude, so as to establish a working mechanism for selecting funding objects
Scientific	The design of each indicator system and the selection of evaluation indicators must be scientific. The indicators selected in this paper should objectively and truly reflect the characteristics of students' consumption behavior and learning attitude at school.The evaluation indicators should be representative, not overlapping, too complicated, but not too simple, to avoid information omission, information error and other phenomena
feasibility	The set indicator system is required to be highly operable, and the selected variables can truly reflect the essence of things, so the collected information should be easy to express and process, and the indicator setting should be concise, practical, and highly operable.The availability of data should be considered, so the selected variables must be easy to obtain to improve the accuracy and feasibility of the operation

(2) Data selection and conversion: The amount of data obtained after data cleaning is large, and not all data are available, so it is necessary to select research related data from them to reduce the complexity of the experiment. In addition, there may be inconsistent measurement standards between different features of the acquired data, such as discrete, continuous and other variables. Therefore, before using data mining algorithms for research, data variables need to be discretized and standardized.

(3) Data mining: mining different knowledge with different relationship types according to different algorithms. The common data mining algorithms include classification, clustering, feature extraction, association analysis, etc. According to the specific analysis of specific problems, select the appropriate algorithm.

(4) Mode evaluation: the validity of the data mining model is evaluated by such indicators as accuracy, accuracy, and sensitivity.

Table 2. Index System for Selection of University Subsidies

Primary indicators	Secondary indicators	Level III indicators
Individual consumption characteristics	Daily living expenses	Total supermarket consumption
		Total consumption of boiled water
		Total shower consumption
	Restaurant consumption	Number of restaurant consumption
		Total Meals
		Maximum single consumption amount
		Single minimum consumption amount
		Average meal amount
	Overall consumption	Total times of school consumption
		Total consumption
		Average consumption
		Remaining amount in the card
Learning attitude	academic record	Score ranking
	Daily performance	Book borrowing type
		Total books borrowed
		Times of entering and leaving the library
		Total amount of dormitory
		Total amount out of dormitory
		Proportion of books borrowed from professional courses
Funding	a grant	Access to financial aid

(5) Knowledge representation: use visual knowledge to show users the results of data mining.

The process of in-depth data mining for selecting indicators of university funding objects is shown in Fig. 1.

The process shown in Fig. 1 is used to mine the index data of university funding object selection. Due to the influence of various factors, there are redundancy, missing and other phenomena in the mining data. If it is directly applied, it will affect the accuracy of the selection of university funding objects. In addition, there are many student consumption scenarios, and consumption data is accumulating at an alarming rate every day. There are sensitive information leakage, redundant and useless data, missing key data, abnormal data and other problems in the data, so it is necessary to establish a unified and deployable

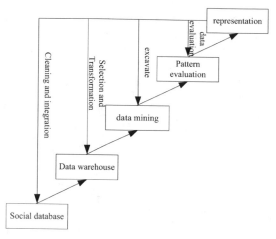

Fig. 1. Schematic diagram of data mining process for selecting indicators of university funding objects

pre-processing framework to apply to the large amount of data. It is necessary and practical to have many data dimensions. The data preprocessing framework in this paper mainly includes four modules: data desensitization, data cleaning, data specification and data integration.The data preprocessing framework is shown in Fig. 2.

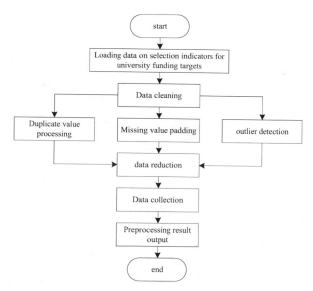

Fig. 2. Framework of data pre-processing for selecting indicators of funding objects

Through investigation and analysis of data, it can be seen that outliers are the abnormal data that has the greatest impact on the selection of university funding targets, so it is necessary to accurately detect and process them. The purpose of calculating the distance between the selection indicators of funding targets is to evaluate the similarity or difference between different targets. By calculating the distance, the degree of difference between different objects in indicator data can be quantified, thereby achieving outlier detection [5]. The formula for calculating the distance between indicator data is

$$d_{ij} = \alpha^o * \sqrt{\sum_{i,j=1}^{n} |x_i - x_j|^{\alpha^o}} \tag{1}$$

In formula (1), d_{ij} It represents the index data of the selection of university funding objects x_i And x_j Minkowski distance. Minkowski distance as a method to calculate the distance between selected index data of funding objects, has the advantages of flexibility, adjustable robustness, interpretability and Foundations of mathematics. These advantages make Minkowski distance a commonly used and reliable distance measurement method, which has wide applications in funding object selection and other data analysis tasks; α^o It represents a variable parameter. The value range is [0, 1]; n It represents the total number of index data for the selection of university funding objects.

According to formula (1), measure the average Minkowski distance between the indicator data x_i for selecting university funding targets and all other data. The calculation formula is

$$\overline{d}_i = \frac{\sum_{j=1}^{n-1} d_{ij}}{n-1} \tag{2}$$

In formula (2), $\overline{d}_i$ It represents the index data of the selection of university funding objects x_i Average distance from all other data.

Calculate the result with formula (2)$\overline{d}_i$ To determine the index data for the selection of university funding objects x_i Whether it is an outlier, the specific determination rules are as follows:

$$\begin{cases} \overline{d}_i \le \beta^* & \text{Normal Value} \\ \overline{d}_i > \beta^* & \text{Outlier} \end{cases} \tag{3}$$

In Eq. (3), β^* It represents the outlier determination threshold, which needs to be set according to the actual selection of indicator data [6].

The outlier detection results of university funding object selection index data obtained according to the above rules are shown in Fig. 3.

As shown in Fig. 3, the data points inside the green dotted circle are outlier detection results, which fully indicates that the outlier detection effect of the selected indicator data of university funding objects is better. Delete them to reduce the proportion of abnormal data in the selected indicator data.

On the basis of the above outlier deletion processing of university funding object selection index data set, each index data is accurately clustered using hierarchical clustering algorithm to provide a basis for subsequent unbalanced data processing.

Hierarchical clustering is mainly divided into top-down hierarchical splitting method and bottom-up hierarchical agglomeration method, which are described in turn below.

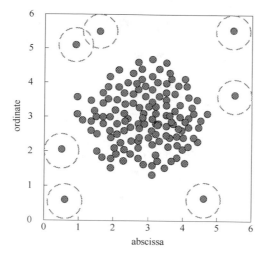

Fig. 3. Example of outlier detection results of university funding object selection index data

The representative of hierarchical agglomeration is AGNES algorithm. For sample size of n Datasets for $\{x_1, x_2, \cdots, x_n\}$ Given the definition of distance and connection mode, the specific steps of the algorithm are as follows:

Initial step: each point is a class, and the dataset is divided into n Classes, namely $C_i \in \{x_i\}$;

Step 2: Calculate the distance matrix between different classes D;

Step 3: find the two nearest classes and merge them into one class.

Step 4: Repeat steps 2 and 3 until all data belong to a class or meet a termination condition[7].

The complexity of AGNES calculation is $O(n^2 \log_2 n)$. (n2logn) is not suitable for large-scale data sets, that is, not suitable for this study. The DIANA algorithm, which is the representative of hierarchical splitting, is completely different from AGNES algorithm in this process. Gradually descending layering[8]is adopted at one point.

First, all data are classified into one category, that is $A_1 = \{x_1, x_2, \cdots, x_n\}$;

Secondly: it is necessary to measure the length between all points, and select the average maximum length leading to different points (mark S), take this point as a new starting point, namely $S \in A_2$;

Then: face every point $x_i \in A_2$, the distance calculation formula is defined as

$$D_i = d_{avg}(x_i, A_1) - d_{avg}(x_i, A_2) \tag{4}$$

In Eq. (4),D_i Represents a class A_1 And A_2 Distance;$d_{avg}(x_i, A_1)$ It represents the selection of indicator data x_i And class A_1 Distance between;$d_{avg}(x_i, A_2)$ It represents the selection of indicator data x_i And class A_2 Distance between.

According to the calculation result of formula (4), select the largest D_i. If D_i If it is greater than 0, then $x_i \in A_2$;

Finally: cycle the previous step until all D_i If it is less than 0, it is regarded as turning a group into two parts.

After all the processes, the DIANA algorithm converts a sample data into X The dataset of is divided into two groups A_1 and A_2, repeat the above links indefinitely, split into different forms, until all the data together or all of them reach the critical value, clustering work is suspended.

Record the index data of college funding object selection after the above processing as $Y = \{Y_1, Y_2, \cdots, Y_n\}$, where, Y_i It refers to the No i The data set of selected indicators provides assistance for subsequent research.

2.3   Unbalanced Data Processing Based on SMOTE

Pre processing results of index data selected by the above university funding objects $Y = \{Y_1, Y_2, \cdots, Y_n\}$ As a basis, through in-depth exploration and analysis, we can see that the data used in the selection of university funding objects is naturally unbalanced, which will have a great adverse impact on the selection results of university funding objects, so it is necessary to deal with it.

SMOTE algorithm is an intelligent oversampling technology proposed by Chawla et al. This method is not to increase the sample size by mechanical copying, but to synthesize new positive samples through certain rules to balance the data. It is a heuristic sampling algorithm. The main idea of this algorithm is to use linear interpolation and K Nearest neighbor method, where the distance from the sample is small, uses linear interpolation to synthesize minority samples, which effectively alleviates the problem of data over fitting [9]. The main steps of SMOTE algorithm are as follows:

Step 1: Select indicator data samples for each Y_i, find it first K Neighbor samples of the same type and the smallest distance from them, so that the upward sampling rate is M, from here K Randomly select samples from neighbor samples with small distance m, and recorded as $\{y_1, y_2, \cdots, y_m\}$;

Step 2: In minority data samples Y_i And $\{y_1, y_2, \cdots, y_m\}$ The random linear interpolation method is used to synthesize new minority data samples, and the expression is

$$Y_{new} = Y_i + rand * (y_i - Y_i) \tag{5}$$

In formula (5),$rand$ It represents any random value between 0 and 1.

Step 3: Insert the new composite sample into the dataset to form a new dataset.

The sampling effect of the index data set of university funding object selection based on SMOTE algorithm is shown in Fig. 4.

With the new dataset above $Y' = \{Y'_1, Y'_2, \cdots, Y'_n\}$ To determine the resampling frequency of each data set, the formula is

$$f_i = \frac{Y'_i - Y_i}{\chi^\Theta} \tag{6}$$

In formula (6),f_i It means to select indicator data set Y_i Resampling frequency of;χ^Θ It represents the auxiliary parameter of resampling frequency calculation, and the value range is [0, 1].

The SMOTE algorithm shown above is used to resample the index data set of university funding object selection, eliminate the adverse effects of unbalanced data, and

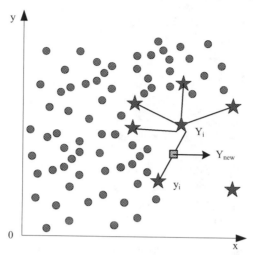

Fig. 4. Sampling Effect Diagram of Index Data Set for Selection of University Funding Objects

make sufficient preparations for obtaining the final results of university funding object selection.

2.4   Obtaining the Results of Selecting University Funding Objects

Record the index data of university funding object selection after unbalanced data processing as $Z = \{Z_1, Z_2, \cdots, Z_n\}$ In combination with the index system of university funding object selection built in Sect. 2.1, a model of university funding object selection is built, with the expression as

$$\psi_i = \frac{7 \times \delta^2 \times \sqrt{Z_i^2}}{\sum\limits_{i=1}^{n} Z_i \times R_i + \varepsilon_i} \tag{7}$$

In Eq. (7),ψ_i It represents the value of the objective function for the selection of university funding objects;R_i It refers to the selection index of university funding objects;ε_i It indicates the adjustment parameters of the selection error of university funding objects;δ^2 It represents the standard factor for the selection of funding objects.

Calculate the result with formula (7)ψ_i Based on, the selection rules of university funding objects are formulated as follows:

$$\begin{cases} \psi_i \geq \varsigma^\tau & \text{Funding Object} \\ \psi_i < \varsigma^\tau & \text{Unfunded Object} \end{cases} \tag{8}$$

In Eq. (8),ς^τ It refers to the decision threshold [10] for the selection of university funding objects.

Through the above process, we obtained the results of the selection of university funding objects, realized the rapid and accurate selection of university funding objects, and provided assistance for the realization of the purpose of funding education.

3 Experiment and Result Analysis

3.1 Select the Indicator Data Set to Determine the Resampling Frequency

The data sources of university funding targets are as follows:

(1) University's own data: Comprehensive universities usually have their own student funding offices or relevant departments responsible for collecting and managing student funding information. These departments may maintain students' personal information, family background, economic status, and other information to determine whether they meet the funding requirements.
(2) Government agencies: Government education departments or related institutions may collect and manage data related to student funding in universities. These institutions may collaborate with comprehensive universities to share student funding information in order to better manage and allocate funding resources.
(3) Survey and research data: Independent survey and research institutions or academic research teams may conduct surveys on university student funding to collect relevant data. These data may include information such as students' economic status, funding needs, and types of funding.
(4) Cooperation with other universities: In addition to comprehensive universities, other universities can also be sources of data on university funding targets. Cooperation and information sharing can be carried out between different universities to better understand and evaluate students' funding needs.

In order to avoid the adverse effects of unbalanced data, a method based on SMOTE algorithm is proposed to resample the selected index data set. Determine and select the resampling frequency of the indicator data set according to Formula (6) to facilitate the subsequent experiments, as shown in Table 3.

As shown in the data in Table 3, the determined resampling frequency is used to collect the required experimental data in the index data set of university funding object selection, so as to maximize the application performance of the proposed method.

3.2 Selection of Evaluation Indicators

In general, there are many situations in the selection of university funding objects, as shown in Table 4.

As shown in Table 4, TP refers to the number of samples originally funded and selected as funded objects;FN represents the number of samples originally funded and selected as non funded objects; FP refers to the number of samples that were originally non funded and selected as funded objects; TN refers to the number of samples that were originally non funded objects but were selected as non funded objects.

According to the needs of university funding object selection, the accuracy rate, recall rate and F value are selected as evaluation indicators, and the expression is

$$
\begin{cases}
Q_1 = \frac{TP}{TP+FP} \\
Q_2 = \frac{TP}{TP+FN} \\
F = \frac{(1+\mu_1^2)Q_1 \times Q_2}{\mu_1^2 Q_2 + Q_1}
\end{cases}
\tag{9}
$$

Table 3. Resampling Frequency of Selected Indicator Data Set

Secondary indicators	Level III indicators	Resampling frequency (Hz)
Daily living expenses	Total supermarket consumption	44
	Total consumption of boiled water	30
	Total shower consumption	50
Restaurant consumption	Number of restaurant consumption	59
	Total Meals	65
	Maximum single consumption amount	42
	Single minimum consumption amount	23
	Average meal amount	20
Overall consumption	Total times of school consumption	47
	Total consumption	58
	Average consumption	59
	Remaining amount in the card	55
academic record	Score ranking	70
Daily performance	Book borrowing type	84
	Total books borrowed	51
	Times of entering and leaving the library	56
	Total amount of dormitory	40
	Total amount out of dormitory	38
	Proportion of books borrowed from professional courses	69
a grant	Access to financial aid	52

Table 4. Selection Results of Funding Objects

	Funding objects	Non funded objects	total
Funding objects	TP	FN	P
Non funded objects	FP	TN	N
total	P'	N'	P + N

In Eq. (9),Q_1 It indicates the accuracy of the selection of university funding objects;Q_2 It refers to the recall rate of university funding object selection;F It represents the F value corresponding to the selection result of university funding objects.F It is a comprehensive evaluation standard,μ_1 In order to adjust the coefficient between recall rate and accuracy rate, it is usually taken as 1.If we can see from the expression,F The criteria can correctly measure the performance of the proposed method, if the recall rate Q_2 And accuracy Q_1 High, description F The value is high, which can better identify minority features. When colleges and universities accurately identify funding objects,F A high value indicates that the precise funding prediction model established can identify more funded students and more accurately identify funding objects.

3.3   Analysis of Experimental Results

According to the determination results of resampling frequency of the above selection index data set and the selection results of evaluation indicators, the quick selection method of university funding objects based on campus all-in-one card and the quick selection method of university funding objects based on poor students' consumption information are set as comparison methods 1 and 2 to carry out a comparative experiment of university funding object selection. The specific analysis process of the experimental results is as follows:

3.3.1   Analysis on the Accuracy of the Selection of University Funding Objects

Table 5 shows the accuracy rate of selecting university funding objects through experiments.

Table 5. Accuracy of Selection of University Subsidies (Q_1/%)

Experimental group	Propose method	Comparison method 1	Comparison method 2
1	89	56	45
2	94	45	56
3	95	52	43
4	98	41	59
5	84	52	53
6	85	58	44
7	78	47	47
8	85	59	59
9	81	41	52
10	76	55	50

As shown in the data in Table 1, after the application of the proposed method, the accuracy rate of university funding object selection is 76% ~ 98%; After the application of comparison method 1, the accuracy rate of the selection of university grants is 41% ~ 59%; After the application of comparison method 2, the accuracy rate of university funding object selection is 43% ~ 59%. Through comparison, it is found that after the application of the proposed method, the accuracy rate of university funding object selection is far higher than that of comparison method 1 and comparison method 2, and the maximum accuracy rate in the fourth experimental group is 98%.

3.3.2 Analysis on Recall Rate of University Funding Object Selection

See Table 6 for the recall rate of university funding objects selected through experiments.

Table 6. Recall rate of university funding object selection (Q_2/%)

Experimental group	Propose method	Comparison method 1	Comparison method 2
1	78	55	36
2	68	45	39
3	89	40	40
4	88	36	41
5	74	45	31
6	71	42	29
7	76	41	45
8	89	48	48
9	90	50	47
10	91	35	59

As shown in the data in Table 6, after the application of the proposed method, the selected recall rate of university funding recipients ranges from 68% to 91%; After the application of comparison method 1, the selected recall rate of university grants is 35% ~ 55%; After the application of comparison method 2, the selected recall rate of university funding objects is 29% ~ 59%. Through comparison, it is found that after the application of the proposed method, the selected recall rate of university funding objects is much higher than that of comparison method 1 and comparison method 2, and the maximum recall rate of the 10th experimental group is 91%.

3.3.3 F-value Analysis of the Selection Results of University Funding Objects

According to Q_1-Q_2 The curve shows that researchers often hope for accuracy Q_1 and recall rate Q_2 Both of them are very high, but in fact they are contradictory. The above two indicators are contradictory and cannot achieve double high. For example, if you want a high accuracy rate, you must sacrifice some recall rates. To get a high recall rate, we must sacrifice some accuracy. But usually, a new indicator can be defined according to the balance point between them F Value, which can consider the accuracy rate and recall rate at the same time, so that both can reach the highest and achieve a balance.

The F value of university funding object selection results obtained through experiments is shown in Fig. 5.

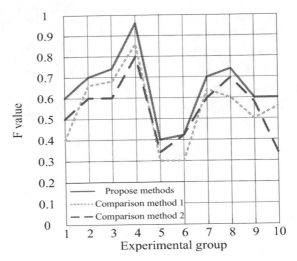

Fig. 5. Schematic diagram of F value of university funding object selection results

As shown in the data in Fig. 5, after the application of the proposed method, the F value range of the selected university funding objects is 0.4 ~ 0.96; After the application of comparison method 1, the F value range of university funding object selection results obtained is 0.3 ~ 0.86; After the application of comparison method 2, the F value range of the selected university funding objects is 0.34 ~ 0.8. Through comparison, it is found that after the application of the proposed method, the F value of the selection result of university funding objects is much higher than that of comparison method 1 and comparison method 2, and the maximum F value of 0.96 is obtained in the fourth experimental group.

4   Conclusion

In the context of big data, introducing big data into the work of funding education, and using data mining methods to analyze students' campus behavior data can objectively provide reliable credit guarantee, reduce human intervention, maximize the accuracy of data, and provide a new method for accurately identifying funding objects. This paper mainly studies the relevant issues when using data mining technology to accurately identify funding objects, and conducts in-depth discussions around the improvement of the target system for accurate identification of funding objects and the imbalance of data when selecting identification methods. First, in view of the status quo of precise identification indicators of university funding objects established in the context of big data, it is proposed that students' daily performance information should be integrated and their learning attitude should be comprehensively considered. Therefore, daily performance indicators are introduced into learning attitude, and a precise identification indicator system of funding objects in line with the background of big data has been established. Secondly, when establishing the prediction model for precise identification of university funding objects, it was found that the unbalanced data made it difficult to accurately identify the behavior characteristics of the funded students. Based on this, this paper proposed to use the SMOTE resampling algorithm to balance the data set in the data pre-processing stage, which enhanced the prediction accuracy of classification, so as to more accurately identify the funding objects. The experimental data shows that the proposed method effectively improves the accuracy, recall rate and F value of university funding object selection, and can provide more effective method support for funding object selection.

Aknowledgement. 2022 University Philosophy and Social Sciences Research Project: Research on the Realistic Dilemma and Technical Demands of Precise Funding in Universities from the Perspective of Big Data (2022SJYB1050).

References

1. Behera, R.K., Jena, M., Rath, S.K., et al.: Co-LSTM: convolutional LSTM model for sentiment analysis in social big data. Inf. Process. Manage. **58**(1), 102435 (2021)
2. Weimar, D., Soebbing, B.P., Wicker, P.: Dealing with statistical significance in big data: the social media value of game outcomes in professional football. J. Sport Manag. **35**(3), 266–277 (2021)
3. Zhang, Z.: Research on the identification of poor college students based on campus card data. Modern Economic Information **12**(13), 12–19 (2017)
4. Cao, L.Z.: Analysis and research on identification of poor students in art vocational colleges based on campus all-in-one card consumption big data. Comput. Knowl. Technol. **16**(30), 3–10 (2020)
5. Guo, W., Gao, J., Tian, Y., et al.: SAFS: object tracking algorithm based on self-adaptive feature selection. Sensors **21**(12), 4030 (2021)
6. Radajewski, D., Hunter, L.: An innovative data processing method for studying nanoparticle formation in droplet microfluidics using X-rays scattering. Lab Chip **21**(22), 4498–4506 (2021). https://doi.org/10.1039/D1LC00545F
7. Juneja, P., Garg, R., Kumar, P.: Uncertain data processing of PMU modules using fuzzy petri net. J. Intell. Fuzzy Syst. **41**(3), 1–13 (2021)
8. Kamiński, M., Szabat, K.: Adaptive control structure with neural data processing applied for electrical drive with elastic shaft. Energies **14**(12), 3389 (2021). https://doi.org/10.3390/en14123389
9. Lee, W.J., Staneva, V., Mayorga, E., et al.: Echopype: enhancing the interoperability and scalability of ocean sonar data processing. J. Acoust. Soc. Am. **149**(4), A63–A63 (2021)
10. Yun, S., We, G.: A python package for intennediate data processing in UAV based plant phenotyping. Remote Sens. **13**(13), 2622 (2021)

Research on Personalized Push of Mobile Education Resources Based on Mobile Social Network Big Data

Huibing Cao[✉]

College of Digital Economics, Nanning University, Nanning 530200, China
caohuibing1979@163.com

Abstract. Nowadays, mobile social networks and mobile educational resources have become two mainstream directions in internet applications. How to combine these two to achieve more personalized and intelligent learning resource push has become a relatively important research direction. Therefore, the research on personalized push of mobile education resources based on mobile social network Big data is proposed. Starting from the mobile social network, the user interest model is constructed by using Big data analysis technology, machine learning, recommendation system and other technical means, and it is matched with the mobile education resource database to achieve personalized recommendation of education resources. The experimental results demonstrate that this method has high recommendation accuracy, with a recommendation accuracy of 95%. It can provide intelligent and efficient learning resource push solutions for mobile learning, promoting the development of mobile learning.

Keywords: Mobile Social Network · Big Data · Educational Resources · Personalized Push · Resource Matching

1 Introduction

Mobile social network Big data refers to massive social network data collected by mobile social applications and platforms. These data include users' personal information, social relationships, social behavior, location information, etc. With the rapid development of mobile Internet, people's lives are increasingly inseparable from mobile devices and social network applications, so the scale of mobile social network Big data is growing and has become one of the most important Big data types at present. Mobile social network Big data has a wide range of application scenarios, such as social analysis, social relationship network analysis, user interest analysis, content distribution optimization, accurate advertising and so on [1]. For example, popular topics on social media can be combined with network analysis and emotional analysis to obtain information such as popularity and emotional tendencies, providing decision-making references for governments and enterprises. Mobile social network Big data can also provide personalized information for recommendation systems and basic data for business models. At the same time, the

L. Yun et al. (Eds.): ADHIP 2023, LNICST 547, pp. 452–465, 2024.
https://doi.org/10.1007/978-3-031-50543-0_31

location information collected in mobile social applications can help merchants carry out geographic positioning marketing and precise location push, strengthening interaction and user stickiness. With the rapid development of mobile internet technology, more and more people are starting to use mobile devices for learning.

Mobile educational resources, as an emerging form of education, not only meet people's personalized learning needs, but also bring unprecedented educational and economic benefits. However, due to the diversity of learners, the diversity of resources, and the uncertainty of the teaching environment, how to achieve personalized push of mobile educational resources has always been a challenge for educators and researchers. Mobile education resources refer to a form of educational informatization achieved by the application of mobile internet technology in the field of education. It includes educational resources accessed and used through various mobile devices, such as course videos, online textbooks, instructional games, educational applications, etc. These resources can be learned and used on students' phones, tablets, or other mobile devices. The emergence of mobile educational resources has enabled learners to access educational resources anytime and anywhere, without being constrained by time and location, effectively improving the accessibility and quality of education. It not only facilitates learning and teaching, but also changes the traditional educational model in the past. Students can learn in a more autonomous, free, and diverse manner through diverse educational resources, enhancing their interest in learning and improving their learning outcomes. In addition, mobile educational resources can also provide teachers with more teaching tools and resources, assist in the teaching process, and reduce the burden on teachers. At the same time, mobile education resources can also provide data support for education managers to better manage teaching and make decisions.

In recent years, many scholars have conducted research on personalized push of mobile education resources. Reference [2] He Ying et al. studied personalized education resource recommendation algorithms based on high-dimensional tensor decomposition, which can preserve the information integrity of high-dimensional space in decomposition and avoid the loss of original information and features in traditional recommendation algorithms, thus providing reference for the research of personalized learning resource recommendation. Reference [3], Li Xiangru et al. proposed the design of a personalized learning resource recommendation system for online education platforms. Through similarity analysis and clustering algorithms, users' learning behavior patterns were discovered, and TF-IDF method was used to mine users' resource preferences, and personalized learning suggestions were given based on this. Although the above two methods can achieve personalized recommendation of educational resources, there are problems with low recommendation accuracy and long recommendation time.

In order to solve the problems in the above methods, this study explores a new personalized push method of mobile educational resources from the perspective of mobile social network Big data. First, this research solved the problem of user interest modeling through Big data analysis, making the recommendation algorithm more accurate. Secondly, in the aspect of recommendation system, this research uses multiple technologies such as Collaborative filtering algorithm, content filtering algorithm and hybrid recommendation algorithm to improve the effect of personalized push and provide a new idea, which is of great significance for the effective use and promotion of mobile education resources.

The innovative technology route of this article is as follows:

(1) Data collection and processing: By utilizing API interfaces provided by mobile social network platforms such as WeChat, Weibo, QQ, etc., users' personal information (such as age, gender, geographical location, etc.), social relationships, and data on their use of mobile devices (such as usage duration, frequency, software and hardware information, etc.) are obtained. At the same time, it is necessary to process these data to analyze users' interests, learning habits, etc., and provide support for subsequent personalized recommendations.

(2) Data Mining and Machine Learning: Utilize data mining and machine learning techniques to extract user interests, habits, and other features, and convert them into mathematical models. Meanwhile, through algorithm training, a user profile and recommendation model are constructed to provide users with more accurate and personalized recommendation content.

(3) Personalized interactive experience: Design and optimize personalized interactive experience to enhance the attractiveness of recommended content and user engagement. For example, establishing social networks for users with similar learning interests, encouraging users to share their learning experiences, notes, and educational resources, thereby enhancing user interaction and learning effectiveness.

(4) Multidimensional content recommendation: Based on user profiles and recommendation models, achieve multi-dimensional content recommendation. This includes content recommendations for different topics, content recommendations for different learning stages, and content recommendations for different subject areas, and provides a variety of recommendation forms (such as text, images, videos, audio, etc.).

2   Integration of Educational Resource Information on Multiple Social Media Websites

Mobile social network is a social network platform for social interaction through mobile devices (such as smart phones, tablets, etc.). With the rapid development and popularization of mobile technology, mobile social network has become one of the important ways for people to socialize daily.

The characteristics of mobile social networks include:

1. Mobile convenience: mobile social networks enable people to interact socially anytime and anywhere through mobile devices such as smartphones and tablets. People can easily send messages, share photos and videos, and post updates, whether on public transport, during travel or during breaks.

2. Multiple social functions: mobile social networks provide a variety of social functions to meet users' different social needs. In addition to the basic functions such as instant chat, sharing photos and videos, and publishing status updates, some platforms also provide creative tools such as live broadcast, short video production, filters and stickers, which enhance the interactive experience between users and friends.

3. Personalized experience: mobile social networks allow users to personalize settings according to their preferences and interests. Users can edit their personal data, select themes and backgrounds, and set privacy settings to show their unique style. In addition, some platforms also recommend interested content and users through algorithms to provide a more personalized social experience.
4. Social media integration: mobile social networks often integrate the functions of other social media platforms. Users can directly access and manage social activities on different platforms in mobile applications without switching applications. For example, users can share photos or tweets on instagram on mobile social networks, or directly initiate Messenger chat from Facebook.
5. Location sharing: mobile social networks usually have location sharing function, allowing users to know their friends or places of interest around them. Users can view the information of nearby people, businesses, scenic spots and so on, and share their location with others. In this way, users can better organize offline social activities, or interact with friends in real time in specific places.

(1) Share via link.

Link sharing has many advantages, making it a popular way of social media sharing. First of all, through link sharing, you can save space and put the details in the link instead of publishing the complete content directly on social media. Secondly, links can be used on multiple platforms, including social media, blogs, forums, e-mail, etc., to facilitate the dissemination of information. In addition, through link sharing, you can use analysis tools to track and analyze the number of clicks, conversion rate and other data of links, so as to understand the audience, optimize the sharing strategy and evaluate the sharing effect. Link sharing can also expand the scope of influence of the content. When others click on your link and share it with their friends or followers, your content has the opportunity to be seen and disseminated by more people. Finally, by adding personal opinions, comments, tags, etc., you can customize and personalize the content to attract readers' attention and promote their interaction with you. In general, link sharing is a convenient, flexible and multiple advantage social media sharing method, which can help you spread content on different platforms, expand the scope of influence, and get more interaction and feedback.

With the user's satisfaction, different social media can form a strategic cooperation relationship to realize the information association of the same user. The feasible ways are as follows.

(2) Share through various social media platforms.

Users can use 'sharing' to quickly publish information from one social media to another. However, sharing behavior is limited by the scope of cooperation on social media sites, which has certain limitations and a relatively small impact.

(3) Internal forwarding on social media.

QQ, WeChat and other social media can achieve information forwarding and sharing, and users can forward their favorite content to friends. The social circle on WeChat is more convenient for users to share their favorite information.

This information is highly representative of user interests and serves as an important data source for analyzing hierarchical models of user demand for mobile education resources. Moreover, friends often share the same interests, and finding the interests of the user's friend can infer and improve the user's interest hierarchy model.

3 Personalized Recommendation of Mobile Education Resources

3.1 Data Collection and Processing

In the research of personalized push of mobile education resources, information collection and processing play (u, i) crucial role in recommendation effectiveness. The user data obtained through the API interface provided by mobile social network platforms can be represented as a binary, where a represents the behavior of user u towards item i. Among them, item i includes educational resources (such as courses, materials, etc.). When user u performs actions on item i (such as browsing, bookmarking, commenting, etc.), the corresponding position can be represented as 1, otherwise it is represented as 0. Based on the binary generated by user behavior, a user behavior matri X can be established as follows:

$$
X = \begin{bmatrix}
x_{11} & \cdots & x_{1i} & \cdots & x_{1j} & \cdots & x_{1m} \\
\cdots & \cdots & \cdots & \cdots & \cdots & \cdots & \cdots \\
x_{u1} & \cdots & x_{ui} & \cdots & x_{uj} & \cdots & x_{um} \\
\cdots & \cdots & \cdots & \cdots & \cdots & \cdots & \cdots \\
x_{n1} & \cdots & x_{ni} & \cdots & x_{nj} & \cdots & x_{nm}
\end{bmatrix}
\tag{1}
$$

where x_{ui} represents the behavior of user u for learning resource i.

In order to improve the accuracy of mobile educational resources recommendation, it is necessary to analyze the user's attitude towards educational resources, so a user's attitude matrix is constructed:

$$
Y = \begin{bmatrix}
y_{11} & \cdots & y_{1i} & \cdots & y_{1j} & \cdots & y_{1m} \\
\cdots & \cdots & \cdots & \cdots & \cdots & \cdots & \cdots \\
y_{u1} & \cdots & y_{ui} & \cdots & y_{uj} & \cdots & y_{um} \\
\cdots & \cdots & \cdots & \cdots & \cdots & \cdots & \cdots \\
y_{n1} & \cdots & y_{ni} & \cdots & y_{nj} & \cdots & y_{nm}
\end{bmatrix}
\tag{2}
$$

where y_{ui} represents user u's attitude towards learning resource i.

After obtaining the user behavior matrix X and user attitude matrix Y, preprocessing is needed to obtain more information and provide support for subsequent recommendations. Pre processing includes removing inactive users, removing a portion of items, and so on. Meanwhile, based on the user's behavior matrix towards the item, the relationship matrix R between the user and the item can be obtained:

$$
R_{ij} = X^T Y^T
\tag{3}
$$

Among them, R_{ij} represents the similarity between item i and item j. At the same time, due to implicit factors such as user preferences and item attributes, it is necessary to decompose the behavior matrix to obtain potential user and item features. By using a recommendation algorithm based on matrix decomposition, a user feature matrix U and an item feature matrix V can be obtained, as shown below:

$$U = \begin{bmatrix} u_{1,1} & \cdots & u_{1,p} \\ \cdots & \cdots & \cdots \\ u_{u,1} & \cdots & u_{u,p} \\ \cdots & \cdots & \cdots \\ u_{n,1} & \cdots & u_{n,p} \end{bmatrix}, V = \begin{bmatrix} v_{1,1} & \cdots & v_{1,q} \\ \cdots & \cdots & \cdots \\ v_{i,1} & \cdots & v_{i,q} \\ \cdots & \cdots & \cdots \\ v_{m,1} & \cdots & v_{m,q} \end{bmatrix} \tag{4}$$

Among them, p and q represent the number of potential features of users and items.

After data collection and processing [4], the user behavior matrix, relationship matrix, user feature matrix, and item feature matrix were obtained, providing a solid foundation for subsequent personalized recommendations.

3.2 Building a Hierarchical User Interest Model

In the personalized recommendation system for mobile education resources, data mining and machine learning techniques can be used to extract meaningful features from users' personal information, social relationships, and data using mobile devices. Based on these features, mathematical models can be constructed, and then user profiles and recommendation models can be trained through machine learning algorithms to further provide personalized recommendation content for users.

Assuming that the user's behavior matrix X has been obtained, the feature matrix U of the user and the feature matrix V of the item can be obtained through the matrix decomposition algorithm. On this basis, user features can be extracted from the following aspects:

(1) Basic information features: This includes basic information such as age, gender, and geographic location, which can usually be obtained from the user's personal information. Therefore, a basic information feature vector u_{basic} can be defined to represent the user's basic information:

$$u_{basic} = \begin{bmatrix} age \\ gender \\ location \\ \cdots \end{bmatrix} \tag{5}$$

(2) Social relationship characteristics: By utilizing users' social relationships (such as friends, followers, etc.), users' i interests, hobbies R, and preferences can be obtained. Assuming the relationship matrix between user and item ac is R, the social relationship feature vector g_i of item i can be defined as;

$$g_i = \sum_{u_j \in F_i} \frac{R_{ij}}{\sqrt{k_{u_j}}} \cdot u_j \tag{6}$$

Among them, F_i represents all users associated with the item, and k_{u_j} represents the number of items that user u_j likes.

(3) Mobile device data characteristics: The usage of mobile devices can reflect users' learning habits and behaviors, such as their device model, usage duration, and application usage. The device data feature vector d_u can be represented as:

$$d_u = \begin{bmatrix} model \\ time \\ apps \\ \dots \end{bmatrix} \qquad (7)$$

Among them, *model* represents the device model, *time* represents the total time the user has used the mobile device, and *apps* represents the application that the user frequently uses.

By decomposing the user behavior matrix X and utilizing the above feature vectors, a user profile and recommendation model can be constructed. By utilizing machine learning algorithms such as clustering, classification, regression, etc., user profiles can be analyzed and learned, helping the system better understand user behavior and preferences, and providing more accurate and personalized educational resource recommendations[5, 6].

3.3   Optimize Personalized Interactive Experience

Suppose there is a set of users, where each user has $I(u_i) = [2, 3, 1, 0][2, 3, 1, 0]$ learning interest, which can be represented as a vector. For example, represents the degree of interest of user u_i in a course with four features, each with k. Referring to the concept of social networks, a learning social network $G = (V, E)$ can be defined, where node v_i represents user u_i, and edge $e_{i,j} \in E$ represents that users u_i and u_j have similar learning interests, which can be represented by cosine similarity:

$$s_{i,j} = \frac{I(u_i) \cdot I(u_j)}{\|I(u_i)\| \|I(u_j)\|} \qquad (8)$$

When a is greater than $s_{i,j}$ certain threshold, connect users u_i and u_j to form an edge $e_{i,j}$.

When it is greater than a certain threshold, connect the user and into an edge.

Furthermore, u recommendation algorithm can be designed to enhance user interaction and learning effectiveness. Taking a user a as an example, first find its adjacent node set $N(u) = \{v_1, v_2, \dots, v_n\}$, and then calculate the average value vector $\bar{I}(N(u)) = \frac{1}{n} \sum_{i=1}^{n} I(u_i)$ of its adjacent users' learning interests. Then, based on cosine similarity, calculate the number of users in their adjacent user set that are most similar to their learning interests, which can be represented as $N_k(u) = \{v_{i_1}, v_{i_2}, \dots, v_{i_k}\}$. Finally, recommend the resources learned by these k users to user u. The recommended algorithm can be represented by the following formula:

$$R(u) = \bigcup_{v_i \in N_k(u)} I_{v_i} \qquad (9)$$

where I_{v_i} represents the set of resources that user v_i has learned. In this way, personalized interaction experiences can be designed and optimized to enhance the attractiveness of recommended content and user engagement.

3.4 Implementing Personalized Recommendations

In the personalized recommendation system for mobile education resources [7], multi-dimensional content recommendation for different themes, learning stages, and subject areas is essential. Based on user profiles and recommendation models, multi-dimensional content recommendations can be achieved based on the following aspects:

(1) Topic recommendation: Based on user interests and search records, educational resources related to topics of interest to users can be recommended [8–10]. Specifically, the vector space model (VSM) can be used to calculate the similarity between the user interest vector and the subject vectors of various educational resources, so as to recommend the most relevant educational resources for users. The relationship matrix between user i and topic j is defined as Q, and the interest of user i in topic j is expressed as q_{ij}. The degree of preference of user i for topic j can be expressed as:

$$s_{i,j} = \sum_{k=1}^{n} q_{i,k} tanh(v_{k,j}) \tag{10}$$

where, $v_{k,j}$ represents the vector representation of the k-th topic. By calculating the similarity between each topic and the user interest vector, the most relevant topics can be obtained and corresponding educational resources can be recommended for users.

(2) Learning stage recommendation: Based on the user's learning stage and the difficulty level of educational resources, educational resources suitable for the user's current learning stage can be recommended. By utilizing information such as the category and difficulty of educational resources, embedded expressions can be made on the resources, and based on this, i recommendation model for users and educational resources can be constructed. If the relationship matrix between user a and educational resource y is P, and the user's preference for educational resources is expressed as $p_{ij}p_{ij}$, then the degree of user i's preference for educational resource y can be expressed as:

$$s_{i,y} = \sum_{k=1}^{m} p_{i,k} tanh(w_{k,y}) \tag{11}$$

Among them, $w_{k,y}$ represents the vector representation of the k-th difficulty level. By calculating the similarity between each difficulty level and the user's preference vector, the most suitable educational resources for the user's current learning stage can be obtained.

(3) Subject domain recommendation: Based on the subject similarity between the user's subject domain and educational resources, relevant educational resources can be recommended to the user. If the relationship matrix between user i and educational resource y is S, user i, preference for subject K is expressed as s_{ik}, and the correlation between educational resource j and subject K is expressed as s_{yk}, then user i, s preference for subject K can be expressed as:

$$p_{i,k} = \sum_{j=1}^{m} s_{ik} s_{yk} \tag{12}$$

Among them, K represents the number of the subject area. By calculating the correlation between each subject area and user preference vectors, corresponding educational resources can be recommended for users.

Through the above recommendation model, diverse content formats such as text, images, videos, audio, etc. can be recommended to meet the different needs of users.

The ultimate goal of establishing a multi-level user interest model is to achieve personalized and precise push of information services, provide users with a better user experience, and also improve the promotion effect of educational resources and students' learning efficiency, achieving a win-win situation for both parties. The specific flowchart of personalized recommendation of mobile education resources is shown in Fig. 1:

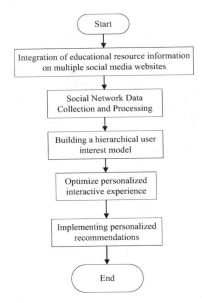

Fig. 1. Flow Chart of Personalized Recommendation of Mobile Education Resources

4 Experiment

4.1 Experimental Environment

By introducing the hardware environment, software environment, and dataset, it is possible to have a more comprehensive understanding of the environment and data sources of this experiment, and to better understand the feasibility and reliability of this experiment.

In terms of hardware environment, using the services of Alibaba Cloud computing platform can ensure the availability and stability of computing resources, provide a good operating environment for data processing and model training, and also avoid a

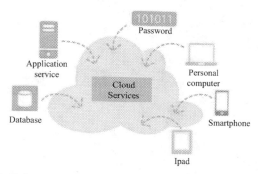

Fig. 2. Schematic diagram of Alibaba Cloud computing platform

large amount of equipment procurement and maintenance costs. The Alibaba Cloud computing platform is shown in Fig. 2.

In terms of software environment, many open source tools such as Python are used, more advanced data analysis and machine learning technologies are used, and a series of functions such as data analysis, data visualization, model training are provided with various Python libraries, providing basic tools and operational support for the experiment. In terms of dataset, a mobile learning dataset from a certain mobile social network platform is adopted, which includes various data such as user learning records, browsing records, and social relationships. By analyzing and processing these data through data mining technology, a user interest model can be constructed and matched with educational resources to achieve personalized recommendation of educational resources. At the same time, desensitizing the data, protecting user privacy, ensuring the legality and morality of the experimental process, provides an effective environment and means for the research of personalized recommendation of educational resources. The operation process of Python is shown in Fig. 3.

4.2 Experimental Process

This experiment is mainly divided into four steps: data processing, interest modeling, resource recommendation, and experimental verification.

(1) In the data processing stage, the collected data is cleaned and preprocessed, including data deduplication, data format conversion, etc.
(2) In the interest modeling phase, the user's learning history and social relations are modeled, the user interest tags and knowledge point tags are obtained through the random walk algorithm, and machine learning technology is used for modeling and training.
(3) In the stage of resource recommendation, collaborative filtering and deep learning methods are used to screen the resources most suitable for users' interests from the educational resource database.
(4) Compare the method proposed in this article with the sampling and statistical analysis of the dataset in references [2] and [3], and evaluate and validate the effectiveness of the proposed recommendation algorithm.

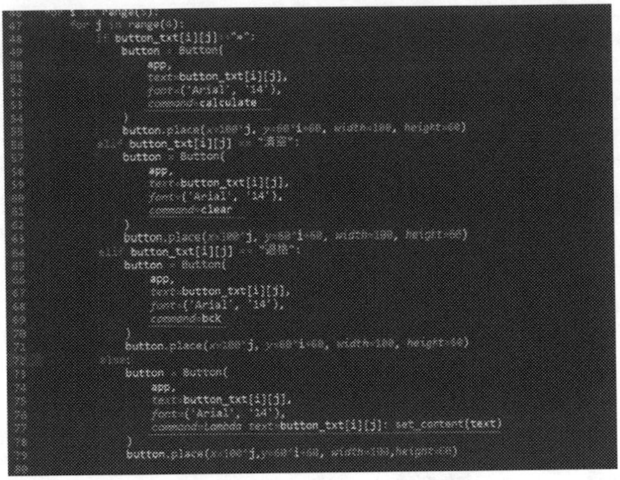

Fig. 3. Python Operation Process

4.3   Experimental Results

In order to verify the reliability of the method proposed in this paper, the dataset was trained using the methods proposed in this paper, as well as those in references [2] and [3]. The accuracy, recall, and F-value were used as indicators to verify the reliability of the method proposed in this paper in personalized recommendation. The specific experimental results are shown in Table 1.

Table 1. Reliability Verification of Three Methods

method	accuracy	recall	Fprice
proposed method	95	99	0.90
Reference [2] Method	70	92	0.85
Reference [3] Method	85	85	0.82

Analysis of Table 1 shows that the method presented in this paper exhibits high scores in accuracy, recall, and F-value. This proves that the method proposed in this article has significant help in improving recommendation accuracy. Specifically, accuracy: This method can accurately recommend the educational resources that students truly need when recommending, and the recommendation results have high accuracy, which is beneficial for improving user experience and satisfaction. Recall rate: This method can accurately discover and recommend educational resources that meet user needs and interests among numerous educational resources, covering user needs more comprehensively, which is conducive to improving the utilization rate and promotion effect of mobile educational resources. F-value: Taking into account both accuracy and recall, it is an important indicator for evaluating the comprehensive performance of personalized recommendation systems. The F-value of this method is higher than the other two methods, representing better recommendation performance and better meeting user needs. In summary, the method presented in this article performs well in terms of recommendation accuracy and can better provide users with suitable educational resources.

In order to verify the personalized recommendation efficiency of the method proposed in this study, the dataset was trained using the method proposed in this paper, as well as the methods in references [2] and [3], respectively. The efficiency of the method proposed in this study was verified using personalized recommendation time as an indicator. The specific experimental results are shown in Fig. 4.

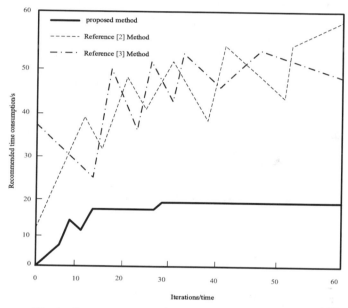

Fig. 4. Comparison of time consumption of three methods

As can be seen from Fig. 4, the method proposed in this paper exhibits a relatively short recommendation time, and as the number of iterations increases, its recommendation time does not fluctuate significantly. This indicates that the method proposed in this article can quickly personalized recommend educational resources and maintain a relatively stable recommendation time. In practical applications, recommendation time is a very important indicator, especially for mobile users. If the recommendation process is too time-consuming, it can easily affect users' purchasing decisions and even reduce their experience and satisfaction. Therefore, the shorter recommendation time of this method can better meet the needs of mobile users. Compared with the other two methods, our method can complete large-scale data processing and computation in a short time, which is related to the cache based optimization strategy proposed in our method. This strategy uses caching technology to preprocess data, improving the efficiency of data queries and reducing the time and resource costs required for computation, thereby improving recommendation efficiency. In addition, the recommendation time fluctuation of the method in this article is relatively small, which can avoid long waits and circles, and increase user satisfaction on the mobile end. In summary, the method proposed in this article performs well in terms of recommendation time for personalized recommendation of educational resources and can meet the needs of mobile users.

5 Conclusion

Personalized push can improve learners' learning effect and satisfaction. By customizing educational resource push according to learners' interests, learning styles and learning needs, it can better meet learners' personalized needs, provide more targeted and effective learning content, and enhance learners' learning motivation and enthusiasm. Data analysis and machine learning technology play a key role in personalized push. Educational institutions and platforms should continuously collect and analyze learners' feedback data to understand the effect of personalized push and the room for improvement. Based on the feedback results, the push strategy can be adjusted and optimized to adapt to the changing needs of learners and provide a better learning experience.

To sum up, personalized push is of great significance in mobile education resources. Through the rational application of data analysis and machine learning technology, the protection of personal privacy, and continuous improvement and feedback, we can achieve more intelligent, accurate and effective personalized push of mobile education resources, and promote the learning effect and satisfaction of learners.

Aknowledgement. The 2021 Social Science Foundation Project of Nanning University, titled "Research on the Factors Influencing the Leadership of the Director of the Teaching and Research Office of Nanning University," with fund number 2021JSGC10.

References

1. Li, Z., Zhou, Y., Liu, Y., et al.: Research on personalized recommendation technology for mobile applications. Cyberspace Secur. 2021, 12 (Z2): 80–84Li, X., et al. Design of personalized learning resource recommendation system in online education platforms. Comput. Technol. Dev. 2021, 31 (02): 143–149
2. Zhipeng, Y., Xuexue, L.: Design of a personalized recommendation system for ideological and political theory resources based on data mining. Autom. Technol. Appl. **42**(01), 93–96 (2023)
3. Shaohua, H., Siqing, Y., Zhiyu, J., et al.: Educational resource recommendation method based on learner model. Comput. Digital Eng. **50**(04), 697–702 (2022)
4. Hui, W., Dehong, S.: Personalized learning resource recommendation algorithm based on improved graph neural network and user preference clustering. J. Heilongjiang Univ. Eng. **36**(06), 30–34 (2022)
5. Xiufeng, Z.: Research on personalized resource intelligent recommendation system based on user profile. J. Libr. Inf. Sci.Libr. Inf. Sci. **3**(12), 17–21 (2018)
6. Xia, Y.: Research on personalized recommendation model of online course resources based on knowledge graph. Electron. Technol. Softw. Eng. **11**, 30–31 (2021)

7. Qin, H.: SQL based automatic recommendation model for educational resource database indexing [J]. Autom. Technol. Appl. 2022, 41 (10): 117–120+136
8. Mingyan, Z.: Research on contextual perceived personalized recommendation model for mobile catering. Inf. Comput. (Theor. Ed.) **34**(24), 95–98 (2022)
9. Junjuan, L.: Personalized recommendation of ideological and political education resources based on K-means clustering. Inf. Comput. (Theor. Ed.) **35**(01), 242–244 (2023)
10. Xinmei, W.: Research on personalized recommendation services for special education resources based on multidimensional and multi-source user characteristics. Educ. Communi. Technol. **06**, 85–91 (2022)

Evaluation Method of Online Education Effect in Colleges and Universities Based on Data Mining

Huibing Cao(✉)

College of Digital Economics, Nanning University, Nanning 530200, China
caohuibing1979@163.com

Abstract. In order to improve the accuracy of online education evaluation results and lay a solid foundation for improving the level and quality of online education, data mining technology is introduced. Based on data mining technology, this paper studies the effectiveness evaluation method of online education in colleges and universities, and puts forward the effectiveness evaluation model of online education in colleges and universities. The decision problem is decomposed into hierarchical structure, the data mining technology is used to analyze the relevant data, and the evaluation table is constructed. The effectiveness of online education is evaluated comprehensively and multi-dimensionally based on association rules. The experimental results show that the method has high feasibility and accuracy, the confidence level is more than 96%, and the response time is less than 10ms, which is superior to the traditional method, proving that the method has significant advantages and potential in practical application.

Keywords: Data mining · Universities · On-line · Education · Effect · Evaluation

1 Introduction

Improving the quality of teaching in colleges and universities is currently the biggest challenge faced by schools due to the wave of higher education popularization. To address this challenge, it's essential for teachers to listen to students' feedback on their teaching methods, course content, and effectiveness. By using student evaluations as a tool to adjust teaching strategies, we can improve the overall quality of education. Ultimately, involving students in monitoring and evaluating the teaching process will ensure that they become active participants in their educational journey.Online education in colleges and universities is a crucial higher education category, providing education to front-line professionals and promoting China's move from elite to mass education. However, the expansion of national college enrollment and a rise in high enrollment rates has led to a need for quality monitoring, particularly for vocational colleges. As education shifts from elite to mass, schools face challenges like resource shortages, declining student quality, and insufficient monitoring of the teaching process. Thus, an effective online education evaluation method is essential [1].

© ICST Institute for Computer Sciences, Social Informatics and Telecommunications Engineering 2024
Published by Springer Nature Switzerland AG 2024. All Rights Reserved
L. Yun et al. (Eds.): ADHIP 2023, LNICST 547, pp. 466–480, 2024.
https://doi.org/10.1007/978-3-031-50543-0_32

Due to various reasons, the effect of teaching monitoring and evaluation in colleges and universities is not ideal. Although some colleges and universities have carried out online evaluation, the large amount of data generated has not been effectively used to improve the teaching quality. For example, the design of a teaching effect evaluation system for accounting major based on big data technology was proposed in literature [2]. It uses MySQL database as the management platform to design four modules: basic information of teaching evaluation, setting of teaching evaluation scheme, online teaching evaluation and statistical analysis of evaluation results. Then it adopts fuzzy comprehensive evaluation and key technologies of data mining, and uses Java language and Android system to make the platform run more efficiently. However, there are still some problems in data processing and it is difficult to apply. Another example is the design of college English teaching effect evaluation method based on intelligent algorithm proposed in literature [3]. This method first collects English teaching evaluation data, intelligently selects evaluation indicators within the demarcated range, establishes a multi-functional intelligent algorithm teaching effect evaluation model, and adopts hierarchical goal evaluation method to realize the design, which also has the problem of difficult data processing.

Data mining is the core step of "knowledge discovery", which is a process of discovering hidden rules or valuable information and knowledge by collecting a large number of random original data samples and using a series of mining tools. Association analysis is to discover association rules between data samples by mining frequent item sets in data sets. More appropriate rules can usually be measured and determined through support, feasibility, relevance, etc. [4, 5]. At present, data mining is widely used in many disciplines and fields, but its workflow tends to be consistent, that is, the process of repeatedly predicting, analyzing, modeling and verifying from massive data through various methods to find interesting, valuable and applicable knowledge or patterns[6]. On this basis, this paper introduces the data mining technology, and makes a deep research on the evaluation method of online education in colleges and universities based on data mining technology.

2   Design of Evaluation Method for Online Education Effect in Colleges and Universities

2.1   Build an Online Education Effect Evaluation System Model

According to the actual education and teaching situation in colleges and universities, the online education effect evaluation system model is constructed to lay the foundation for the subsequent education effect evaluation.

First of all, it is necessary to clarify the correct orientation of the construction of the evaluation system model of online education effect in colleges and universities, including two orientations, namely, diversification orientation and differentiation orientation.

Diversified orientation. The personnel involved in the evaluation of teachers' teaching should be diversified so as to make a more comprehensive, objective and scientific evaluation of teachers' teaching activities from multiple aspects and perspectives [7].Diversified teaching evaluation should be composed of the following parts: first, expert evaluation; Second, peer evaluation; Third, the evaluation of students; Fourth, the evaluation of teaching management personnel; Fifth, teachers' self-evaluation.

Differentiation oriented. The so-called micro differentiation means that the evaluation content of teaching should have a high degree of clarity, which should be easy to master standards and collect information. The first level indicators of the teaching evaluation content of university teachers generally include: teaching attitude, teaching design, teaching content, teaching methods, teaching effects, teaching characteristics and other major aspects, while the contents of the second level indicators and the third level indicators must be differentiated and refined according to the actual situation of the university and the differences in the course.

Through the research of practical problems, the decision-making problems are decomposed into different hierarchies according to the overall evaluation objectives, and a hierarchical structure is constructed to represent the characteristics and functions of the system. Secondly, using the same method as the judgment solution matrix vector, the judgment matrix of each element at each level to the related elements at the previous level is constructed to determine the weight of each element. For the decision problems whose target value can only be qualitatively described and the target system with hierarchical and staggered evaluation indicators, it is more appropriate to use the analytic hierarchy process [8].The process steps of constructing the model of online education effect evaluation system in colleges and universities based on analytic hierarchy process are as follows.

1. Establishment of evaluation system (index system).

To realize the establishment of the evaluation system, it is necessary to carry out a systematic analysis, including the scope of the system, the factors involved and the relationship between these factors. Through this analysis, the system objectives, evaluation criteria and index system are established, and the corresponding hierarchical structure is established. The hierarchy structure required by AHP usually consists of three levels: the highest level is the target set of the problem -- the target level; The middle level is the standard level which affects the realization of the goal. The lowest level is the measure that promotes the realization of the goal - the measure level. As shown in Fig. 1, this is a hierarchical diagram.

The application of analytic hierarchy process has different purposes, and the hierarchy can be selected and adjusted accordingly. For example, when selecting the evaluation index system to determine the ranking of index weights, the first two or three hierarchies can be used, and the alternative scheme layer should also be added when selecting the decision scheme.

2. Construct judgment matrix.

This paper constructs the judgment matrix according to the hierarchical structure. The specific method to construct the judgment matrix is to take the criterion, that is, each element with downward subordinate relationship as the first element of the judgment matrix, place it in the upper left corner, and then arrange the elements belonging to it in the first column and the first row [9] in order.

Usually, the method used to fill in the judgment matrix is to repeatedly ask the experts who fill in the judgment matrix; According to the criteria of the judgment matrix, compare the elements in pairs, select the important elements, evaluate their

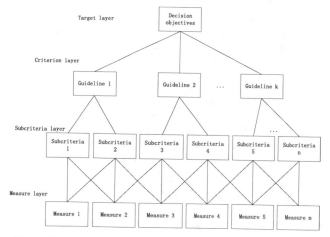

Fig. 1. Hierarchy Diagram of AHP Decision Analysis Method

importance, and use the importance scale value table to assign the importance by 1–9. The importance scale values are shown in Table 1.

Table 1. Importance Scale Values

Factor ratio factor	Quantized value
Equally important	1
Slightly important	3
Strongly important	5
Strongly important	7
Extremely important	9
Intermediate value of two adjacent judgments	2,4,6,8

3. Computational weight vector

To calculate the weight vector, appropriate mathematical methods should be used. The essence of single-layer hierarchical ranking is to compute the weight vector that indicates the relative weights of each factor in its standard under each judgment matrix. The computation of the weight vector includes methods such as sum method, power method, and root method. The sum method principle is to normalize each column in a consistent judgment matrix to obtain the required weights. In an inconsistent judgment matrix, the results obtained by normalizing each column only approximate the corresponding weights. The arithmetic mean of the column vectors is taken as the final weight:

$$W_i = \frac{1}{n} \sum_{i=1}^{n} \frac{a_{ij}}{\sum_{k=1}^{n} a_{k1}} \tag{1}$$

Among them, a_{ij} represents the elements in the judgment matrix under the criteria; a_{k1} represents the lower level element of the judgment matrix. In some special cases, the judgment matrix can be consistent and transitive. In general, this property does not require the judgment matrix to be strictly satisfied. However, a correct judgment matrix importance ranking requires certain logical rules from the perspective of human cognition rules. For example, if A is more important than B and B is more important than C, A comparison can be made between A and C. [10] Therefore, in practice, the judgment matrix generally meets the requirements of consistency, and consistency inspection is required. Only if the judgment matrix passes the test can it be explained that it is logically reasonable, and then continue to analyze the results.

4. Verify consistency. First, CI (consistency index) is calculated according to the one-time index, and the calculation formula is:

$$CI = \frac{\lambda_{max} - n}{n - 1} \tag{2}$$

Among them, λ_{max} represents the maximum eigenvalue of the judgment matrix. According to the judgment, different orders in the matrix can be obtained. Again, the consistency ratio of the evaluation index CR (consistency ratio) is calculated and judged. The formula is:

$$CR = \frac{CI}{RI} \tag{3}$$

where, RI is the average random consistency index corresponding to the evaluation index. When $CR < 0.1$, the judgment matrix consistency meets the requirements; When $CR > 0.1$, the judgment matrix is unacceptable and does not meet the consistency requirements, the judgment matrix should be modified.

2.2 Data Mining for Online Education Effect Evaluation

After building the online education effect evaluation system model based on the above, next, use data mining technology to comprehensively mine and analyze the relevant data of online education effect evaluation of colleges and universities. The data mining system is generally composed of various databases, mining pre-processing modules, mining operation modules, pattern evaluation modules, and knowledge output modules. The achievements of these modules constitute the architecture of data mining, as shown in Fig. 2.

As shown in Fig. 2, the data mining architecture of online education effect evaluation for colleges and universities designed in this paper. The data mining process is as follows:

1. Define the problem.

The primary factor to realize data mining is to clarify the problems that users need to solve. However, we do not know what kind of pattern structure the knowledge data to be mined is, but we can know the specific objects that users want to mine. Because we do not know what is meaningless to mine, we must understand users' mining needs, combine specific needs, and study the feasibility of data implementation to finally determine users' business problems.

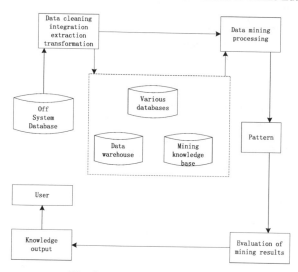

Fig. 2. Data Mining Architecture

2. Prepare data.

Data exists in different ways, and can be distributed in the same database or under different departments of the enterprise. Because different storage media often have inconsistent data formats, and a large number of erroneous data will accompany. Therefore, it is necessary to properly delete or complete data at a stage, then smooth the noisy data, and then convert the data format into a unified format to facilitate data processing; Finally, all the data are put together to find useful potential data in the integrated data, and then generate useful test data sets. In the actual teaching evaluation process, it is inevitable that due to the influence of personal subjective emotions, the evaluation data is not credible. Therefore, it is necessary to screen out the teaching evaluation data that deviates excessively from the actual situation for data cleaning to ensure the objectivity of the subsequent model construction. In data cleaning, regression uses a function to fit data to smooth data and identify noise; However, the density based outlier detection method has a good effect on anomaly data detection for data sets with uneven distribution.

3. Browse data.

The purpose of browsing data is to test and evaluate the dataset generated in the previous step to determine whether the dataset meets the user's requirements. However, data correction cannot be successful once. It requires repeated experimental correction until the tested data set can map out the real data.

4. Generate model.

Select a mathematical model that can match the data mining algorithm. The adaptability of the model and algorithm is an important factor for success. If the adaptability is high, then the model is successful. The design model of this paper is the entity relationship model of education effect evaluation, as shown in Fig. 3.

As shown in Fig. 3, it is an individual for teachers and a whole for schools. When building the entity relationship model of teaching evaluation, we can only build it for a single teacher, thus improving the matching degree of the model.

5. Browse and verify the model.

Before bringing the above model to the user environment, relevant tests should be carried out on the model, and the evaluation should be carried out according to relevant data indicators. If the model can be used, it can be directly used in the user environment; If it is not good, it needs to be modified repeatedly until the model produces the best data results.

6. Deploy and update models.

After the model is established, the new data will be described under the new model, and the data model that is least difficult to understand will be used to realize visual expression for users. At the same time, the model will be adjusted and changed immediately according to the specific effects or needs in the actual process.

After the completion of data mining for online education effect evaluation of colleges and universities, the evaluation table for online education effect of colleges and universities is constructed based on the above weight calculation results, and the corresponding weight distribution is divided, as shown in Table 2.

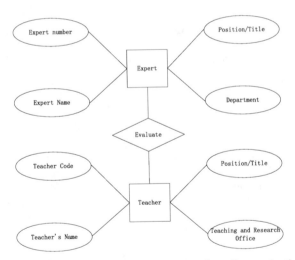

Fig. 3. Entity relationship model of education effect evaluation

For the above five first level indicators, the corresponding second level indicators have the same weight. In data mining, data preparation includes all activities of constructing the final dataset from the original data. These data will be input values to the model. Because the data stored in the computer system database has problems such as noise, missing and redundancy, it is not suitable for direct data analysis and mining. Therefore, it is necessary to further preprocess the original data. In the data preprocessing stage, data cleaning, integration, reduction, transformation and other steps need to be carried out. The final data after preprocessing should be accurate, complete and consistent. The

Table 2. Evaluation of online education effect in colleges and universities

Primary indicators	Secondary indicators	Weight distribution
Teaching quality U1	U11	0.05
	U12	0.03
	U13 (well prepared, skilled in teaching, patient in answering questions)	0.04
	U14 (Reasonable teaching design, clear organization and sufficient teaching preparation)	0.07
	U15	0.01
Teaching Methods U2	U21 (Rational use of modern information technology and digital teaching resources such as the Internet and artificial intelligence)	0.09
	U22 (pay attention to the cultivation of students' ability and encourage innovation)	0.04
	U23 (pay attention to individual differences of students and give guidance to students' learning methods)	0.08
	U24 (active classroom atmosphere, good interaction between teachers and students)	0.06
	U25 (pay attention to the cultivation of students' practical ability and innovative consciousness)	0.017
Teaching content U3	U31 (clear teaching purpose and reasonable time arrangement)	0.15
	U32 (earnestly practice the teaching plan to make it meet the requirements of the syllabus)	0.03
	U33 (brief explanation, prominent key and difficult points, and standard demonstration operation)	0.08
	U34 (organic integration of cultural education, curriculum ideological and political, professional ideological and political, professional and creative integration and other elements)	0.07
	U35 (combine and extend classroom knowledge and extracurricular knowledge)	0.09
Teaching effect U4	U41 (student participation in class)	0.23
	U42 (improvement of students' practical ability)	0.16
	U43 (Student Quality Development)	0.19
Teaching Features U5	U51 (innovative teaching mode, both scientific and artistic)	0.11

values of the five first level indicators are equal to the sum of the scores of each second level indicator. After conversion, we converted the scores of the secondary indicators into the relevant data of the five primary indicators, and sorted out the collected data as the original data samples to cluster the five primary indicators. After data collection, preprocessing and transformation, data samples for data mining can be obtained.

In the clustering analysis of data mining, we use K-Means algorithm, which usually selects K data from n data as the initial cluster center. Then, the remaining objects are allocated to the corresponding clusters according to the similarity with the selected initial cluster center. Finally, the corresponding average value of each cluster center is obtained through calculation, and the task is completed by repeating this process to obtain the final convergent standard measure function. In the actual operation process, we divide the sample data into three clusters, and randomly select three data as the center point of the initial cluster analysis. Then the number of three clusters represents the clustering results of "excellent", "qualified" and "unqualified" teaching evaluation results respectively.

2.3 Evaluating the Effect of Online Education in Colleges and Universities

After the above data mining of online education effect evaluation is completed, the initial clustering evaluation results of online education effect are obtained. On this basis, the effect of online education in colleges and universities is evaluated comprehensively and multi-dimensionally based on association rules.

The data of the five primary indicators in Table 1 are converted through data discretization. In this paper, we convert the five first level indicators in the above table into four levels: "excellent", "good", "medium" and "poor". See Table 3 for the specific conversion criteria.

Table 3. Discrete conversion standard of primary indicators

Primary indicators	excellent	good	in	difference
Teaching quality	Score ≥ 22	19 ≤ score < 22	16 ≤ score < 19	Score < 16
teaching method	Score ≥ 22	19 ≤ score < 22	16 ≤ score < 19	Score < 16
content of courses	Score ≥ 26.5	23 ≤ score < 26.5	19.5 ≤ score < 23	Score < 19.5
teaching effectiveness	Score ≥ 13	11 ≤ score < 13	9 ≤ score < 11	Score < 9
Teaching characteristics	Score ≥ 4.5	4 ≤ score < 4.5	3.5 ≤ score < 4	Score < 3.5

According to the data discretization conversion standard in Table 3, the primary evaluation indicators are converted to obtain the evaluation table of online education effect of colleges and universities shown in Table 4.

Table 4. Evaluation of online education effect in colleges and universities

Teaching quality	teaching method	content of courses	teaching effectiveness	Teaching characteristics	Evaluation grade
excellent	excellent	excellent	excellent	good	excellent
excellent	excellent	good	good	good	excellent
excellent	excellent	excellent	excellent	excellent	excellent
good	in	In	in	good	qualified
good	good	In	good	good	qualified
excellent	good	good	good	good	qualified
good	in	difference	difference	difference	unqualified
in	in	difference	difference	difference	unqualified
…	…	…	…	…	…

Through the online education effect evaluation table in Table 4, we can obtain the education effect evaluation grade, and then achieve the goal of online education effect evaluation of colleges and universities based on data mining.

3   Experimental Analysis

3.1   Experiment Preparation

The above content is the whole design process of the online education effect evaluation method for colleges and universities based on data mining technology proposed in this paper. Before putting forward the method for online education effect evaluation in colleges and universities, the feasibility and effectiveness of the method need to be objectively tested to avoid abnormal problems in direct use, which is not conducive to improving the level and quality of online education in colleges and universities.

In order to verify the effectiveness of relevant algorithms, the course Fundamentals of Computer Application for the automobile maintenance specialty of X University is selected as the experimental object, and the first 497 (38250 in total) data extracted from the data set are used as the data source for example analysis. Each record contains 35 attributes, The data mining algorithm in this paper is analyzed by observing the change of the algorithm execution time and the time required under different minimum support states from the fixed minimum support. The software and hardware environment of the experiment is shown in Table 5.

According to the software and hardware environment configuration shown in Table 5, the online education effect evaluation test environment of colleges and universities was built. The online education effect evaluation index system used in this experiment consists of multiple sub indicators. In view of the needs of X university teachers and students for online education, as well as the various factors that affect the education effect, the rating indicators are mainly divided into two parts: teacher quality indicators and teaching

Table 5. Software and Hardware Environment Configuration for Experimental Test

Environmental Science	project	to configure
Hardware	CPU	Intel (R) Core (TM), 17
	Memory	4G
	Hard disk	500G
Software	CPU	WindowsServer 2019
	Memory	SQL SERVER 2019
	Hard disk	/

quality indicators. Among them, teacher quality indicators include teacher's education background, professional title, teaching achievements, etc. The teaching quality indicators are divided into four items: classroom driving, classroom participation, curriculum difficulty and comprehensive teaching evaluation. In the whole data evaluation indicator system, the data information of teachers' quality indicators is reflected, as shown in Table 6.

Table 6. Data Information of Teacher Quality Indicators

Field Name	Field Description	Field Type	length
JS-GH	Teacher ID	INT	8
JS-MM	password	VARCHAR	64
JS-XM	full name	VARCHAR	20
JS-ZGXL	Highest education	VARCHAR	20
JS-ZC	title	VARCHAR	20
JS-XB	Gender	VARCHAR	20
JS-GL	working years	DATETIME	20
JS-NL	Age	INT	4

The above table shows the composition of teachers' basic information, from which the irrelevant items, such as gender, age, reserved job number, educational background and other attribute items that are closely related to teaching work, are excluded. Among them, each attribute item corresponds to several different attribute sectors. For example, the education level includes four levels: junior college, bachelor, master and doctor. The rating information of teaching quality comes from the students' evaluation and rating of each teacher on the school system after each semester. It is also the main part of the data set in this teaching evaluation system. Different attributes of teaching evaluation are also divided into different rating levels. For example, classroom driving includes four levels: excellent, good, medium and poor. There are also Boolean variables, such as whether comprehensive evaluation has two attribute levels: yes and no. In the scoring system, different attribute levels correspond to different score ranges. The teaching evaluation information table is shown in Table 7.

Table 7. Evaluation Information of Online Education Effect

Field Name	Field Description	Field Type	length
PJ-ID	Course No	INT	8
PJ-JSGH	Teacher ID	VARCHAR	8
PJ-KTDDX	Driving power of online classroom	VARCHAR	20
PJ-KTHDX	Online classroom interaction	VARCHAR	20
PJ-KCJSD	Online classroom acceptance	VARCHAR	20
PJ-YYBDNL	Online classroom expression ability	VARCHAR	20
PJ-XSCYD	Student participation in online classes	VARCHAR	20
PJ-KHFS	Online classroom assessment method	VARCHAR	20
PJ-ZHPJ	Comprehensive evaluation of online classroom	VARCHAR	20

The teacher education evaluation information is obtained through the online education effect evaluation information of colleges and universities in Table 7. On this basis, According to the online education effect evaluation method designed in this paper, the online education effect of X University is objectively evaluated. The flow of this experimental analysis is shown in Fig. 4.

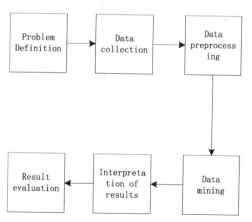

Fig. 4. Schematic Diagram of Experimental Analysis Process

According to the experimental analysis process shown in Fig. 4, the online education effect evaluation test of colleges and universities was carried out to test the feasibility and application effect of the proposed evaluation method.

3.2 Result Analysis

In order to make the experimental analysis results more intuitive and clear, the method and principle of comparative analysis are introduced. Set the above online education effect evaluation method based on data mining proposed in this paper as the experimental group, and set the traditional online education effect evaluation method as the control group, and compare and analyze the evaluation effects of the two methods respectively. The credibility of the evaluation results of online education in X colleges and universities is selected as the evaluation index of this experimental analysis. The higher the confidence of the evaluation results, the higher the feasibility of the online education effect evaluation method, the higher the accuracy of the evaluation results, and vice versa.

Randomly select some items in the online education effect evaluation information table, and map the centroid to the score table of the percentile system. Starting from five dimensions of online education quality, online education content, online education method, online education effect, online education management, the above two methods are used to evaluate the online education effect of X university from five different dimensions. SPSS statistical analysis software was used to measure the confidence of the results of the two methods and compare them. The results are shown in Fig. 5.

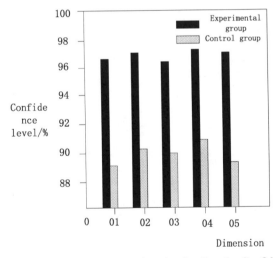

Fig. 5. Comparison Diagram of Evaluation Results Confidence

It can be seen from the comparison results of evaluation indicators in Fig. 5 that after the application of the online education effect evaluation method for colleges and universities based on data mining proposed in this paper, the confidence level of evaluation results is significantly higher than that of traditional methods, reaching more than 96%. It is not difficult to see that the online education effect evaluation method proposed in this paper has high feasibility, high accuracy of evaluation results, significant advantages in application effect, and can be put into practical use.

Next, the evaluation time of the two methods is compared, and the comparison results are shown in Fig. 6 below.

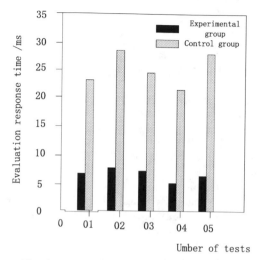

Fig. 6. Evaluation response time comparison

As shown in Fig. 6, the evaluation response time of the proposed data mining based online education effectiveness evaluation method for universities is less than 10ms, while the evaluation response time of traditional methods is above 20ms. Therefore, the proposed method is better and can be better applied to online education effectiveness evaluation.

4 Conclusion

In view of the problems of low accuracy and poor effect of traditional online education evaluation methods in practical application, this paper introduces data mining technology and puts forward an online education evaluation method based on data mining technology. This method establishes the evaluation model of the effectiveness of online education in colleges and universities, decomposes the decision-making problem into a hierarchical structure, analyzes the relevant data by using data mining technology, constructs an evaluation table, and evaluates the effectiveness of online education comprehensively and multidimensional based on association rules. The experimental results show that compared with the traditional method, the evaluation results obtained by the proposed method are more than 96% confident, the evaluation response time is less than 10ms, and the performance is better. It can improve the technical level of teaching evaluation, science, objectivity and justice, and improve the teaching quality in colleges and universities.

Aknowledgement. The 2021 Social Science Foundation Project of Nanning University, titled "Research on the Factors Influencing the Leadership of the Director of the Teaching and Research Office of Nanning University," with fund number 2021JSGC10.

References

1. Koprda, S., Magdin, M., Reichel, J., et al.: Time efficiency of online education in technical subjects without decreasing didactic effectiveness during the COVID-19 Pandemic. Int. J. Eng. Educ. **6**, 37 (2021)
2. Kok, K., Paterakis, N.G., Giraldo, J.S.: Development, application, and evaluation of an online competitive simulation game for teaching electricity markets. Comput. Appl. Eng. Educ. **30**(3), 759–778 (2022)
3. Li, X.: Design of accounting teaching effect evaluation system based on big data technology. Chin. New Technol. New Prod. **449**(19), 136–138 (2021)
4. Sun, Y.: Design of college English teaching effectiveness evaluation method based on intelligent algorithm. Inf. Comput. (Theor. Ed.) **34**(14), 243–245 (2012)
5. Apriyanto, R., Adi, S.: Effectiveness of online learning and physical activities study in physical education during pandemic COVID 19. Kinestetik J. Ilmiah Pendidikan Jasmani **5**(1), 64–70 (2021)
6. Sia, C.H., Ng, S., Hoon, D., et al.: The effectiveness of collaborative teaching in an introductory online radiology session for master of nursing students. Nurse Educ. Today **1**, 105033 (2021)
7. Lu, J., Sun, T., Lou, J., Yang, G., Lu, Z.: Effect evaluation of teaching quality evaluation system for university teachers. Chin High. Med. Educ. **297**(09), 41–42 (2021)
8. Xue, Y, Wang, T. online and offline mixed teaching effect evaluation of big data analysis. Inf. Technology, 2021, 357(08):70–74+80
9. Song, H., Lv, X.: Evaluation method of classroom teaching effect of military academy based on KPI. J. Air Force Early Warning Acad. **35**(04), 295–298 (2021)
10. Yang, Y., Zhang, M., Wu, B., et al.: Design of simulation system for discontinuity network in rock mass and its application in teaching. Comput. Simul. **39**(9), 5 (2022)

Author Index

Printed in the United States
by Baker & Taylor Publisher Services